Enemy Civilian Casualties

Enemy Civilian Casualties

Politics, Culture, and Technology

Ofer Fridman

LEXINGTON BOOKS
Lanham • Boulder • New York • London

Published by Lexington Books
An imprint of The Rowman & Littlefield Publishing Group, Inc.
4501 Forbes Boulevard, Suite 200, Lanham, Maryland 20706
www.rowman.com

6 Tinworth Street, London SE11 5AL

British Library Cataloguing in Publication Information Available

Library of Congress Cataloging-in-Publication Data

Names: Fridman, Ofer, 1979- author.
Title: Enemy civilian casualties : politics, culture, and technology / Ofer Fridman.
Description: Lanham : Lexington Books, [2019] | Includes bibliographical references and index.
Identifiers: LCCN 2018057329 (print) | LCCN 2018060786 (ebook) | ISBN 9781498586924 (electronic) | ISBN 9781498586917 (cloth : alk. paper)
Subjects: LCSH: Civilian war casualties—Government policy. | Nonlethal weapons—Government policy. | Casualty aversion (Military science)—Government policy. | United States—Military policy. | Russia (Federation)—Military policy. | Israel—Military policy. | Politics and war. | War and society. | War—Technological innovations. | War—Moral and ethical aspects.
Classification: LCC UA11 (ebook) | LCC UA11 .F75 2019 (print) | DDC 363.34/98—dc23
LC record available at https://lccn.loc.gov/2018057329

∞™ The paper used in this publication meets the minimum requirements of American National Standard for Information Sciences Permanence of Paper for Printed Library Materials, ANSI/NISO Z39.48-1992.

Printed in the United States of America

To my grandfather, who taught me by example
to strive for best in everything I do.

Contents

Acknowledgments

I am indebted to many great scholars whose works are dutifully noted and who, of course, might not subscribe to the argument I present and definitely bear no responsibility for errors and mistakes in the text. The responsibility for all views and mistakes in this research is mine and mine alone.

This project would not have been possible without generous and priceless assistance from many people, to whom I would like to express my debt of gratitude. While the meritable advice from scholars, the moral endorsements from friends and the unconditional support from the family—were all equally invaluable for the successful accomplishment of this research, I believe that due to the different nature of these contributions, it will only be right to phrase my gratefulness separately.

On the academic front, special thanks are due to my professor and mentor Beatrice Heuser from the University of Reading. I owe her very much. I was lucky to benefit from her open-door policy and encyclopedic knowledge, her inputs, remarks, and continuous support throughout all stages of this research were absolutely invaluable. For their comments and advice during the initial stages of my research (despite the fact that most of them were done at Park House), I am especially grateful to two former members of the Department of Politics and International Relations at Reading: Patrick Porter and Burak Kadercan. Thanks are also due to other researchers in the department, whose support, knowledge and friendship I have benefited from over the years of my research, specifically Corinne Heaven, Nadya Ali, Ben Whitham, Mona Jibril, Zoi Verdanika, Jan Claassen, Mark Jakeman, Christos Vrakopoulos, Alex McLaughlin, Moeen Koa, and Ibrahim Muhammad Aziz.

In the process of this project I was fortunate to spend a very productive time at the Moscow State Institute of International Relations (MGIMO). To making my time especially fruitful, I owe my appreciation to Igor Istomin, whose organizational skills and readiness to help were beyond expectations; to Yulia Nikitina, whose wise advice and assistance I benefited from enormously; and to Vitaliy Kabernik, from the Military-Political Research Centre at MGIMO, whose valuable inputs were always useful. A special word of thanks is due to Denis Levin, from the Bauman Moscow State Technical University and Artem Muranov, from the Research & Development Institute of Applied Chemistry, who were always there for me, as friends and as professionals.

I own a great deal of appreciation to Dima Adamsky from the Interdisciplinary Centre Herzeliya, Israel, whose enthusiasm during the initial stages of this project was invaluable for its final success. Similarly, I am indebted to Randolf Cooper, whose comments and readiness to help in the final stages were very beneficial and encouraging, as well as to Professor Theo Farrell from King's College London and Professor Dominik Zaum from University of Reading, who both contributed their time and efforts helping me to hone my arguments.

On the friendship front, there were too many people, who had probably never quite understood why and what I was doing, but (for some unknown reason) had always believed in me and in my project. I owe my special gratitude to Vicki Clayton, who, despite her own ups and downs, was always there for me, and to Eviatar Perkal and Moshe Katz, whose long and unconditional friendship had always inspired me. I want to extend my appreciation to Asher Elankry, who believed in me through and through. I also owe a special appreciation to Simon Dunstan and Jenny Spencer-Smith, who had turned my stay in the UK into an exciting experience, opening their hearts (and Tanqueray) and providing an invaluable insight into British culture.

I also want to thank my family, who has always believed in me, even when I did not. To my parents and grandparents, whose moral support was unconditional and to my sister and brother-in-law, whose living room sofa was occupied on too many occasions during my research trips to Israel.

Last, but definitely not least, I wish to acknowledge Sagit Carmeli for her devoted and excellent editing that turned my scrambled thoughts into a readable text—there are not enough crisps and pies in the world that would help me repay my debt.

Introduction

This book has two main purposes. First, it offers a more comprehensive understanding of the relations between society and its military in the twenty-first century. Civilian casualties and collateral damage have long been considered as an undesired outcome of military activity that has to be reduced. While most of the contemporary discourse on this topic has primarily concentrated on three main factors: the legal aspects of causing civilian casualties, the impact of war on local populations, and different factors of military professionalism required to avoid disproportional harm to civilians, this book addresses the topic from an entirely different perspective. As the subject of civilian casualties during military operations seems to be highly politicized, this book takes this discourse out of its usual niches and suggests that the indirect responsibility rests with the politicians and the public that they represent. In other words, this book makes an attempt to answer a very controversial question: When a society, at the beginning of the twenty-first century, sends its troops to war, does it really care about the enemy civilian casualties? In answering this question, this book intersects with a growing body of work on the relationship of warfare and violence. One of the most fundamental works, recently written on this subject, is Steven Pinker's *The Better Angels of Our Nature: A History of Violence and Humanity*, which suggests that violence as a phenomenon of social behavior in general, and in making war in particular, is in decline.[1] Another good example is Christopher Coker; in his *Humane Warfare* he discusses the idea that in the twenty-first century "Western Societies can only now fight wars which minimize human suffering"[2] and in his *The Future of War* he proposes that "developed societies are likely to continue with war but in a form that is more rational and optimal . . . [and that] we may be able to humanize war, or make it more humane."[3] Having these opinions in mind, this book would make an attempt to provide an alternative, though not necessarily a contradicting, interpretation and approach to Pinker's and Coker's arguments, discussing political and cultural aspects that have sustained the violence in twenty-first century military operations at a significantly higher level than it could possibly be. While the analysis of how different societies deal with the problem of enemy civilian casualties is the main purpose of this book, it addresses this problem through the prism of military technological transformation, or, in fact, the lack of it.

This directly leads to the second purpose of this book. Non-Lethal Weapons (NLW)—weapons that are intended to minimize civilian casualties and collateral damage—are based on technology that, during the 1990s, was expected to revolutionize the conduct of warfare. As Alvin and Heidi Toffler suggested in 1993:

> Non-lethality emerges not as a simple replacement for war, or an extension of peace, but as something different. It is something radically different in global affairs—an intermediate phenomenon, a pausing place, an arena for contests where more outcomes are decided bloodlessly.[4]

The rapid rise of interest in NLW sparked an entirely new military, as well as academic, discourse concerning their potential contribution to military success on the twenty-first century battlefields. It seems, however, that except for this debate, very little has been done within the military forces themselves. This book aims to address this problem analyzing three different countries—the United States, Russia, and Israel—that were chosen not only because of their differences, but also due to their similarities. In the past decades, all three were involved in armed conflicts: Israel in Lebanon and Gaza, the United States in Iraq and Afghanistan, Russia in Chechnya and Georgia. All three have armed forces with a track record as organizations capable of revolutionary changes.[5] All three have strong defense industries that are able to support military demands for NLW.[6] And, finally, all three seek to be military powers (whether regional or global), and therefore invest significant efforts (political and economic) to produce as effective armed forces as possible. Given this last similarity between the American, the Russian, and the Israeli military forces, it seems vital to understand the social-political rationale that prevents them from employing NLW that "may provide more effective power" than their lethal counterparts.[7] Due to the combination of these characteristics, each one of these three states represents political, military, and technological prowess, and, therefore, is expected to implement revolutionary changes in military affairs ahead of any other country. Investigating the reasons behind the absence of the RMA (Revolution in Military Affairs) of NLW in these three particular cases, this book addresses the political, social and cultural factors that prevent the employment of NLW on the battlefield. It presents the path that has led three completely different militaries, from completely different societies and political realms, to the same outcome: rejection of the technologies intended to minimize civilian casualties during military operations.

WHAT THIS BOOK IS ABOUT

While this book analyzes the absence of the RMA of NLW, its main focus is not about NLW per se. Instead, it focuses on how different societies approach the problem of enemy civilian casualties. The argument developed throughout this book has two main parts. Firstly, it rethinks the validity of the claim that harming civilian populations during military operations has high significance in international relations. Indeed, this claim is frequently used to criticize political adversaries in the international arena. Moreover, different non-state international institutions, such as the Red Cross, Amnesty International, Human Rights Watch, and others, repeatedly use this argument to criticize countries for the civilian casualties caused by their militaries. This book, however, asks to rethink the effectiveness of this criticism, arguing that countries that are involved in military operations have developed political mechanisms to fend such criticism off. While it is generally argued that international political pressure is a crucial factor that leads to the significant reduction of harmed civilians and destroyed civilian infrastructure, the relevant foreign policies have already been crafted in the context of international criticism, allowing the fending off any criticism related to military activity, including the one related to civilian casualties.

Each one of the analyzed countries has developed its own foreign policy mechanisms that allow them to resist the international criticism raised before, during and after military operations. Using very effective foreign policy narratives, political decision-makers can be relatively relaxed about their military performance regarding civilian casualties and they feel no need to press their military to apply fundamental changes that would be required to make operations less lethal than is politically required (i.e., to employ NLW on a large scale).

This leads to the second main argument of this book, which challenges us to rethink the weight that is given to the value of human life in the twenty-first century, and its universality. Each nation has its own "way of war," its "strategic culture," influenced by its geography, society, history, religion, and other factors, and, therefore, it is assumed that different societies would value the life of the enemy civilian population differently. Yet in all three analyzed cases, the societies showed very similar high levels of tolerance toward enemy civilian casualties during military operations. Indeed, each culture has gone its own way in shaping its unique attitude towards the enemy. With regard to enemy civilian casualties, however, it seems that the outcome is surprisingly similar.

In the twenty-first century people, indeed, value human life significantly more highly than before. This appreciation, however, has been translated largely into general unwillingness to sacrifice one's own life and the lives of one's relatives, rather than a universal value. In other words, in the twenty-first century, societies do not send their sons and husbands to war as easily as they did in the past; and if they do, they demand that their leadership (and therefore their military) minimizes the casualties of their own military personnel to the operationally possible and politically defendable minimum. However, when it comes to enemy civilian casualties, this narrow appreciation of human life has contributed to a general apathy among public opinion toward enemy civilian population in conflict zones, due to the fact that society's relatives (military personnel) sacrifice their lives there, and by this justifying enemy civilian casualties.

In the last decade, all three countries analyzed in the following chapters were involved in armed conflicts, i.e., Israel in Lebanon and Gaza, the United States in Iraq and Afghanistan, Russia in Chechnya and Georgia. On the one hand, it seems safe to assume that some countries paid more attention and invested more efforts to minimize enemy civilian casualties, than the others. On the other hand, as this book will show, they all were reducing enemy civilian casualties to a politically acceptable minimum (which differs from society to society and from conflict to conflict), rather than minimizing them through fundamental technological innovation. This book suggests that the roots of this situation are much deeper than the simple professional misconduct of the military establishment, or the poor political behavior of political leaders who had sent them to fight. While the technology that allows the lowest possible minimization of civilian casualties, rather than just politically acceptable reduction, has been around for several decades, without extensive political pressure, the required military transformation just cannot occur. Therefore, despite the seemingly bloodless images of contemporary warfare produced by precision-guided munitions, drones, and selective strikes, war remains a very lethal affair, especially for enemy civilian populations.

The history of previous military transformations shows that facing unmet challenges, the military knows how to perform fundamental revolutionary changes to confront them. This book argues that NLW are not employed because, due to political and cultural aspects, the challenge of minimizing civilian casualties has never been really put to the military.

POLITICAL ASPECTS OF RMA: WHY RMAS OCCUR (OR NOT)

During the last several decades, different historians and strategists have suggested different ways to explain the interconnection between technology, military, and society that produces a fundamental, even revolutionary, change in the ways of war.[8] While analyzing this phenomenon, different scholars use different terminology, but it seems right to argue that the two most acceptable terms are: *Military Revolution* (MR) and *Revolution in Military Affairs* (RMA).

The term *Military Revolution* was first introduced by historian Michael Roberts in 1956. He used his inaugural lecture at Queen's University Belfast to expound on an early modern "military revolution." In his later publications, he continued to claim in favor of this term and finally determined it in his prominent article "The Military Revolution, 1560–1660," first appearing in his *Essays in Swedish History* in 1967. The main idea of Roberts's military revolution was "that purely military developments, of strictly technological kind, did exert a lasting influence upon society at large."[9] In other words, the main idea of MR is that fundamental changes in the means of warfare can lead to the dramatic changes in society and its social institutions, and vice versa, dramatic political-societal changes can lead to revolutionary changes in the ways of warfare.

The concept of RMA appeared independently and almost at the same time as the concept of MR did. First invented by Soviet military thinkers during the Cold War as Military-Technological Revolution, the idea was "Americanized" in the early 1990s by U.S. strategists, such as Andrew Krepinevich and Andrew Marshall.[10] Pioneering the theoretic-conceptual debate on RMA, they established the conceptual framework of an RMA that included three basic dimensions: doctrinal, organizational, and technological. According to Krepinevich, an RMA is "when the application of new technologies into significant military systems combines with innovative operational concepts and organizational adaptation in a way that fundamentally alters the character and conduct of war."[11] In fact, compared to Roberts' MR, RMA is a much more limited phenomenon. While MR claims to explain the relations between military innovations and social structure, RMA is restricted only to the military field. In other words, RMA theory tries to explain the very specific chain between military innovations and military institutions.

There is no doubt that these two independently developed concepts, MR and RMA, share a mutual nature. The very first and most comprehensive attempt to integrate them was made by the historian Williamson Murray in 1997 in his article "Thinking about Revolution in Military Affairs"[12] and later in his book, coedited with another military historians MacGregor Knox, *The Dynamics of Military Revolution, 1300–2050*. Murray's main breakthrough is rooted in his attempt to take the idea of MR out of its context of

Early Modern Europe and determine a conceptual definition of this phenomenon and the relation between it and RMA. According to Murray, MRs are "earthquakes" that bring "systematic changes in politics and society . . . [and] recast society and the state as well as military organization"; and RMAs are "less-embracing changes" that appear within military institutions and "require the assembly of a complex mix of tactical, organizational, doctrinal, and technological innovations in order to implement a new conceptual approach to warfare or to a specialized sub-branch of warfare."[13] Murray claims a very interesting connection between MRs and RMAs: "If Military Revolutions are compared with earthquakes, we can think of RMAs as pre- and aftershocks."[14] In other words, RMAs can be considered as traceable parts of the military transformations that constitute MRs. Since this book focuses on the RMA of NLW, a better explanation of this phenomenon is required.

Since its introduction, the concept of RMA has attracted many advocates,[15] as well as critics.[16] It seems, however, right to argue that at the core of the theory of RMA rests an attempt to conceptualize military transformation by providing clearly defined analytical framework, i.e., a simultaneous and interdependent adaptation of new technologies, concepts of operations, and organization. In other words, while the term "military transformation" "should simply be understood to mean 'profound change' in military affairs" that "need not imply rapid or across-the-board change" as a "transformation is a process with no simple end point"[17]; the concept of RMA seems to provide more rigid methodological tools for the analysis of the military transformations, as it has clearly defined variables (technology, concepts of operations, and organization) that not only to define the beginning and the end of military transformation, but also its pace.[18]

While one of the main arguments of the opponents of the RMA theory is that it is too technologically deterministic and it "separated war from its political, cultural and psychological context,"[19] this doesn't seem to be the case, as some scholars explicitly analyzed the RMA phenomenon from the cultural perspective.[20] Moreover, there is an extensive literature that discusses cultural, political, social, and other factors of military transformations. For example, in analyzing military transformations, Theo Farrell and Terry Terriff argue that politics and culture are amongst the main factors behind fundamental military changes[21]; Colin Gray, Jeremy Black, and Theo Farrell claim that national norms and culture shape and transform strategic behavior[22]; Stephen Rosen suggests that social structures transform national styles in generating military power[23]; and Elizabeth Kier and Beatrice Heuser argue that cultural predispositions of specific nations shape their military doctrines.[24] And since an RMA is just more narrowly defined military transformation, the observations and conclusions proposed by these scholars also apply to the RMA phenomenon.

A combination of this extensive literature on military transformation with the conceptual original literature on RMA allows us to construct a more complete analytical framework for the analysis of fundamental changes in military affairs.[25] First, it is important to note that while technology is, undoubtedly, an essential component of the RMA phenomenon, it does not ignite a revolution by itself and, in fact, history provides a large number of cases when a novel military technology failed to ignite an RMA.[26] Moreover, when Richard Hundley argues that "RMAs are often adopted and fully exploited first by someone other than the nation inventing the new technology,"[27] underlining the fact that the "novel" technology behind an RMA is usually not exactly novel. The story of the RMA of armored warfare offers one of the best examples of a technology (i.e., tanks) that was developed by the British military in World War I, but the RMA itself occurred first in Nazi Germany, 20 years later.[28]

The second important element of RMA, widely discussed in the literature, is an "unmet military challenge." While Hundley's argument that "unmet military challenges are an essential element" that drives the breakthrough process of RMA[29] might sound too technologically deterministic, Colin Gray elaborates on this idea, arguing that the implementation of RMA demands that "some lawful authority" will "decide that there is a challenge, a threat, or an opportunity, in need of RMA solution."[30] Combining this understanding with the conclusion of Farrell and Terriff that "strategic pressures operate through political processes in shaping military change,"[31] it seems possible to argue that a successful RMA requires the political leadership to realize that there is a challenge that demands a fundamental military transformation. In other words, for an RMA to occur, a military organization has to realize that a challenge that had been put in front of it by its political leadership cannot be met by existing capabilities and, therefore, a change is required, and the bigger the challenge, the bigger the change.

While this combination highlights the importance of politics for fundamental military transformations (i.e., RMAs), it also emphasizes the importance of different cultural factors, as, according to Farrell and Terriff: "cultural continuity can shape military change by defining organizational responses to strategic, political, and technological developments."[32] As "under the impact of [different] cultural factors countries interpret differently the changing character of war,"[33] it seems right to argue that culture shapes the way of the political-military decision-making process, as well as the way a novel technology is perceived by the military. Therefore, culture influences not only political leadership in their definition of challenges and threats, but also the military in its interpretation of these challenges and finding possible solutions.

These combinations ultimately lead to the conclusion that an RMA, as a form of military transformation, is not technology per se, as suggested by the critics of this theory, but it is a result of continuous interaction between politics, the military, and the cultural context that shapes them both. Combining these three main factors that stand behind the occurrence of RMA, it seems right to propose a conceptual framework that explains the occurrence of RMA: *an RMA occurs when a political leadership, shaped by certain socio-cultural circumstances, defines to its military a challenge that military culture interprets as unmet and dealing with it requires fundamental change that combines novel, but already available technology, operational concept, and organization.*[34]

Moreover, this conclusion allows us to establish a very clearly defined analytical framework for the analysis of RMAs, as according to it an RMA occurs when:

> (1) a political leadership, shaped by certain socio-cultural circumstances, defines to its military a challenge;
> (2) the military, shaped by its military culture, interprets this challenge as unmet and dealing with it requires fundamental change that combines novel, but already available technology, operational concept, and organization;
> (3) in return, the political leadership decides to supports the military fundamental transformation.[35]

In other words, the first and foremost reason for the occurrence of an RMA is a political environment that creates its necessity, or, as Williamson Murray and MacGregor Knox put it:

> [RMA] always occur within the context of politics and strategy . . . New Zealand will not realize a revolution in military affairs in the foreseeable future: its strategic environment is too benign to inspire one.[36]

The importance of the political aspects for the occurrence of an RMA can be easily explained in the conceptual context of a strategy-making process. Strategy is the "direction and use made of means by chosen ways in order to achieve desired goals,"[37] and, therefore, a fundamental change of "desired goals" would require a revolutionary change in the means and ways to use them (i.e., RMA). Thus, the way to understand why certain means are not used is to examine whether there are political goals that require these means, and, therefore, an analysis of the RMA in a "certain affair" that did not occur has to follow the following questions:

1. Did the political leadership define this "certain affair" as a military challenge? If not, then what was the socio-cultural context that influenced the political-military leadership in its decision-making process?
2. If the challenge was defined, how did the military, shaped by its military culture, interpret this challenge? If it was interpreted as a chal-

lenge that the military can deal with, then what was the military cultural context that influenced the military leadership in its decision-making process?

3. If the challenge was defined by the military as unmet, did the political leadership support its military through the transformation? And if it did not, what was the socio-cultural context that prevented the political leadership from supporting it military?[38]

While these three questions seem to be straightforward, it will be wrong to assume that any of them will provide a clear-cut answer. Human nature is a too-complex phenomenon, let alone a combination of culture, politics, and military affairs. It seems, however, that through these questions it would be possible to shed light on the reasons behind the absence of an RMA, an understanding that might contribute no less than the existing analysis of successful RMAs, because, as Richard Hundley put it: "there are probably as many failed RMAs, as successful."[39]

THE PUZZLE OF THE REVOLUTION IN MILITARY AFFAIRS OF NON-LETHAL WEAPONS

Since the last decade of the twentieth century, different scholars and political analysts have claimed that human life is valued more and more among the general international community and, particularly, within Western societies. This argument has led many researchers to assume that democratic culture and traditions, modern ethical and moral issues have created a desire for a world without war or, at least, a demand that contemporary armed conflicts, if unavoidable, at least have to be far less lethal, forcing the military to seek new technologies that have "greater precision, shorter duration, less lethality, and reduced collateral damage . . . [as these technologies] may provide more effective power than their larger and more destructive, but also more inexact and crude, predecessors."[40]

Military operations, during the last several decades, have deviated from previous wars, where victory was considered solely by military terms. Winning the "hearts and minds" has become a vital factor in achieving the political aims of military operations.[41] Hence, the achievement of a comprehensive victory demands support from the local population[42]; and therefore, the unlimited harming of civilians, in most cases, will increase their mobilization against the military and delay the successful end of operations. Moreover, it is assumed that in the era of mass media and especially the Internet, harming local populations significantly decreases domestic and international support for the continuing of military operations.[43]

While enemy civilian casualties are considered as an unnecessary and an undesirable outcome of warfare, an attempt to reduce the number of non-combatant casualities raises the problem of distinguishing between combatants and non-combatants. This problem has become more challenging due to two major changes that have occurred in the last twenty years. The first one has a military nature and constitutes the transformation of warfare—from traditional-conventional warfare to the new concept of war, frequently called Military-Operations-Other-Than-War (MOOTW) (such as, *asymmetrical war*, *hybrid warfare*, *war on terror*, *small war*, *peacekeeping operations*, etc.). In this type of warfare the distinguishing between combatants and civilians is even more complex than it had been in the past.[44]

The second major change has occurred within society. Today, there is no doubt that the Internet has transformed the social landscape of human life. One of the outcomes of this transformation is the ability of non-state actors to organize civilians to participate in conflicts, which has risen significantly after the appearance of the Internet.[45] Moreover, the development of new social media Internet platforms, such as Facebook, Twitter, and YouTube, took these capabilities a few steps further, providing civilians with the ability to organize themselves.[46] Civilians have always taken part in wars and armed conflicts. Their direct participation, as part of an underground movement, or a partisan unit (or even as a terror organization), as well as their indirect involvement by providing an essential support—were always a problem for military activity. While it seems that social-political reality in the beginning of the twenty-first century demands from the military a significant decrease in the lethality of conflicts,[47] the increase of civilian participation in zones of conflicts, triggered by social media, makes this demand even more challenging.[48] Consequently, it seem right to assume that the creation of less-lethal capabilities would be a military challenge that has to be met.[49]

There are many different terms that describe weapons that are intended to incapacitate people without causing death or permanent injury—"non-lethal," "less-lethal," "less-than-lethal," etc.[50] Different organizations in different countries have adopted different terminologies. Nevertheless, all these terms are very similar in their referring to a group of weapons that are "explicitly designed and primarily employed so as to incapacitate personnel or material immediately, while minimizing fatalities, permanent injury to personnel, and undesired damage to property, facilities, materiel and the environment."[51] It is important to emphasize that while the term "non-lethal" suggests that no fatalities can ever occur through an employment of NLW, this is not the case. By definition, NLW "do not, and are not intended to, eliminate risk of those actions entirely."[52] Similar to lethal weapons that do not always fulfil their purpose, their lethality depends on many factors (the level of the soldiers' professionalism, the state of the target, environmental conditions, etc.); NLW are also victims of these factors, and, despite their non-lethal nature, can, under certain circumstances, cause permanent damage or even death.

During the past two decades, much literature has been written about NLW that can be divided into three major groups. The first group describes NLW and non-lethal technologies; these works discuss the technological developments of NLW and/or the history of their usage.[53] The second group advocates NLW and emphasizes the revolutionary promise of NLW and their potential to ignite new RMA and alter the nature of the way of conducting conflicts.[54] The third group questions the perspective of NLW, and claims that NLW have a place in the future, but also cautions against bestowing any special status to NLW and discusses the problems of the ethical and legal aspects of their employment.[55] While all literature comes mostly from academia, there is also a noteworthy interest in NLW within military research publications. Interestingly, this attention to the possible implementation of NLW has appeared, almost simultaneously, in military organizations with completely different cultural backgrounds, such as the United States,[56] Russia,[57] and Israel.[58] And yet, despite this extensive discussion on the possible perspectives of NLWs, and the significant investments in technological development that has occurred in this area,[59] it seems that not much has happened in the last thirty years, as "the list of non-lethal weapons that are currently available does not differ greatly from similar compilations 30 years previously."[60]

As it will be shown, current non-lethal technologies are able to provide a vast spectrum of non-lethal effects that are ready to be adopted by the military. There are no technological obstacles that can prevent efficient integration of existing effective non-lethal agents within existing conventional military platforms. As it will be shown throughout the following chapters, different industries around the globe already have introduced effective NLW that can be integrated within military forces. The fact, however, that this integration has not occurred yet, points to the existence of other aspects that prevent the vast adaptation of NLW within military.

WHAT THIS BOOK IS NOT ABOUT

While this book is formed around Non-Lethal technologies, it does not mean that it accepts the approach of technological determinism, as Max Boot argued:

> Technology sets the parameters of the possible; it creates the potential for a military revolution. . . . No technical advance by itself made a revolution; it was how people responded to technology that produces seismic shifts in warfare.[61]

Non-Lethal technologies have been around for many years,[62] in fact, certain non-lethal chemical agents have been widely employed during different military conflicts in the past hundred years, as far back as World War I.[63] While technology is vital, "a true revolution in the way military institutions organize, equip and train for war, and in the way war is itself conducted, depends on the confluence of political, social, and technological factors."[64] Technology is not the reason behind the absence of the RMA of NLW, at least not in the United States, Russia, and Israel, and therefore it is not the main focus of this book. Political and social factors, by contrast, seem to provide a more adequate explanation for the absence of NLW on the modern battlefield.

The research presented in this book also does not try to predict the future of NLW. The future is a kind of a "black box" and an attempt to open it would ultimately fail this research by a gargantuan number of assumptions and tendencies that have to be considered during forecasting efforts. Instead, this research analyzes the past, focusing on the reasons that have led to the failure of the RMA of NLW so far. While the analysis of three different countries examined by this research might help the reader glance at a possible future for non-lethal technologies, it is at the reader's peril only.

While focusing on the problem of civilian casualties, this book does not deal with the definition of "civilian casualties." It does not try to define or redefine who should, or should not, be defined as an "innocent" or an "undesired" casualty of war. Existing literature suggests two different ways to devise a definition. The first one is based on more traditional dichotomous distinctions between combatants and non-combatants in a war zone, which is based on the four Geneva Conventions of 1949 and the two Additional Protocols of 1977. The second way suggests that this dichotomy is rooted in regular warfare and is no longer suitable when applied to contemporary counter-insurgencies, and, therefore, the distinguishing aspects should be not between combatants and not-combatants, but rather between different levels of participation in war. One of the best examples for this type of distinction is given by the Israeli High Court of Justice that suggests a more general division between "direct" and "indirect" participation in hostilities. While "direct" participation includes transporting militants, operating weapons, or supervising the operations; "indirect" participation consists of providing food, medicine, and shelter to combatants, monetary aid, logistical support, and distributing propaganda.[65] As it will be discussed throughout this book, its focus is not on the legal issues regarding the harming of enemy's population, and, therefore, the reader can adopt any of the existing ways to define "civilian casualties."

Moreover, this book does not engage in the debate on the morality of harming civilians. It does not try to claim that harming enemy civilian population is immoral, unethical, or illegal; this job is preserved for philosophers and lawyers. Instead, the purpose of this work is a much more constructive one. As discussed previously, the main argument of this book is with those who assume that democratic culture and its traditions, modern ethical and moral issues, have been restricting the level of violence during military operations in the twenty-first century. Indeed, it will be unfair to claim that the American, Russian and Israeli militaries do not understand the consequences of disproportionally harming civilians in the political environment of the twenty-first century and try to reduce the collateral damage caused by their activities. It seems that the World War II–style city bombings, as practiced against Hamburg, Dresden, or Tokyo, indeed belongs to the past. However, there is a difference between the politically acceptable level of civilian casualties and the minimum possible one, between the "reduction by avoiding" and "minimization by preventing." This book tries to explain why, while the technology of NLW allows the latter, the military organizations still follow the former.

Finally, while those who are interested in the foreign policy narratives or cultural attitudes toward war and violence of each one of the analyzed nations might find this book interesting, they also might find that the discussions of each one of these aspects are quite limited. While in analyzing the lack of RMA of NLW in the United States, Russia, and Israel, the following chapters will discuss the foreign policies, national characters, and military cultures of these countries, this book is not explicitly about these aspects. Instead, this book constructs three different puzzles regarding the different factors that have prevented the employment of NLW in three different countries. In an attempt to assemble these puzzles of the failure of the RMA of NLW in the United States, Russia, and Israel, this book discusses political and cultural narratives of these countries as general cohesive complexes, rather than separate specific phenomena consisting of many distinctive patterns, which they obviously are. Those who are explicitly interested in these specific aspects are strongly advised to use this book only as a source-full starting point, and not as a detailed analysis.

METHODOLOGICAL REMARK

As shown above, RMA is a very complex phenomenon that involves a nexus of relationships between different variables from technological, political, cultural, and military spheres. In the light of this multifaceted reality that does or does not produce revolutionary changes in the character and conduct of warfare, only a finely-honed methodology would be able to ensure the analytical empirical nature of the following discussion. As this book focuses on the RMA of NLW that did not occur, this ultimately creates an important methodological question: is it possible to analyze something that did not occur? To address this question, it is important to clarify that, as discussed previously, an RMA is a finely defined concept; thus, this research analyzes a finely defined scientific phenomenon. As with any scientific phenomenon, it is possible to establish a list of the required components, a combination of which leads to its creation. Moreover, an RMA, as a defined phenomenon, occurred many times in the past, proving the fact that when the required components were there, it did occur, and when one of them was missing, the RMA failed.[66] Consequently, it seems right to argue that the absence of an RMA can be explained by the analysis of the required components pointing to those, which failed to create the desired result—and this book does exactly that.

This book is based on data and evidence gathered from a vast spectrum of sources. The chapter that discusses the RMA of NLW that did not occur utilizes different primary sources, such as governmental reports, international laws, archival documents, industry reports, etc. The case of the United States is constructed from unclassified dissertations and articles written by senior American officers, as well as on academic publications produced by leading scholars and documents published by different political establishments. The Russian case utilizes declassified scientific studies published in leading military periodicals, different governmental documents, and products of scientific studies conducted in leading Russian universities. The Israeli case is based on combined data gathered from curricula of academics, leading research institutions in the military and political spheres, and professional journals. Where primary sources are unavailable, the book relies on secondary literature utilizing the indigenous languages of each one of the cases: English, Russian, and Hebrew. This multi-linguistic approach allows the enhancing and accentuating of the different sources, especially in the Russian and Israeli cases, which have not been translated into English before.

THE STRUCTURE

Following this introduction, the book consists of four chapters and, a conclusion:

Chapter 1: Non-Lethal Weapons: The Revolution in Military Affairs That Did Not Occur

The purpose of this chapter is threefold. The main subject of this book is the RMA of NLW, and as such, the first part of this chapter examines the general background of these weapon systems. It discusses the perplexity of the definition of NLW, what they are, and what they are not. It provides a historical brief discussing the implementation of NLW since their appearance in the beginning of the twentieth century and focuses on different technological aspects of NLW and their effects. Finally, it examines the complex, and to some degree paradoxical, relations between NLW and modern international laws of armed conflict. The second part of this chapter reproduces a theoretical depiction of the potential RMA of NLW. While the main purpose of this book is to explain the reasons behind the failure of the RMA of NLW, this part provides an essential understanding of what this RMA could or should have been, as a reference for the analysis of its absence. The third part utilizes this picture of the potential RMA of NLW to frame the examination of the following American, Russian and Israeli cases.

Chapter 2: From the Frontier to the "Protector of the Free World": The United States and Enemy Civilian Casualties

This chapter examines the reasons behind the failure of the RMA of NLW in the U.S. military. Despite the fact that the very idea of military employment of NLW was a product of the American military thinking, they have hardly been ever used. American military regards technology as a lodestar of military success. However, without political pressure that defines the minimization of civilian casualties as an integral part of the military success, non-lethal military technologies remain underdeveloped and idle. American political leadership has developed strong foreign policy mechanisms to reduce the influence of international criticism on their decision-making process. The rhetorical narratives, such as *the protector of the free world against a great evil*, *the promoter and the defender of democracy*, and the ability to create *coalitions of willing nations*, have been allowing American leadership to minimize international criticism in general, and criticism regarding enemy civilian casualties during U.S.-led military operations in particular. Moreover, American public opinion has been showing very little interest in enemy civilian casualties. The American mind-set, shaped by different historical and cultural predispositions to disregard everything that is not American and a general belief that war is a temporary unfortunate phenomenon that has to be resolved as quickly and aggressively as possible, leaves very little space for appreciation of enemy population. Without political domestic demand to minimize civilian casualties, and with a foreign policy that successfully fends off international criticism, the American political leadership does not pressure its military to revolutionize its ways of conduct and fully integrate NLW in military organizational and operational concepts. Consequently, this chapter demonstrates how NLW remain a marginal and oxymoronic phenomenon in the history of the technological adventures of the U.S. military.

Chapter 3: "Besieged Fortress" and the Politics of "Double Standards": Russia and Enemy Civilian Casualties

This chapter examines the reasons behind the failure of the RMA of NLW in the Russian military. While the growth of interest toward NLW in the Russian military has been directly influenced by American military attitudes that emerged in the mid-1990s, the initial enthusiasm has been successfully buried by the Russian bureaucracy machine. The types of non-lethal technologies and the level of their employment in the Russian military basically does not differ from its American counterpart. Despite a very productive professional debate on the possible usefulness of NLW on the battlefield, the Russian political leadership has never felt pressured to minimize enemy civilian casualties. Building on its status as a nuclear power and permanent member of the UN Security Council, its growing economy, and broad domestic political support, the Kremlin has been able to withstand international criticism and accusations of violations of international law in general, as well as during military operations in particular. Using the foreign policy narrative of *double standards* (i.e., the West has no right to criticize Russia, as it operates in the same manner), Russian leadership feels no pressure to make its military less lethal than its American counterpart is. Moreover, on the level of the unconscious cultural context, the Russian people are ready to suffer and sacrifice by fighting to protect their country, and by demonstrating levels of endurance that cannot be grasped by Western mind. If readiness for such self-victimization is so high, it subsequently leaves no place for the concern for enemy civilian casualties. In other words, the suffering (and casualties) of the enemy's population during military operations is generally approved in the Russian cultural context, just as the Russians are ready to suffer themselves. Without internal political pressure, and with a resilient foreign policy, NLW do not pass this test of political necessity in Russia, and are consequently doomed to remain a thought-provoking, yet unused tool in the Russian military arsenal.

Chapter 4: "Never Again Shall Masada Fall!": Israel and Enemy Civilian Casualties

This chapter examines the reasons behind the failure of the RMA of NLW in the Israeli military. The Israeli Defense Forces (IDF) had been employing NLW significantly before the American conceptualization of these weapons in the mid-1990s. Since the end of the Six Day War, the IDF has been forced to fulfill law enforcement duties in the Territories (Judea, Samaria, and Gaza) and deal with civilian disturbances, massive riots, and other forms of civilian violence organized by the local population. Despite this long history of NLW employment, which was highlighted especially during the First and Second

Intifadas, the IDF has always thought of NLW as a tool for riot and crowd control only, and the idea of their employment on the battlefield has never been considered as a serious option. Facing international criticism, the Israeli political leadership has been insisting that international opinion is dominated by anti-Israeli countries, and, therefore, Israel's performance is irrelevant—it does not matter what Israel does (or does not do); it will be widely criticized. Israeli political decision-makers have successfully played *(good) few against many (bad)* card, arguing that the IDF is *the most ethical military in the world* and it does the utmost to reduce enemy civilian casualties, thus, fending off international criticism and tolerating the IDF's refusal to fundamentally change their practices. Moreover, the Israeli social and cultural context, based on the long Jewish history of persecution, the Holocaust heritage, and the history of the Arab-Israeli confrontation, pays little regard to enemy population life. The Israeli *siege mentality*, the *existential anxiety*, and the cultural narrative that *Never again shall Masada fall!* constitute the Israeli national character, shaping its intolerance toward enemy population and, therefore, stimulating no reason to minimize enemy civilian casualties less than it is politically required. With successfully defied international pressure, and with no internal political demand to minimize civilian casualties, the employment of NLW in the IDF remains limited to riot control missions with no role on the "real" battlefields.

Conclusion: Enemy Civilian Casualties: Do We Really Care?

The concluding part of the book examines why the impression of minimizing enemy civilian casualties in the contemporary political-cultural environment is more important than their actual minimization. This chapter discusses two different views on the phenomenon of declining violence in the twenty-first century, trying to explain the failure of military organizations to adopt NLW. The first is the idea presented by Steven Pinker, who argues the general decline of violence as a consequence of the development of human nature. The second is the concept presented by Ian Morris, who argues that the decline of violence is a pragmatic outcome of successful wars that helped to create bigger societies. While neither of these views suggest a hopeful interpretation of the failure of militaries to employ weapons that significantly minimize civilian casualties and collateral damage, the last part of the chapter offers a ray of optimism based on the analysis of the last military campaigns in Afghanistan, Iraq, and Syria.

NOTES

1. Steven Pinker, *The Better Angels of Our Nature: A History of Violence and Humanity* (London: Penguin Group, 2011).

2. Christopher Coker, *Humane Warfare* (London: Routledge, 2002), 2.

3. Christopher Coker, *The Future of War: The Re-Enchantment of War in the Twenty-First Century* (Oxford: BlackWell, 2004), 141.

4. Alvin Toffler and Heidi Toffler, *War and Anti-War: Survival at the Dawn of the 21st Century* (Boston: Little, Brown and Company, 1993), 134.

5. Dima Adamsky, *The Culture of Military Innovation: The Impact of Cultural Factors on the Revolution in Military Affairs in Russia, the US, and Israel* (Stanford: Stanford University Press, 2010).

6. See Homeland Security Research Cooperation (HSRC), *Non-Lethal Weapons: Technologies and Global Market—2012–2020*, 2011, http://www.homelandsecurityresearch.com/2011/10/non-lethal-weapons-technologies-global-market-2012-2020/ [accessed: 10 May 2018].

7. David Koplow, *Death by Moderation: The U.S. Military's Quest for Useable Weapons* (New York: Cambridge University Press, 2010), ix–x.

8. For example, Toffler and Toffler, *War and Anti-War*; Max Boot, *War Made New: Technology, Warfare and the Course of History, 1500 to Today* (New York: Gotham Books, 2006); Williamson Murray and MacGregor Knox (ed.), *The Dynamics of Military Revolution 1300–2050* (Cambridge: Cambridge University Press, 2001); Theo Farrell and Terry Terriff (ed.), *The Sources of Military Change: Culture, Politics, Technology* (London: Lynne Rienner, 2002).

9. Michael Roberts, "The Military Revolution, 1560–1660," in Clifford Rogers (ed.), *The Military Revolution Debate: Readings on the Military Transformation of Early Modern Europe* (Oxford: Westview Press, 1995), 13.

10. Williamson Murray and MacGregor Knox, "Thinking About Revolutions in Warfare," in Williamson Murray and MacGregor Knox (eds.), *The Dynamics of Military Revolution*, 1–10; Dima Adamsky, "Through the Looking Glass: The Soviet Military-Technical Revolution and the American Revolution in Military Affairs," *Journal of Strategic Studies* 31, no. 2 (2008): 257–94; Stephen Rosen, "The Impact of the Office of Net Assessment on the American Military in the Matter of the Revolution in Military Affairs," *Journal of Strategic Studies* 33, no. 4 (2010): 569–82.

11. Andrew Krepinevich, "Cavalry to Computer: The Pattern of Military Revolutions," *National Interest*, no. 37 (1994): 30.

12. Williamson Murray, "Thinking about Revolution in Military Affairs," *Joint Force Quarterly* 16 (1997): 69–76.

13. Murray and Knox, "Thinking about Revolutions in Warfare," 6–7, 12.

14. Murray, "Thinking about Revolution."

15. For example, Colin Gray, *Strategy for Chaos: Revolutions in Military Affairs and the Evidence of History* (London: Frank Cass Publishers, 2002); Richard Hundley, *Past Revolutions, Future Transformation: What Can History of Revolutions in Military Affairs Tell Us about Transforming the US Military?* (Santa Monica: RAND, 1999); Boot, *War Made New*; Adamsky, *The Culture of Military Innovation*; Murray and Knox, *The Dynamics of Military Revolution*.

16. For example, Jeremy Black, *War and Technology* (Bloomington: Indiana University Press, 2013); Martin Van Creveld, *The Age of Airpower* (New York: Public Affairs, 2011); H. R. McMaster, "On War: Lessons to be Learned," *Survival* 50, no. 1 (2008): 19–30; Tim Benbow, "Talking 'Bout Our Generation? Assessing the Concept of 'Fourth-Generation Warfare,'" *Comparative Strategy* 27, no. 2 (2008): 148–63.

17. Paul Davis, "Military Transformation? Which Transformation, and What Lies Ahead?," in Stephen Cimbala (ed.), *The George W. Bush Defense Program: Policy, Strategy, and War* (Washington: Potomac Books, Inc., 2010), 11.

18. For example, Stephen Metz and James Kievit, *Strategy and the Revolution in Military Affairs: From Theory to Policy* (Carlyle: US Army Strategic Studies Institute, 1995); Gray, *Strategy for Chaos*; Hundley, *Past Revolutions, Future Transformation*.

19. McMaster, "On War: Lessons to be Learned," 20.

20. For example, Adamsky, *The Culture of Military Innovation.*

21. Farrell and Terriff, *The Sources of Military Change*.

22. For example, Theo Farrell, *The Norms of War: Cultural Beliefs and Modern Conflict* (London: Lynne Rienner Publishers, 2005); Colin Gray, *Out of the Wilderness: Prime Time for Strategic Culture* (Fort Belvoir: Defence Threat Reduction Agency, 2006); Jeremy Black, *War and the Cultural Turn* (Cambridge: Polity Press, 2012).

23. For example, Stephen Rosen, "Military Effectiveness: Why Society Matters," *International Security* 19, no. 4 (1995): 5–31.

24. For example, Elizabeth Kier, *Imagining War: French and British Military Doctrine between the Wars* (Princeton: Princeton University Press, 1997); Beatrice Heuser, *Nuclear Mentalities? Strategies and Beliefs in Britain, France and the FRG* (London: MacMillan Press, 1998).

25. Ofer Fridman, "Revolutions in Military Affairs That Did Not Occur: A Framework for Analysis," *Comparative Strategy* 35, no. 5 (2016): 388–406.

26. Gregory Wilmoth, "False-Failed Innovation," *Joint Force Quarterly* 22 (1999): 51–57.

27. Hundley, *Past Revolutions, Future Transformation*, xiv.

28. John Stone, "The British Army and the Tank," in Farrell and Terriff (eds.), *The Sources of Military Change*; Williamson Murray, "Contingency and Fragility of the German RMA," in Murray and Knox (eds.), *The Dynamics of Military Revolution*; Robert Larson, *The British Army and the Theory of Armoured Warfare, 1918–1940* (London: Associated University Press, 1984).

29. Hundley, *Past Revolutions, Future Transformation*, xv.

30. Gray, *Strategy for Chaos*, 76.

31. Farrell and Terriff, *The Sources of Military Change*, 10.

32. Ibid., 7.

33. Adamsky, *The Culture of Military Innovation*, 141.

34. Fridman, "Revolutions in Military Affairs That Did Not Occur," 397.

35. Ibid., 402.

36. Williamson Murray and MacGregor Knox, "Future behind Us," in Murray and Knox (eds.), *The Dynamics of Military Revolution*, 180.

37. Colin Gray, *Perspectives on Strategy* (Oxford: Oxford University Press, 2013), 2.

38. Fridman, "Revolutions in Military Affairs that Did not Occur," 402–3.

39. Hundley, *Past Revolutions, Future Transformation*, 15.

40. Koplow, *Death by Moderation*, IX–X.

41. Tracy Tafolla, David Trachtenberg, and John Aho, "From Niche to Necessity: Integrating Non-Lethal Weapons into Essential Enabling Capabilities," *Joint Force Quarterly* 66 (2012): 72.

42. See Robert Mandel, *The Meaning of Military Victory* (London: Lynne Rienner Publishers, 2006); Roger Barnett, *Asymmetric Warfare* (Dulles: Brassey's, 2003); Michael Gross, *Moral Dilemmas of Modern War: Torture, Assassination, and Blackmail in an Age of Asymmetrical Conflict* (New York: Cambridge Press, 2010).

43. See Gil Merom, *How Democracies Lose Small Wars* (Cambridge: Cambridge University Press, 2003); Ivan Arreguin-Toft, *How the Weak Win Wars: A Theory of Asymmetric Conflict* (Cambridge: Cambridge University Press, 2005); Andrew Mack, "Why Big Nations Lose Small Wars: The Politics of Asymmetric Conflict," *World Politics* 27, no. 2 (1975): 175–200.

44. Eric Patterson, *Just War Thinking: Morality and Pragmatism in the Struggle against Contemporary Threats* (Lanham, MD: Lexington Books, 2007), 4–8.

45. Peter Van Aelst and Stefaan Walgrave, "New Media, New Movements? The Role of the Internet in Shaping the 'Anti-Globalization' Movement," *Information, Communication & Society* 5, no. 4 (2002): 465–93.

46. Ofer Fridman, "'Are We Ready for the Revolution of Nonlethal Weapons?': Using a Comprehensive RMA Model to Examine the Current Strategic Situation," *Comparative Strategy* 32, no. 3 (2013): 192–206.

47. Sjef Orbons, "Non-Lethality in Reality: A Defence Technology Assessment of its Political and Military Potential" (PhD Dissertation, The University of Amsterdam, 2013), 29.

48. Fridman, "'Are We Ready for the Revolution of Non-Lethal Weapons?'"

49. Tafolla, Trachtenberg, and Aho, "From Niche to Necessity."

50. Neil Davison, *"Non-Lethal" Weapons* (Eastbourne: Palgrave Macmillan, 2009), 1–3.

51. US Department of Defense, *Policy for Non-Lethal Weapons, Directive 3000.3E*, Washington, April 25, 2013.

52. Ibid.

53. For example, Davison, *"Non-Lethal" Weapons*; Gross, *Moral Dilemmas of Modern War*; James Carafano, "The Future of Anti-Terrorism Technologies," *Heritage Lectures*, no. 885 (2005); Orbons, "Non-Lethality in Reality"; Charlie Mesloh, Mark Henych, and Ross Wolf, *Less Lethal Weapon Effectiveness, Use of Force, and Suspect & Officer Injuries: A Five-Year Analysis. Report to the National Institute of Justice* (Florida Gulf Coast University, 2008).

54. For example, John Alexander, *Future War: Non-Lethal Weapons in 21st Century Warfare* (New York: St. Martin's Press, 1999); Nick Lewer and Steven Schofield, *Non-Lethal Weapons: a Fatal Attraction? Military Strategies and Technologies for 21st Century Conflict* (London: Zed Books, 1997); Koplow, *Death by Moderation*.

55. For example, David Koplow, *Non-Lethal Weapons: The Law and Policy of Revolutionary Technologies for the Military and Law Enforcement* (New York: Cambridge Press, 2006); Nick Lewer (ed.), *The Future of Non-Lethal Weapons: Technologies, Operations' Ethics and Law* (London: Frank Cass Publishers, 2002); Robert McNab and Scott Richard, "Non-Lethal Weapons and the Long Tail of Warfare," *Small Wars & Insurgencies* 20, no. 1 (2009): 141–59; David Fidler, "Non-Lethal Weapons and International Law—Three Perspectives on the Future," *Medicine, Conflict and Survival* 17, no. 3 (2001): 194–206.

56. For example, Tafolla, Trachtenberg, and Aho, "From Niche to Necessity"; Richard Scott, "Nonlethal Weapons and the Common Operating Environment," *Army Magazine* 50, no. 4 (2010): 21–26; Frank Siltman and John Frisbie, "Fire Support Just Got Harder: Adding Nonlethal Fires as a Core Competency," *Fires Bulletin* (2008, July–September): 6–8.

57. For example, V. Moiseev and V. Orlyanskii, "Oruzhiye neletal'nogo deystviya i printsipy taktiki [Weapons with Non-Lethal Action and the Principles of Tactics]," *Voennaia Misl'* 6 (2011): 26–33; A. Nagovitsyn, A. Grudzinskii', and A. Sporykhin, "Oruzhiye neletal'nogo deystviya i perspektivy yego ispol'zovaniya v interesakh sil Organizatsii Dogovora o kollektivnoy bezopasnosti [Weapons with Non-Lethal Action and the Perspectives of Their Employment in the Interest of Collective Security Treaty Organization]," *Voenaya Misl'*, no. 3 (2011): 51–59; V. Snigur, "Ovladet' gorodom bez zhertv i razrusheniy [To Capture a City without Casualties or Damage]," *Armei'skii' Sbornik*, no. 12 (2006): 50–52; A. Pronin, A. Leonov, and L. Kaplyarchuk, "Osnovnyye kriterii voyenno-ekonomicheskoy otsenki oruzhiya neletal'nogo deystviya [The Major Criteria of Military-Economic Evaluation of Weapons with Non-Lethal Action]," *Voennaia Misl'*, no.10 (2012): 43–50; V. Antipov and S. Novichkov, "K voprosu o razrabotke i primenenii neletal'nykh sredstv porazheniya na khimicheskoy osnove [Regarding the Question about the Development and Employment of Non-Lethal Means Based on Chemicals]," *Voenaya Misl'*, no. 9 (2009): 54–61.

58. For example, Uzi Ben-Shalom, "'Faser lematzav ham'em': mah'shevot al neshek pahot katlani be'sde hakrav ["Faser switched to Shock": Thoughts on Less Lethal Weapons on the Battlefield]," *Be'yabasha*, no. 15 (2010, September); Yitzhak Ben-Israel, "Neshel al-heger [Non Lethal Weapon]," *Ma'arachot* 363 (1999): 20–24; Guy-Zahar Shtultz, "Haf'alat koah medureget be'imutim besviva ezrahit [The Employment of Graduate Force in Conflicts within Civilian Environment]," *Ma'arachot*, vol. 438 (2011): 8–17.

59. Koplow, *Death by Moderation*, 198–99.

60. Davison, *"Non-Lethal"Weapons*, 209.

61. Boot, *War Made New*, 10.

62. Davison, *"Non-Lethal" Weapons*, 12–40.

63. See Stockholm International Peace Research Institute (SIPRI), *The Problem of Chemical and Biological Warfare. Volume I: The Rise of CB Weapons*, 1971.

64. Col. Richard H. Witherspoon, "Foreword," in Earl Tilford (ed.), *The Revolution in Military Affairs: Problems and Cautions* (Carlisle, PA: Strategic Studies Institute, US Army War College, 1995).

65. *Public Committee against Torture in Israel and Palestinian Society for the Protection of Human Rights and the Environment vs. The Government of Israel* (High Court of Justice 769/02, December 11, 2005), Paragraphs 34–37 (Hebrew).

66. Fridman, "Revolutions in Military Affairs That Did Not Occur."

Chapter One

Non-Lethal Weapons

The Revolution in Military Affairs That Did Not Occur

NLW—THE BACKGROUND

Non-Lethal Weapons—A Perplexing Definition

Defining "Non-Lethal Weapons" is not a simple task. The very idea of non-lethality sounds like an oxymoron, when applied to weapons. Different terms have been suggested to avoid this semantic contradiction: "less-lethal," "less-than-lethal," "disabling," "sub-lethal," "soft-kill," and others. While all of these refer to the same family of weapons, none can escape criticism.[1] This, in fact, is one of the main reasons for the variety of terms that have been used by different institutions in different countries. In the United States, the two most prevalent terms are "Non-Lethal Weapons" and "Less-Lethal Weapons," while in Russian military literature, the dominant term is *oruzh'ie neletal'nogo deistviya* (Weapons with Non-Lethal Action)[2]; the Russian Ministry of Interior has decided to avoid any connection with lethality, defining devices with non-lethal effects employed by the Russian Police and other Ministry of Interior Forces, by adopting the term *special'nie sredstva* (Special Means).[3] In Israel, there are three terms, used concurrently: *neshek pahot katlani* (Less-Lethal Weapons), *neshek al-hereg* (Non-Kill Weapons), and simply *emtzaim lepizur hafganot* (Riot Control Means).[4]

This variety of terms directly projects the perplexities of the attempt to fit together the concept of weapons with the idea of non-lethality. Nonetheless, this book adopts the term "Non-Lethal Weapons" (NLW),[5] as it focuses on the military (and not law enforcement) implementation of these weapons, and the professional military literature in all three countries of interest titles them in that way. The interesting reason for this similarity is the fact that the U.S. Department of Defense (DOD) was the first military organization to distinguish NLW from other armaments, starting to pay more coherent attention to NLW in the early 1990s and igniting the intensive discourse about the military implementation of NLW.[6] Since then, the DOD's definition states that NLW are weapons "that are explicitly designed and primarily employed to incapacitate personnel or materiel immediately, while minimizing fatalities, permanent injury to personnel, and undesired damage to property, facilities, materiel, and the environment."[7] This definition serves as a starting point in official and professional military literature not only in the United States, but also in Russia and Israel.

This commonly used definition, however, requires several clarifications. First, unlike their lethal counterparts that are designed to eliminate the threat through gross physical destruction caused by blast, penetration, or fragmentation, NLW prevent targets from functioning through temporary and reversible incapacitation.[8] Secondly, though the term "non-lethal" suggests that no fatalities can ever occur through an employment of NLW, this is not the case. By their definition, NLW are developed with "the intent to minimize the probability of producing fatalities, significant or permanent injuries . . . but do not, and are not intended to, eliminate risk of those actions entirely."[9] Similar to lethal weapons that do not always fulfil their purpose, their lethality depends on many factors (the level of the soldiers' professionalism, the state of the target, environmental conditions, etc.); NLW are also victims of these factors, and, despite their non-lethal nature, can, under certain circumstances, cause permanent damage or even death. Third, NLW do not exclude the probability of damage to property and the environment. In fact, there are certain systems that are deliberately designed to cause property damage in an attempt to reduce the probability of causing serious injuries or fatalities.[10] Fourth, unlike lethal weapons that are designed to cause irreversible damage to the target, NLW "are intended to have relatively reversible effects on personnel or materiel."[11] And finally, while there are many military actions that aim to reduce fatalities during military confrontations, such as Psychological Operations (PSYOPs) or cyber warfare (except for non-lethal "electronic attack"),[12] NLW are physical weapons, rather than activities, and therefore have to be considered accordingly.

A Brief History of Non-Lethal Weapons

The definition of NLW, as discussed above, is the fruit of American military thought of the early 1990s; however, the history of weapons, defined today as non-lethal, is long and complex. Probably the first known application of NLW can be dated to over 2,000 years ago, when the Chinese forces used ground pepper to temporary blind their enemies.[13] The modern history of

NLW can be traced as far back as 1912, when the French Police for the first time employed hand-grenades that dispersed an irritant chemical agent (ethyl-bromoacetate) to subdue "criminals and riotous individuals or groups."[14] During World War I, similar non-lethal irritant agents were the first chemical weapons to be used, before they were precipitously replaced by more lethal gases.[15] These two examples from the early history of NLW already point to the fact that non-lethal technologies and their employment policies have two different, but interconnected, story-lines—military and law enforcement.[16] While it is obvious that these two complement each other, creating the most comprehensive history of NLW development, the principal focus of this book is military, thus, its main emphasis is on the military record of NLW. Moreover, as this book is about the RMA of NLW, its main interest is the cases of the military implementation of NLW on the battlefield and during military operations, rather than civil disorder and riots, e.g., the British military employment of NLW in Northern Ireland or the IDF's usage of NLW in the West Bank.[17]

The history of the military implementation of NLW on the battlefield can be generally divided into four main periods: (1) World War I, (2) the inter-war period and World War II, (3) the Vietnam War and other military interventions during 1960s and 1970s, and (4) Post Cold War conflicts. It is important to clarify that until 1970s, the NLW employed by the different militaries were mainly irritant non-lethal chemical agents, that, on the one hand, increased the effectiveness of conventional lethal weapons;[18] but on the other, (due to their non-lethal nature) allowed bypassing legal issues related to the implementation of chemical weapons.

The first appearance of non-lethal chemical agents on the battlefield was very short. An interesting fact is that, while the question of which nation initiated chemical warfare in World War I ignited an acrimonious debate among French and German commentators in the early 1920s, very little attention was paid to the French employment of tear-gas hand-grenades. The use of irritant chemical agents on the battlefields of World War I was apparently initiated by a conscripted Parisian policeman who, upon returning from leave, brought back to the war-front some tear-gas hand-grenades. Though the significance of this incident is unclear, non-lethal irritants, undoubtedly, were the first chemical agents to attract the imagination of military commanders on both sides of the front. As war progressed, more effective ways to deploy gases were invented (e.g., artillery shells) and more lethal agents developed. The introduction of chlorine gas[19] in the Second (1915), and the later employment of mustard gas in the Third (1917) Battles of Ypres were milestones in the rapid military adaptation of more lethal chemicals that replaced previous non-lethal irritants.[20]

After their short appearance on the battlefields of World War I, chemical NLW remained in the focus of military research and development as integrated major parts of the inter-war chemical and biological warfare (CBW) programs.[21] For example, during the 1920s the U.S. Army Chemical Warfare Service (CWS) performed more research on CN (chloroacetophenone) than any other chemical agent, promoting civilian use of this irritant agent as an anti-riot police weapon. The broadness of the 1925 Geneva Protocol, which aimed to limit the usage of chemical weapons as a method of warfare, did not stop either development, or employment of chemical weapons, including non-lethal ones. Two such examples were the employment of artillery shells with tear-gas by Spanish governmental forces in 1936, during the Spanish Civil War; and the employment of artillery shells, air-bombs, and toxic candles consisting of CN by the Japanese forces in China in 1937 (although, CN was only a part of the vast variety of chemical agents employed by the Japanese, including highly lethal hydrogen cyanide, mustard gas, and others). Toward, and during, World War II large arsenals of chemical weapons (including chemical NLW) were produced and stockpiled by both the Allies and the Axis powers. These arsenals, however, remained unused for fear of retaliation and the bewildering lack of certainty in their military effectiveness.[22]

The period of the 1960s and the 1970s was, probably, the period of the greatest military employment of non-lethal chemicals. While CS (2-chlorobenzalmalononitrile) was invented and researched in the United States during the 1930s and the 1940s, it was the British military that in the mid-1950s concentrated its efforts on developing weapons based on this more effective non-lethal agent. The incomparable effectiveness of CS over CN was demonstrated by the British police as early as 1958–1959, however, both agents remained in the service of the British colonial police and military and were employed on, at least, 124 occasions during the short period between 1960 and 1965. CS munition became an integral part of the weapons arsenal of British troops sent abroad; for example, the British contingent deployed to Anguilla in 1969 was armed (in addition to conventional weapons) with some 2,000 CS cartridges and grenades.[23] While the British were the first to employ CS, it was the U.S. military in Vietnam that realized the concept of massive employment of this non-lethal agent. Despite the initial intention of the U.S. military to use CS only for riot control, CS based weapons were, almost immediately, widely employed in the form of various munitions, including hand-grenades, mortar and artillery shells, and even helicopter and aircraft canister dispensers.[24] The second part of the 1960s was the period with the most massive employment of CS in regular combat operations; between 1964 and 1969 the U.S. forces employed almost 7,000 tons of different forms of CS.[25] It is important to note that while the initial explanation of American officials for the use of chemical non-lethal agents was the humanitarian nature of these weapons and their ability to harmlessly separate between combatants and non-combatants,[26] CS based weapons very quickly became the normative choice among available weapons and were frequently used to make other fires more lethal.[27]

Due to the widespread criticism of the American employment of CS in Vietnam and uncertainty regarding the effectiveness of CS against troops equipped with gasmasks,[28] the 1980s were a period of relative stagnation regarding NLW in the military. Despite the fact that most NLW available to law enforcement at this time were the result of military-funded research, "[NLW] were given no real priority in context of Cold War military planning."[29] With the end of the Cold War there were two main factors that reshaped the attitude toward NLW. The first was the reevaluation of new political-military environments and the emerging understanding of top level Washington decision-makers regarding the enormous potential of NLW in future military conflicts. The second was a bottom-up process formulated within the growing demand for NLW from the U.S. operational commanders. The culminating point of this demand in the early 1990s was the operation "United Shield," the U.S. assistance to the withdrawing UN peacekeepers from Somalia, when the I Marine Expeditionary Force, due to a specific requirement "to pursue less-lethal alternatives in dealing with unarmed hostile elements," became a test-bed for military employment of new technologies designed "to fill the gap between verbal warning and deadly force."[30]

These two vectors drove the DOD to start paying more attention to NLW. The process of the institutionalization of NLW in the DOD started in 1990 when Richard Cheney, then the U.S. Secretary of Defense, became interested in the subject and, in March 1991, commissioned the Nonlethal Strategy Group (NSG) to examine it.[31] Based on the NSG report, the process was led by a Non-Lethal Weapons Steering Committee (NLWSC) established in 1994 and the Council on Foreign Relation; the process was finalized in 1996 with the establishment of the Joint Non-Lethal Weapons Program (JNLWP).[32]

While it is obvious that this new policy toward NLW, led by the United States, was almost immediately adopted by NATO;[33] more interestingly, the military institutions of other countries started to developed their own policies echoing the post–Cold War political military environment dominated by low intensity asymmetric conflicts involving civilian population and growing international concern regarding harming innocent civilians. The best example of this is the Russian "Conception of the Development and Employment of NLW" approved by the Military-Industrial Commission of the Russian Federation in 2005.[34]

The rapid rise of interest toward NLW in the early 1990s ignited an entirely new military, as well as academic, discourse, an outcome for which different centers of knowledge and research were established, such as the Institute for Non-Lethal Defense Technologies (INLDF) at the University of Pennsylvania,[35] the Bradford Non-Lethal Weapons Research Project

(BNLWRP) at the University of Bradford,[36] and the European Working Group Non-Lethal Weapons (EWG-NLW).[37] There is little doubt that in the last 20 years NLW have been at the center of a vast professional and academic debate, concerning their potential contribution to military success on the twenty-first century battlefield. It seems, however, that except for this debate, very little has been done within the military forces themselves. From the technological point of view, it is notable that "[NLW] that have been recently adopted by the military are primarily commercial off-the-shelf technologies, rather than the product of military sponsored research and development." Moreover, the spectrum of NLW that are currently employed "does not differ greatly from similar compilations 30 years previously."[38]

Furthermore, from the implementation point of view, the "operational use of available NLW by the military has been limited."[39] For example, the employment of NLW by the U.S. forces in Iraq was mostly in two very limited areas: compliance tools for controlling prisoners and the stopping of vehicles at check-points. The reports on, and the analyses of, two other military operations—the Russian forces in the Russia-Georgia War (2008) and the Israeli forces in Operation Cast Lead (2008–2009)—that were greatly criticized for the disproportional use of force and harming enemy civilian populations, mention nothing of NLW. Not only had NLW not been used, but there were not even plans to use them.[40] In 2004, an Independent Task Force sponsored by the U.S. Council on Foreign Relations stated that "currently, both military and civil leadership remain insufficiently familiar with the capabilities and limitations of NLW."[41] It seems that ten years later on, this situation has not been changed.

Non-Lethal Weapons—A Technological Brief

As it was mentioned in the introduction, this book does not accept the idea of technological determinism. However, organized around NLW, it has to provide a basic technological background for these types of weapons and their capabilities. As the main purpose of NLW is the temporary and reversible incapacitation or neutralization of the target, the main focus in the development of NLW is the Non-Lethal Effects (NLE) that these weapons cause to the target. The following discussion focuses on seven main types, or "families," of NLW according to their NLE: kinetic, chemical, biological, electrical, optical, acoustic, and directed energy. Another important division of NLW is the nature of the target, and includes two main categories: counter-personnel—NLW that affect people, and counter-materiel—NLW that affect machines of war and, related to them, equipment. In other words, each family of NLW can be divided to two: counter-personnel NLW and counter-materiel NLW (e.g., chemical counter-personnel NLW and chemical counter-materiel NLW).

Kinetic NLW

"Kinetic," "Low-Kinetic Impact," "Blunt Impact," or "Traumatic"—all these terms describe a family of impact projectiles that are intended to "induce pain, irritation and minimal injury to the subject without causing any life-threatening injuries."[42] First introduced in 1960s in the form of baton rounds made of teak wood, kinetic NLW experienced significant and rapid development creating vast variety in calibers (from 5.56 mm and 12 ga to 40 mm and above) and impact power (from 30 Joules to 100 Joules and even more); projectiles' material (plastic, rubber, foam, beans bags, and others) and shape (batons, rounds, and bullets); and the methods to shoot them (from the direct use of lethal platforms such as rifles, shotguns, and grenade launcher, through different attachments and adaptations to these lethal platforms, to the specially designed non-lethal rifles, pistols, and launchers).[43]

The NLE of kinetic NLW are directly related to the blunt trauma created by an immediate impact of the projectile. While this family of NLW is designed to merely cause superficial hurt rather than penetrate, causing more serious injury or even death, there are no assured guaranties, as the non-lethality (or lethality) of the kinetic NLW depends on many different variables. For example, striking an especially vulnerable person (young, old, or sick), or in a very sensitive or defenseless body part (eye, head, or neck) can cause permanent damage or even death. Another very influential variable is the difference between the effective distance of the projectile and the actual distance to the target (each kinetic NLW has a very specifically defined effective range, i.e., when the impact occurs at a shorter distance it might be lethal, and when it occurs at a longer distance it is not effective).[44] This operational disadvantage, however, does not stop kinetic NLW from being widely employed, as a way to overcome this disadvantage has been found in suitable training and appropriate rules of engagement.

The second part of kinetic NLW include the various counter-materiel technologies, which consists of different stopping devices (explicitly designed barriers, nets, etc.) that are intended to stop the vehicles and vessels containing kinetic power and, in this way, minimizing damage to their operators.[45]

Chemical and Biological NLW

Chemical, Biological, and Biochemical NLW are different non-lethal agents commonly dispersed in form of gases, smokes, aerosols, powders, and liquids that cause lacrimation, irritation, and inflammation or other types of temporary incapacitation.[46] While the most known agents are chloroacetophenone (CN), 2-chlorobenzalmalononitrile (CS) and oleoresin-capsicum (OC),[47] there are many other malodorants, calmatives, and other substances, such as Skunk, a liquid that creates a terrible odor of rot and sewage on whatever it touches.[48]

The spectrum of the possible delivery methods of chemical and biological NLW is vast. It can be generally divided into two main groups. The first is special methods, designed explicitly as NLW; these include water cannons, aerosol sprays, canisters and dispensers, a variety of non-lethal munitions and projectiles that consist of non-lethal chemical agents (hand grenades, pepper-balls, pneumatic and pyrotechnic explicitly designed pistols and rifles). The second group is non-lethal munition that consist of non-lethal chemical agents fired from conventional lethal weapon systems (rifles, shotguns, grenade launchers, mortars, artillery guns, and even tanks, helicopters, and jets).[49] The NLE of chemical and biological NLW are directly related to the spectrum of symptoms that these non-lethal agents cause. On the one hand, the more common CS, CN, and OC cause direct irritation to eyes, skin, and mucous membranes, creating burning and tearing of the eyes, flushing, tingling, and intense burning sensation of the skin, and the feeling of burning in the mouth, increased salivation, gagging, nausea, and vomiting.[50] On the other hand, there are other calmatives that cause completely different symptoms varying from unconsciousness to hallucinations.[51]

As most chemical and biological NLW are delivered by kinetic methods (i.e., bullets, rounds, grenades), their NLE might be complemented by blunt trauma (for example, FN-303).[52] Most chemical and biological NLW, however, are not designed for a kinetic impact with the target, but only for delivering the agent into the area. In the case of direct hitting, these munitions suffer from the same disadvantages as Kinetic NLW. Together with these counter-personnel chemical and biological NLW, there is a large group of counter-materiel agents that are intended to disable vehicles and prevent effective infrastructure functioning. These NLW include combustion modifiers, fuel and lubricant contaminants, and other substances that disable engines and vehicles. Against infrastructure might be used corrosives, abrasives, and depolymerization agents.[53] Despite the fact that theoretical literature considers these agents as NLW, most of them barely answer the criteria of the reversibility required of NLW and, in many cases, they might even be lethal to personnel. Consequently, despite the vast research programs in this field, there is no official employment of such counter-materiel chemical or biological agents in any modern military or law enforcement agencies.

Electrical NLW

Unlike many others, Electrical NLW do not originate from the military or law enforcement, but rather from farming and torture. The first stunning device was the *picana electrica*, an electrical version of the barbed cattle prod, introduced in Argentina in the 1930s and rapidly adopted by local police as a torture device during interrogations.[54] Since then the family of electric NLW has significantly expanded, including different types of electro-shocking devices, stun batons, and stun guns.[55]

The NLE of electrical NLW depend on implemented technology and can be generally divided into two groups: neuromuscular incapacitation and muscular pain. The first group includes different stun guns that fire harmless projectiles that deliver an electrical charge by means of very thin wires that temporary immobilize the target, as "the affected person normally falls to the ground due to inability to operate his or her legs."[56] The second group includes different "touch stun" devices (i.e., batons) with effects ranging from significant painful muscular sensations up to unconsciousness, but have no neuromuscular paralyzing effects. Despite this general division, there are certain stun guns that project electrical charge with wires (as in the first group), but create only muscular effect similar to "touch stun."[57]

The counter-materiel counterparts of these electrical NLW are different electrical devices for the temporary and reversible disabling of vehicles by targeting their electrical circuits and components (e.g., the Pre-Emplaced Electric Vehicle Stopper that stops and disables vehicles by providing an electrical pulse through deployed contacts that shut down power overloading electrical circuits and components).[58]

Optical and Acoustic NLW

While Acoustic NLW produce a direct beam of powerful acoustic energy, and Optical NLW create a light powerful enough to dazzle or temporary blind the target,[59] the NLE of both of these families are similar—disorientation and temporary incapacitation of the target. Acoustic NLW aim to employ powerful audio beams to create effects ranging from significant discomfort and nausea to a tickling in the mouth and a disturbance of physical equilibrium, without causing permanent damage to the hearing mechanism.[60] Despite significant research efforts from the early 1960s to the late 1990s, the technological difficulties of producing an effective, but also non-lethal, Acoustic NLW led researchers to generally dismiss this concept.[61] Only two non-lethal acoustic technologies have been developed. The first one is Acoustic Hailing Devices (AHD),[62] which, in fact, are not NLW, but devices that "provide scalable, directional warning tones or intelligible voice commands beyond 500 meters."[63] The second type of acoustic technology is a range of underwater acoustic guns. Interestingly, as acoustic beams behave differently under water than in the air, it seems more feasible to implement this technology in a non-lethal way. These systems, however, are still under development.[64]

Optical NLW aim to produce a powerful light that targets the eyes, creating their reflexive closure and saturating the vision cells, leading to a temporary loss of vision.[65] Despite significant improvement in optical technologies during the last decades, optical NLW proved inefficient, especially after the appearance of directed energy technologies.

While acoustic and optical NLW have struggled to find their ways as stand-alone technologies, their combination, in one NLW, proved very effective. Different "flash-bang" devices that combine a very powerful beam of light (a flash) and a painful, but short, sound pressure (a bang) have been in use for more than 30 years. The most common devices are hand grenades; there are, however, other different types of air-burst munitions, intended to produce these effects above the target.[66]

Directed Energy NLW

While Directed Energy is a general term that covers different types of technologies, which create "a beam of concentrated electromagnetic energy or atomic or subatomic particles," the directed energy weapon is "a weapon or system that uses directed energy to incapacitate, damage, or destroy enemy equipment, facilities, and/or personnel."[67] In general, the technology of directed energy weapons embraces two main areas: lasers (laser is an acronym for Light Amplification by Stimulated Emission of Radiation) and devices that produce radio frequency, microwave, or millimeter wave beams.[68] While the major military interest in directed energy weapons was in their potential as lethal weapons, different technological developments in the field have found their way to NLW.

The NLE of directed energy NLW, based on laser technology (so called "dazzling" lasers), are "temporary blindness (flash-blindness) or visual disturbance without permanent adverse effects on the eyes."[69] In the early 1990s, when "dazzling" lasers were first introduced, the immature technologies struggled to find the right balance between the type and the power of the devices and the safety concerns about possible permanent damage. The fruit of this labor is the fact that from the mid-2000s, different types of "dazzling" lasers have been successfully employed in different countries.[70]

Today, directed energy NLW, based on laser technology, is solely counter-personnel.[71] The research and development process, however, has not stopped; and there are several very promising developments in this field (e.g., the Pulse Energy Projectile, Wireless Electrical Weapons, and Thermal Lasers) that might expand the future arsenal of counter-personnel, as well as counter-materiel, NLW. Though these programs receive their funding mainly because of their potential as lethal weapons, there is still time for their non-lethal implications to be studied.[72]

The development of directed energy weapons that produce radio frequency, microwave, or millimeter wave beams also started as extremely classified military programs in the second part of the twentieth century. Only in the late

1980s and early 1990s had the concepts of possible non-lethal implementation emerged.[73] While today this technology is "perhaps the most intriguing new nonlethal weapons technology,"[74] and there are already different concepts and prototypes, the United States remains the only country that has ongoing (or, at least, unclassified) research programs in this field.

The counter-personnel NLW that is based on this technology is the Active Denial System (ADS) that produces an invisible and inaudible millimeter wave energy beam that causes targets to feel "as if they are being intensely burned."[75] The NLE of the ADS are based on beams that have a frequency of 95 gigahertz (GHz) that instantly heat the skin of the target at a depth of 0.3–0.4 millimeter to the temperature of 45–55 degrees Celsius. Permanent damage and burns that might be caused by these beams are prevented by very limited duration of exposure.[76] Although the ADS has already been proven safe in more than 11,000 tests and demonstrations on more than 700 humans, it has never been employed as "the newness of the technology, coupled with concerns over its mischaracterisation as a 'microwave weapon that fries, cook, or sterilize its targets,' resulted in a lack of willingness to employ it."[77]

The counter-materiel NLW that produce radio frequency, microwave, or millimeter wave beams are the Multi-Frequency Radio-Frequency Vehicle Stopper (MFRFVS) and the Radio Frequency Vessel Stopper (RFVS) that produce high power microwaves to disrupt the vehicle's or vessel's electrical components and cause their engines to stall.[78] These two systems are currently under development and there is no indication as to when they will be ready to be employed.

Another possible technology for counter-materiel directed energy NLW is Electromagnetic Pulse (EMP)—"the electromagnetic radiation from a strong electronic pulse, most commonly caused by a nuclear explosion that may couple with electrical or electronic systems to produce damaging current and voltage surges."[79] The potential of EMP to damage and destroy electrical equipment has attracted the military from the first time it was observed in high-altitude airburst nuclear tests. Since then, extensive work has been done in different countries on nonnuclear EMP[80]; however, to this day much of this work is classified and there is no evidence of emerging NLW based on EMP.

Other Non-Lethal Technologies

While these seven families embrace the most prevalent modern NLW, there are certain devices, systems, and technologies that are not covered by this umbrella. The first group of such NLW includes technologies and concepts that had been developed and even employed in the past, but proved to be inefficient. The best examples of these are different net-guns that were intended to capture and restrain individuals by deploying a net, and different foam-based weapons that were intended to create barriers by instantly employing very sticky foam.[81] Even though some of these systems had actually been employed on several occasions, they proved themselves as tactically impractical.[82]

The second group of NLW that are not included in this brief consists of different features or activities that despite their non-lethal nature, are not physically perceptible weapons. While some researchers include these features and activities within their discussion of NLW due to cultural, legal, or linguistic differences in the definition of NLW, others do it due to the lack of conceptualization. An example for the first type is the fact that the term *special'nie sredstva* ("special means"—the equivalent of NLW used by the Russian Ministry of Interior) does not comprise the word "weapons," which allows it to include, for example, specially trained police dogs.[83] An example for the second type is the misleading attempt to place Information Warfare under the definition of NLW.[84]

The last group includes a very limited list of very different, mainly counter-materiel, niches of NLW that due to their unique and original nature cannot be referred to by any of the families listed above. The best example of this type of NLW is cluster bombs (in which each bomblet is stuffed with 147 tiny spools of glass-like and highly conductive aluminum fibers) designed to cause reversible damage to electricity lines and transformation stations, taking them out of service for many hours or even days.[85]

NLW and International Law

While the idea that armed hostilities can be regulated by rules and laws is a part of "almost all societies, without geographical limitations,"[86] the history of the current, applicable, laws of war starts from the mid-nineteenth century when unilateral codes and bilateral agreements[87] started their transformation into open-ended multilateral treaties (i.e., open to later accession by other states).[88] The modern law of armed conflict (LoAC) is based on four cardinal principals: arms control regulations on specific weapons, discrimination, military necessity, and proportionality.[89] While the first principle deals with the prohibitions or restrictions on the development, production, stockpiling, transfer, or use of certain military technologies and weapons (Arms Control Regimes), the next three focus on the regulations and restrictions on the employment of permitted weapons (international humanitarian law).

As NLW are weapons by definition, they have to be subject to the full corpus of the LoAC and answer to all its principals. Consequently, it is not surprising that any comprehensive analysis of NLW includes their legal implications.[90] Moreover, the relations between the concept of "non-lethality" and the LoAC have inspired a vast debate in the field of international law.[91] Today, there are 19 different international treaties, laws, protocols and regulations that concern NLW[92] and the following discussion is divided into three main parts: NLW and international arms control regimes; NLW and international humanitarian law; and the impact of the LoAC on the employment NLW.

NLW and International Arms Control Regimes

The seven main families of NLW can be regrouped into three main categories according to international arms control regimes: chemical, biological and conventional.

Chemical NLW and International Arms Control Regimes

While the 1925 Geneva Protocol generally prohibited the use of "asphyxiating, poisonous or other gases, and all analogous liquid materials or devices,"[93] the more contemporary and detailed 1993 Chemical Weapons Convention (CWC) prohibits developing, producing, acquiring, stockpiling, retaining, transferring, or using chemical weapons.[94] It defines chemical weapons as "toxic chemicals and their precursors . . . which through [their] chemical action on life processes can cause death, temporary incapacitation or permanent harm to humans or animals."[95] Consequently, the counter-personnel chemical NLW, designed to incapacitate enemy troops, fall under this treaty. It is important to note that counter-materiel chemical NLW are not directly restricted by CWC, as long as their employment does not affect humans or animals (or the environment, as covered by international environmental laws), and, hence, these types of chemical weapons are not forbidden.[96]

The CWC also differentiates Riot Control Agents (RCA) from other chemical weapons. It defines RCA as "any chemical not listed in a Schedule, which can produce rapidly in humans sensory irritation or disabling physical effects which disappear within a short time following termination of exposure,"[97] and it permits their use for "law enforcement including domestic riot control purposes,"[98] but prohibits them "as a method of warfare."[99] Thus, it seems that while chemical counter-personnel NLW are permitted for law enforcement purposes, they are banned from the battlefield.

Interestingly, the CWC contains no definition for "methods of warfare" and while it undeniably bans counter-personnel NLW during inter-state conflicts, the legitimacy of their employment during Military Operations Other Than War (MOOTW)—peacekeeping operations, counter-terror operations, hostage rescue operations, etc.—is debatable. For example, the position of both the United States and Russia is that during MOOTW the use of RCA, and therefore counter-personnel chemical NLW, is unaffected by the CWC.[100]

Biological NLW and International Arms Control Regimes

While the same 1925 Geneva Protocol generally prohibited "bacteriological methods of warfare,"[101] the contemporary 1972 Biological and Toxin Weapons Convention (BTWC) bans developing, producing, stockpiling, acquiring, retaining or using biological weapons—"microbial or other biological agents, or toxins whatever their origin or method of production, of types and in quantities that have no justification for prophylactic, protective or other peaceful purposes."[102]

Unlike CWC, BTWC does not distinguish between different groups of biological weapons and its ban is absolute. Consequently, despite the promising potential of biological NLW described above, international law unconditionally prohibits their development and employment in any circumstances.

Conventional NLW and International Arms Control Regimes

In an addition to CWC and BTWC, both of which restrict or ban chemical and biological weapons, the LoAC includes the 1980 UN Conventional Weapons Convention[103] with its four Protocols. Protocol I prohibits the use of "any weapon the primary effect of which is to injure by fragments which in the human body escape detection by X-rays."[104] Protocol II prohibits use of land mines, body-traps and other devices that are designed "to kill, injure or damage and which are actuated by remote control or automatically after a lapse of time"[105] (this field is also regulated by the 1997 Land Mine Convention, which bans anti-personnel mines designed "to be exploded by the presence, proximity or contact of a person and that will incapacitate, injure, or kill one or more persons").[106] Protocol III prohibits making civil population the object of attack by incendiary weapons ("any weapon or munition which is primary designed to set fire to object or to cause burn injury to persons").[107] And Protocol IV bans the employment of "laser weapons specifically designed . . . to cause permanent blindness."[108]

These international legal regulations, written to regulate, restrict, or prohibit certain military technologies, concern the development and employment of NLW in various ways. For example, Protocol I has to be considered in the development of different kinetic counter-personnel NLW (this is one of the reasons for the metal core of many rubber, plastic, and other bullets and rounds); and Protocol IV had vast influence on the development of "dazzling" lasers.[109]

While these treaties refer only to very specific military technologies (including non-lethal ones), all others remain generally permitted for use. For example, there are no prohibitions on electrical, acoustic, optical, and electromagnetic NLW.[110] The employment, however, of all these permitted or regulated NLW (exactly as their lethal counterparts), has to be done in accordance with international humanitarian law.

NLW and Humanitarian Law

In addition to arms control regimes as mentioned above, there are three additional principles of the LoAC: discrimination, military necessity, and proportionality. These principles regulate the ways in which military force can be employed and are defined by different international treaties, conventions, and protocols. These documents, however, do not distinguish between different types of weapons, but set concrete principles by which any weapon can be employed. Consequently, NLW are subject to these international laws, just as any other lethal weapon or military device or platform.

Although there is no international law that directly uses the terms "proportionality," and "indiscriminate harm of civilians," the 1977 Additional Protocol I to the 1949 Geneva Convention is considered the main reference to the legal debate in this area.[111] The protocol states that:

> those who plan or decide upon an attack shall . . . take all feasible precautions in the choice of means and methods of attack with a view to avoiding, and . . . minimizing, incidental loss of civil life . . . [and] refrain from deciding to launch any attack which may be expected to cause incidental loss of civilian life . . . which would be excessive in relation to the concrete . . . military advantage anticipated.[112]

While this statement can be interpreted in different ways, it is clear that it argues against the unnecessary harming of civilians caused by indiscriminate military activity or disproportional use of force even in the case where there is a legitimate military necessity. In other words, anything beyond the minimum required force directed exclusively toward military objectives in a way that best minimizes possible damage to civilians, will be unnecessary, disproportional, or indiscriminate, or all of the above.

In fact, NLW seem to have no contradictions with these international humanitarian law principles, as they offer a solution placed somewhere "between bullhorns and bullets."[113] By their definition and essence, NLW are designed as weapons that minimize "fatalities, permanent injury to personnel, and undesired damage to property, facilities, materiel, and the environment."[114] And, while there are understandable legal problems with the indiscriminant characteristics of chemical NLW[115] (usage of which is anyway regulated by CWC), all other legitimate families of NLW seem to offer the required solutions to the demands of humanitarian law for selective and proportional attacks directed toward legitimate targets, according to legitimate military necessity.

The Impact of the LoAC on the Employment of NLW

While it seems that NLW are "new weapons coming in scope with the law of war,"[116] in fact, some of them have become the victims of this law. On the one hand it seems that "it is more humane to incapacitate enemy than kill him,"[117] on the other hand, a legitimate enemy can be legitimately killed.[118] The main purpose of NLW is not to prevent the legitimate killing of a legitimate enemy (although they are able to); their main purpose is to offer a flexibility of action in situations where lethal alternatives are less suitable for legal, military, economic, or political reasons. On the one hand, NLW are designed to minimize civilian casualties and property losses, especially in combat situations, where it is difficult to distinguish between combatants and non-combatants (e.g., urban warfare),[119] or in MOOTW, where the objectives contradict the harming of people that the military was sent to protect.[120] On the other hand, by minimizing collateral damage, NLW "deliberately target civilian non-combatants"[121] and, thus, are automatically exposed to legal criticism.

The legal paradox of NLW was best formulated by former U.S. Secretary of Defense Donald Rumsfeld, who complained to the House Armed Services Committee that "in many instances our forces are allowed to shoot somebody and kill them, but they're not allowed to use a non-lethal riot control agent."[122] Although his argument regards non-lethal riot control agents, the military employment of which is banned by CWC, it emphasizes the main legal problem of NLW: from the legal perspective it is sometimes simpler to wait for a situation that will allow the use of lethal force and legitimately kill combatants, as well as a number of non-combatants, than to employ NLW from the onset and accidentally injure or kill a single non-combatant. There are many explanations for this awkward and paradoxical relation between the LoAC and NLW, and there are certain scholars in this field who advocate radical changes in the existing system concerning NLW.[123] Until then, these changes will occur; however, the development and employment of NLW have to comply with the existing laws, or find a way to bypass them by the implementation of new technologies that are not regulated (e.g., electromagnetic), or by fully utilizing gaps in the current legislation (e.g., the fact that CWC has no definition for "methods of warfare" and, de facto, legitimizes[124] the usage of chemical counter-personnel NLW during MOOTW).

THE REVOLUTION IN MILITARY AFFAIRS OF NLW (THAT DID NOT OCCUR)

While the main purpose of this book is to explain the reasons behind the absence of the RMA of NLW, it is essential to understand what this RMA could or should be. In other words, it is important to create a general picture of the potential RMA of NLW as a reference for the analysis of its absence. This theoretical manipulation, however, might attract extensive criticism, as it attempts to portray phenomenon that did not occur. To avoid this criticism, the construction of this hypothetical RMA of NLW has to be based on three main pillars: the existing knowledge about the RMA process; extensive literature, written mainly in the late 1990s, that explicitly predicted the RMA of NLW and its characteristics[125]; and other, mainly official, documents that define a possible future military employment of NLW. Although these three allow creating a more or less detailed portrait of the RMA of NLW, it is important to keep it in very general lines to avoid the discussion, unnecessarily and unproductively, sliding into fiction.

As discussed in introduction, the RMA phenomenon consists of changes in three dimensions—military technology, military organization, and the concept of operations. Each of these dimensions is "a necessary, but not a sufficient, condition"[126] for an RMA, because only changes in all three have a potential to revolutionize military affairs. Thus, portraying the theoretical RMA of NLW has to concentrate on these three dimensions.

The Concept of Operations (CONOP)

CONOP is a general, but clear and concise, statement that expresses what the military intends to accomplish and how it will be done.[127] From the historical analysis of past RMAs it seems right to argue that the development of a new CONOP, and not the military technology in itself, is the first main step toward the RMA's implementation. For example, the tank, as an innovative military technology, initially appeared during World War I; however, it was the theory of the armored warfare that propelled the tank towards the RMA.[128]

While the theoretical and conceptual debate within the U.S. military started in the early 1990s,[129] the first official military document that attempted to define the CONOP of NLW was probably the U.S. Army's "Concept for Nonlethal Capabilities in Army Operations," published in 1996.[130] The main force behind this quest for new non-lethal technologies and the concepts of their employment was the emerging understanding that the post–Cold War security environment would demand military solutions that lie between diplomacy and force "where the destructiveness of conventional weaponry is too much and diplomacy is not enough."[131] In other words, due to the changes in global security and the political system, the military "must be capable of attaining its objectives and be politically usable," and therefore "armed forces equipped with both lethal and nonlethal weapons [have] greater political utility than those which only have lethal means."[132]

Defining the CONOP of NLW, the literature of the 1990s and early 2000s explicitly emphasized that NLW are equally applicable in MOOTW and major theatre wars, and they are capable in fulfilling a spectrum of missions, minimizing civil casualties and collateral damage. This early literature clearly stated that while the military would intend to accomplish its mission while avoiding any unnecessary damage (minimizing unacceptable political costs and pressures), NLW would offer a solution that is operationally attractive, legally defensible, ethically and morally favorable, technically feasible, and economically affordable.

Organization

An analysis of past RMA suggests two possible ways in which militaries fundamentally change their organizations. The first is the way of "reorganization," a process that reshapes the existing forces by building their capabilities according to the new CONOP. The second way emphasizes the establishment of new types of forces that, according to the new CONOP, are intended to fight on their own, fulfilling specific missions on the main battlefield, or fighting within an independent dimension.

Consequently, these two ways suggest two different possible changes in the organization of the military that implements the RMA of NLW: (1) the reorganization of the existing forces that integrate non-lethal capacities, and (2) the establishment of a new branch of non-lethal forces. While it is obvious that any new type of organization depends very much on the nature of military technology, the new CONOP plays a significantly greater role, as it defines how this technology is going to be employed. Consequently, the answer to the required change in military organization lies in the CONOP of NLW.

Early architects of the CONOP of NLW formulated two main characteristics that suggest the way in which military forces that intend to employ NLW successfully should be organized. The first is the fact that "nonlethal capabilities do not replace or diminish the role of lethal capabilities for the force . . . [they] simply provide the commander additional options for applying military force."[133] The second is the expectation from NLW to be employed by military forces simultaneously with their lethal counterparts, especially in such tactically challenging scenarios as urban warfare in major theater wars.[134]

Consequently, it seems right to assume that the new military organization demanded by the RMA of NLW is not the establishment of independent non-lethal forces, but an integration of non-lethal capabilities within existing conventional forces. After the implementation of the RMA of NLW, military units should be able to accomplish their stated or directed objectives exploiting the minimum required, but maximum necessary, force from the whole available arsenal of capabilities (from non-lethal to lethal).

Technology

An analysis of previous RMAs suggests that the technology that rested at the core of these RMAs usually was around for many years, and it was a new CONOP that ignited the process of change. The same analysis shows that with the implementation of a new CONOP and a new organizational structure, the technology also experienced significant transformation, adopting and improving according to the challenges created by the conceptual and the organizational changes. This process of technological change is understandable, as the specific characteristic of the new CONOP and the organizational challenges direct the R&D efforts for the existing technology.

The previous discussion on the non-lethal technologies focused not on the specific existing weapons system employed (or not) by the military, but rather on the technological nature of NLW and their capabilities. It is interesting, however, how the presented CONOP of NLW and the integrated organization (that suits that CONOP) could influence these technologies. The best vision of these new-generation NLW is, in fact, provided by the 2008 U.S. Army TRADOC Pamphlet *Force Operating Capabilities*: "[the military] must be provided with organic nonlethal capabilities . . . with multi-functional/multirole lethality options in integrated multipurpose system configurations."[135] In other words, the implementation of the RMA of NLW will demand integrated weapons and weapon systems that are capable of creating inbuilt lethal as well as non-lethal effects. While the potential integration of non-lethal capabilities within existing fire support conventional platforms, such as aircraft, artillery, and precision-guided munitions, has been vastly debated during the last decade,[136] the RMA of NLW demands more comprehensive integration of non-lethal capabilities, even within the weapon systems and firearms used by combatants on the ground (i.e., assault rifles).[137]

THE ABSENCE OF RMA OF NLW: THE FRAMEWORK FOR THE ANALYSIS

While technological determinism would immediately suggest that it is the immature technology of NLW that holds the RMA back, this argument can be easily fended off for two reasons. First, the process of taking a military technology and making it a revolutionary one is not a matter of science alone, and it involves a wide range of factors. For example, the tank technology of the 1920s was also immature for the RMA of armored warfare performed by the German military, and it was a combination of the German military culture, German national-cultural context, and Adolf Hitler who "recognized the potential of tank warfare more rapidly than some other national leaders of the day"[138] and pressed its military and industry to raise the tank into its adulthood. The second reason is that, as it will be shown in the following chapters, military industries have already demonstrated several developments that can answer the CONOP and organization of the RMA of NLW—the fact that this weaponry is not employed has not technology to blame. Therefore, the technological immaturity of NLW seems to be a consequential factor, and it appears that there is something much more fundamental that has led to the failure of the RMA of NLW.

The conceptual framework for the analysis of an RMA suggests three main questions that should be asked during the analysis of a failed RMA in a "certain affair":

1. Did the political leadership define this "certain affair" as a military challenge? If not, then what was the socio-cultural context that influenced the political-military leadership in its decision-making process?
2. If the challenge was defined, how did the military, shaped by its military culture, interpret this challenge? If it was interpreted as a challenge that the military can deal with, then what was the military cultural context that influenced the military leadership in its decision-making process?
3. If the challenge was defined by the military as unmet, did the political leadership support its military through the transformation? And if it did not, what was the socio-cultural context that prevented the political leadership from supporting it military?[139]

It is important to understand the nature and character of this "certain affair." In other words, it is important to define the character of the challenge that the political leadership put (or did not put) in front of its military. Consequently, the first step in the analysis of the RMA of NLW has to be the definition of the challenge that this RMA supposedly is intended to overcome. According to the definition of NLW, these weapons are intended to minimize civilian casualties, therefore the challenge that these weapons are intended to deal with is *a minimization of enemy civilian casualties*. Following the conceptual logic of the RMA's occurrence, it seems right to assume that the RMA of NLW would occur when:

1. A political leadership, shaped by certain socio-cultural circumstances, defines *the minimization of enemy civilian casualties during military operations* as a significant challenge.
2. The military, shaped by its military culture, interprets this challenge as unmet and dealing with it requires the RMA of NLW.
3. In return, the political leadership decides to supports this fundamental military transformation.

Keeping this understanding in mind, each one of the following analyses of the RMAs of NLW will be divided into three parts. The first part will analyze the status of NLW in each military, determining whether the RMA of NLW occurred or not. Only when the absence of this RMA is determined, will the analysis focus on the factors that had led to its absence.

The second part will discuss the political leadership and national socio-cultural predispositions that shaped the political attitude toward enemy civilian casualties. This part will consist of two different but highly interconnected areas: national foreign policy, and national character. In an attempt to analyze whether the political leadership defines *the minimization of enemy civilian casualties during military operations* as a significant challenge or not, it is important to understand the degree to which this leadership is pressed to do so. In other words, to what degree the political leadership is pressed (internationally as well as domestically) to demand from its military a significant reduction of civilian casualties and collateral damage. While there is broad agreement among scholars that international political pressure presents a crucial factor that leads to the reduction of killed or injured civilians and destroyed civil infrastructure,[140] an analysis of the existence of this pressure, its strength and roots, is beyond the scope of this book and therefore it will assume its existence, yet not delve into it. It seems very important, however, to understand how policy-makers comply with or fend off this pressure. In other words, to understand whether political leadership defines *the minimization of enemy civilian casualties during military operations* as a significant challenge, it is vital to understand how foreign policy is crafted within the context of international criticism, in general, and the criticism regarding civilian casualties during military operation, in particular.

An additional factor that influences political leadership in its decision-making process is, obviously, domestic political factors. Existing research clearly shows that each nation has its own "way of war" or "strategic culture" and that a nation's cultural context affects the political decision-making process.[141] An understanding of *what people think about war and enemy civilian casualties* will shed a light on the level of domestic political pressure (or lack thereof) to minimize collateral damage during military operations, consequently endorsing (or not) the political leadership to define *the minimization of enemy civilian casualties during military operations* as a significant challenge. While the analysis of this cultural context will focus on several narratives relevant to the idea of reducing violence applied toward enemy populations; it is important to note that this book analyzes a general national cultural character, i.e., the cohesive complex of specific social-physiological aspects of a given nation, rather than specific characters of individual groups or elites in each nation.

The third and the last part of the analysis will focus on military culture and the idea of NLW. The analysis of past RMAs shows that one of the major factors that produces an actual innovation is the extent to which militaries recognize and exploit the opportunities inherent in new tools of war.[142] Consequently, the military approach toward NLW seems to be an important factor in the analysis of the RMA of NLW, or its absence.

CONCLUSION

The purpose of the first part of this chapter was to shed light on the perplexing nature and history of NLW. It aimed to provide a general perspective on the idea of NLW and their technological roots, as well as on the legal implications and complications concerning their employment on the battlefield. While this historical, technical, and legal background was essential for the general understanding of NLW, the second and third parts were critical for the further analysis of the absence of RMA of NLW in the United States, Russia, and Israel.

The theoretical construction of the RMA of NLW (i.e., the RMA of NLW that could be) is crucial as a reference for the further analysis of its absence. Since the main objective of this book is to explain the failure of this RMA in the United States, Russia, and Israel, the first step in the analysis of each one of the cases is to establish that the RMA of NLW indeed did not occur. The CONOP, technology, and organization of the RMA of NLW, generated in this chapter, make the analysis of the argument that *the RMA of NLW did not occur* methodologically possible.

The last part of this chapter paved the methodological way for the analysis of the absence of the RMA of NLW. This process highlighted three main aspects that are relevant for the implementation of the RMA of NLW: (1) foreign policy and its attitude toward international criticism regarding collateral damage and civilian casualties; (2) domestic national cultural attitudes toward enemy civilian populations; (3) the military culture concerning novel technologies in general, and NLW in particular. These three elements are vital to the analysis of the required causal chain that could lead to the RMA of NLW, as only their combination directed toward the minimization of civilian enemy casualties during armed conflicts can create a required process that will ignite the RMA of NLW. In other words, only in cases where (1) *the state complies with the international criticism that requires reducing civilian casualties during conflicts*; (2) *domestic society demands reducing enemy civilian casualties*; and (3) *the military of this state has a culture that is able to promote the idea of NLW*—can the RMA of NLW be implemented.

NOTES

1. See Neil Davison, *"Non-Lethal" Weapons* (Basingstoke: Palgrave MacMillan, 2009), 1; Nick Lewer and Steven Schofield, *Non-Lethal Weapons: A Fatal Attraction?* (London: Zed Books, 1997), 5–6.

2. V. Selivanov, "Oruzhiye neletal'nogo deystviya kak sredstvo bor'by s terrorizmom, obespecheniya mirotvorcheskikh operatsiy i operatsiy po pravoprinuzhdeniyu [Non-Lethal Weapons as Means for Struggle against Terrorism, Provision on Peacekeeping and Law Enforcement Operations]," *Vestnik Rossii'skoi' Akademii Estestvennih Nauk*, no. 4 (2005): 1–11.

3. The Federal Law of the Russian Federation № 3-ФЗ, *O Politsii [About Police]*, Moscow, 1 March 2011.

4. See Roi Ben-Horin, "Non-Lethal Weapons: Theory, Practice, and What Lies Between," *Strategic Assessment* 2, no. 4 (2001): 21–27; Guy-Zahar Shtultz, "Haf'alat koah medureget be'imutim besviva ezrahit [The Employment of Graduate Force in Conflicts within Civilian Environment]," *Ma'arachot*, no. 438 (2011): 8–17.

5. Interestingly enough, the term Non-Lethal Weapons was probably adopted by the U.S. Military due to its political attractiveness, rather than the non-lethal nature of these weapons; see Steven Aftergood, "The Soft-Kill Fallacy," *The Bulletin of the Atomic Scientists* 50, Issue 5 (1994): 40–45.

6. Charles Swett, *Strategic Assessment: Non-Lethal Weapons*, Office of the Assistant Secretary of Defense for Special Operations and Low-Intensity Office, November 9, 1993.

7. U.S. Department of Defense, *Policy for Non-Lethal Weapons, Directive 3000.3E*, Washington DC, September 27, 2017.

8. John Alexander, *Future War: Non-Lethal Weapons in Twenty-First-Century Warfare* (New York: St. Martin's Press, 1999), 5.

9. U.S. Department of Defense, *Policy for Non-Lethal Weapons.*

10. Erik Nutley, *Non-Lethal Weapons: Setting Our Phasers on Stun? Potential Strategic Blessings and Curses of Non-Lethal Weapons on the Battlefield*, Center for Strategy and Technology Occasional Paper No. 34 (Maxwell Air Force Base, Alabama: Air University, 2003), 2–3.

11. U.S. Department of Defense, *Policy for Non-Lethal Weapons.*

12. Electronic attack is "Division of electronic warfare involving the use of electromagnetic energy, directed energy, or anti-radiation weapons to attack personnel, facilities, or equipment with the intent of degrading, neutralizing, or destroying enemy combat capability and is considered a form of fires." See U.S. Joint Chiefs of Staff, *Joint Publication 1–02, Department of Defense Dictionary of Military and Associated Terms*, 8 November 2010 (as amended through October 15, 2016), 87.

13. Alexander, *Future War*, 76–77.

14. Carol Ackroyd et al., *The Technology of Political Control*, Second Edition (London: Pluto Press Limited, 1980), 197–98.

15. Stockholm International Peace Research Institute (SIPRI), *The Problem of Chemical and Biological Warfare, Volume 1: The Rise of CB Weapons* (1971), 131–34.

16. See Davison, *"Non-Lethal" Weapons*, chapters 2, 3, 4.

17. See Sjef Orbons, "Non-Lethal Weapons: Peace Enables or Troublesome Force? Assessing the Role of CS and Baton Rounds in the Northern Ireland Conflict," *Small Wars & Insurgencies* 22, no.3 (2011): 467–94; Sarit Michaeli, *Crowd Control: Israel's Use of Crowd Control Weapons in the West Bank* (Jerusalem: B'tselem, 2013).

18. Davison, *"Non-Lethal" Weapons*, 17–18.

19. Chlorine gas is also an irritant, but it is significantly more lethal than previously employed agents and it is not considered a NLW.

20. SIPRI, *The Problem of Chemical and Biological Warfare, Volume 1*, 129–35.

21. Ronald Sutherland, "Chemical and Biochemical Non-Lethal Weapons,' *Stockholm International Peace Research Institute (SIPRI) Policy Paper*, no. 23 (2008): 1.

22. SIPRI, *The Problem of Chemical and Biological Warfare, Volume 1,* 59–60, 147, 332–35.

23. Ibid., 69–70, 212–13.

24. Dando, Malcolm, *A New Form of Warfare: The Rise of Non-Lethal Weapons* (London: Brassey's Ltd, 1996), 78–79.

25. SIPRI, *The Problem of Chemical and Biological Warfare, Volume 1,* 194.

26. Dando, *A New Form of Warfare*, 76–77.

27. Davison, *"Non-Lethal" Weapons*, 18.

28. Dando, *A New Form of Warfare*, 78–81.

29. Lewer, Schofield, *Non-Lethal Weapons: A Fatal Attraction?*, 34.

30. F. M. Lorenz, "Non-Lethal Force: The Slippery Slope to War?," *Parameters* XXVI, no. 3 (1996): 52.

31. Steve Metz, "Non-Lethality and the Revolution in Military Affairs," in Malcolm Dando (ed.), *Non-Lethal Weapons: Technological and Operational Prospects*, Jane's Special Report, November 2000, 13.

32. Davison, *"Non-Lethal" Weapons*, 52.

33. NATO, *NATO Policy on Non-Lethal Weapons*, Brussels, October 13, 1999.

34. A. Nagovitsyn, A. Grudzinskii', and A. Sporykhin, "Oruzhiye neletal'nogo deystviya i perspektivy yego ispol'zovaniya v interesakh sil Organizatsii Dogovora o kollektivnoy bezopasnosti [Weapons with Non-Lethal Action and the Perspectives of Their Employment in the Interest of Collective Security Treaty Organization]," *Voenaya Misl'*, no. 3 (2011), 52.

35. The Institute for Non-Lethal Defense Technologies (INLDF), https://www.arl.psu.edu/content/institute-non-lethal-defense-technologies-0 [accessed May 10, 2018].

36. Bradford Non-Lethal Weapons Research Project (BNLWRP), https://bradscholars.brad.ac.uk/handle/10454/3960 [accessed May 10, 2018].

37. European Working Group Non-Lethal Weapons (EWG-NLW), http://www.non-lethal-weapons.com/index.html, [accessed May 10, 2018].

38. Davison, *"Non-Lethal" Weapons,* 86, 209.

39. Ibid., 86.

40. For Israel, see Anthony Cordesman and Arleigh Burke, *The "GAZA WAR": A Strategic Analysis* (Washington D.C.: Center for Strategic & International Studies [CSIS], 2009); UN Human Rights Council, *Human Rights in Palestine and Other Occupied Arab Territories: Report of the United Nations Fact Finding Mission on the Gaza Conflict*, 2009; For Russia, see: the Independent International Fact-Finding Mission on the Conflict in Georgia, *Report*, 2009.

41. Graham Allison, Paul Kelley, and Richard Garwin, *Nonlethal Weapons and Capabilities: Report of an Independent Task Force* (New York: Council on Foreign Relations Press, 2004), 33.

42. Gary Vilke and Theodore Chan, "Less Lethal Technology: Medical Issues," *Policing: An International Journal of Police Strategies & Management* 30, no. 3(2007): 342.

43. See Davison, *"Non-Lethal" Weapons*, 29–21; Alexander, *Future War*, 89–93; David Koplow, *Death by Moderation: The U.S. Military's Quest for Usable Weapons* (New York: Cambridge University Press, 2010), 192; Joint Non-Lethal Weapons Directorate, *Non-Lethal Weapons (NLW): Reference Book*, June 30, 2011; A. Silnikov, "Vidy spetsial'nykh sredstv sostoyashchikh na vooruzhenii OVD, i ikh klassifikatsiya [Types and Classification of Special Equipment Adopted by the Russian Interior Ministry Units]," *Vestnik Sankt-Peterburgskogo universiteta MVD Rossii*, no. 3(51) (2011): 66–70; Michaeli, *Crowd Control*; National Research Council, *An Assessment of Non-Lethal Weapons: Science and Technology*, Washington, 2003, 24–26.

44. Koplow, *Death by Moderation*, 192

45. Joint Non-Lethal Weapons Directorate, *Non-Lethal Weapons (NLW): Reference Book*, 63–79.

46. Vilke and Chan, "Less Lethal Technology," 343; Sutherland, *Chemical and Biochemical Non-Lethal Weapons*, 9–10; Davison, *"Non-Lethal" Weapons*, 105–6; National Research Council, *An Assessment of Non-Lethal Weapons*, 26–28.

47. For more detailed information about these and other non-lethal chemical agents see: Eugene Olajos and Woodhall Stopford (ed.), *Riot Control Agents: Issues in Toxicology, Safety, and Help* (Boca Raton: CRC Press LLC, 2004).

48. Michaeli, *Crowd Control*, 35–37.

49. See Davison, *"Non-Lethal" Weapons*, 118–33; Joint Non-Lethal Weapons Directorate, *Non-Lethal Weapons (NLW): Reference Book*; Silnikov, "Vidy spetsial'nykh sredstv [Types and Classification of Special Equipment]"; Michaeli, *Crowd Control*; Michael Crowley, *Drawing the Line: Regulation of "Wide Area" Riot Control Agent Delivery Mechanisms under the Chemical Weapons Convention* (Manchester: Omega Research Foundation, 2013).

50. See Vilke and Chan, "Less Lethal Technology," 343–45; N. Chuvatkin, "Irritanty v sredstvakh neletal'nogo deystviya [Irritant in Non-Lethal Means]," *Proceeding to the International Conference "Special Means of Non-Lethal Action,"* Moscow, 2012, 39–42.

51. Sutherland, *Chemical and Biochemical Non-Lethal Weapons*, 18–19.

52. Joint Non-Lethal Weapons Directorate, *Non-Lethal Weapons (NLW): Reference Book*, 13.

53. Sutherland, *Chemical and Biochemical Non-Lethal Weapons*, 9–19; Alexander, *Future War*, 71–76.

54. See Darius Rejali, *Torture and Democracy* (Princeton: Princeton University Press, 2008).

55. Davison, *"Non-Lethal" Weapons*, 21–23; Alexander, *Future War*, 67; Koplow, *Death by Moderation*, 192–93; National Research Council, *An Assessment of Non-Lethal Weapons*, 32–33.

56. Alexander, *Future War*, 67; See also Davison, *"Non-Lethal" Weapons*, 22–23, 60–62; Koplow, *Death by Moderation*, 192–93; the official website of Taser International Inc.: http://www.taser.com [accessed May 10, 2018].

57. For example, see the official website of MARCH GROUP LLC, http://www.shoker.ru/ (Russian), [accessed May 10, 2018].

58. Joint Non-Lethal Weapons Directorate, *Developing Non-Lethal Weapons: The Pre-emplaced Electric Vehicle Stopper*, http://jnlwp.defense.gov/DevelopingNonLethalWeapons/PreEmplacedElectricVehicleStopper.aspx [accessed May 10, 2018].

59. Davison, *"Non-Lethal" Weapons*, 186 ; Alexander, *Future War*, 67.

60. Davison, *"Non-Lethal" Weapons*, 186–87.

61. Davison, *"Non-Lethal" Weapons*, 188–95; National Research Council, *An Assessment of Non-Lethal Weapons*, 31–32.

62. Or by another name: Long Range Acoustic Devices (LRAD).

63. Joint Non-Lethal Weapons Directorate, *Developed Non-Lethal Weapons: Acoustic Hailing Devices*, http://jnlwp.defense.gov/CurrentNonLethalWeapons/AcousticHailingDevices.aspx [accessed May 10, 2018].

64. Davison, *"Non-Lethal" Weapons*, 201–2; Joint Non-Lethal Weapons Directorate, *Non-Lethal Weapons (NLW): Reference Book*, 46–47.

65. Alexander, *Future War*, 67.

66. Davison, *"Non-Lethal" Weapons*, 97; Joint Non-Lethal Weapons Directorate, *Mission Payload Module—Non-Lethal Weapons System (MPM-NLWS) Fact Sheet*, October 2011; Silnikov, "Vidy spetsial'nykh sredstv [Types and Classification of Special Equipment]," 69.

67. US Joint Chiefs of Staff, *Joint Publication 1–02, Department of Defense Dictionary*, 79.

68. Davison, *"Non-Lethal" Weapons*, 143.

69. Ibid., 149.

70. Joint Non-Lethal Weapons Directorate, *Non-Lethal Weapons (NLW): Reference Book*; Silnikov, "Vidy spetsial'nykh sredstv [Types and Classification of Special Equipment]."

71. However, it is important to note that one of the most prevalent implementations of "dazzling" lasers is as a tool to stop a vehicle at a checkpoint by dazzling the driver. Consequently, these NLW might be considered as counter-materiel.

72. Davison, *"Non-Lethal" Weapons*, 154–62.

73. Ibid., 162–67.

74. Koplow, *Death by Moderation*, 196.

75. Ibid.

76. Davison, *"Non-Lethal" Weapons*, 167.

77. Tracy Tafolla, David Trachtenberg, and John Aho, "From Niche to Necessity: Integrating Non-Lethal Weapons into Essential Enabling Capabilities," *Joint Force Quarterly* 66 (2012): 76.

78. Joint Non-Lethal Weapons Directorate, *Future Non-Lethal Weapons,* http://jnlwp.defense.gov/FutureNonLethalWeapons.aspx [accessed May 10, 2018].

79. US Joint Chiefs of Staff, *Joint Publication 1-02, Department of Defense Dictionary*, 86.

80. Alexander, *Future War*, 65.

81. Ibid., 81–83.

82. Martin Stanton, "What Price Sticky Foam?," *Parameters* XXVI, no. 3: 63–68.

83. For example, Silnikov, "Vidy spetsial'nykh sredstv [Types and Classification of Special Equipment]."

84. For example, Alexander, *Future War*, 103–15.

85. William Arkin, "Smart Bombs, Dump Targeting?," *The Bulletin of the Atomic Scientists* 56, no. 3 (2000): 46–53.

86. Adam Roberts and Richard Guelff, "Introduction by the Editors," in Adam Roberts and Richard Guelff (eds.), *Documents on the Laws of War*, 3rd edition (New York: Oxford University Press, 2002), 3.

87. For example, the "Articles of War" decreed by King Gustavus II Adolphus of Sweden in 1621 for his army departing to fight Russians in the Baltic provinces included at least 7 articles (from 150) that can be considered as humanitarian laws. See Kenneth Orgen, "Humanitarian Law in the Articles of War Decreed in 1621 by King Gustavus II Adolphus of Sweeden," *International Review of the Red Cross*, no. 313 (1996): 438–42.

88. Roberts and Guelff, "Introduction," 4–7.

89. Koplow, *Death by Moderation*, 55–78.

90. For example, Alexander, *Future War*, 189–99; Koplow, *Death by Moderation*, 205–8; Nutley, *Non-Lethal Weapons*, 37–43; Sutherland, *Chemical and Biochemical Non-Lethal Weapons*, 5–7; Davison, *"Non-Lethal" Weapons*, 37–38, 68, 101–2.

91. See Kenneth Merget, "Non-Lethal Weapons and the Possibility of Radical New Horizons for the Laws of War: Why Kill, Wound and Hurt (Combatants) at All?," *Social Science Research Network*, July 1, 2008, http://papers.ssrn.com/sol3/papers.cfm?abstract_id=1295348 [accessed May 10, 2018]; David Fidler, "The International Legal Implications of 'Non-Lethal' Weapons," *Michigan Journal of International Law* 21, no. 51 (1999): 51–100; Eve Massingham, "Conflict without Casualties . . . A Note of Caution: Non-Lethal Weapons and International Humanitarian Law," *International Review of the Red Cross* 94, no. 886 (2012): 673–85; David Fidler, "Non-Lethal Weapons and International Law: Three Perspectives on the Future," *Medicine, Conflict and Survival* 17, no. 3, 194–206; Julian Robinson, "Non-Lethal Warfare and the Chemical Weapons Convention," *A Further HSP Submission to the OPCW Open-Ended Working Group on Preparations for the Second CWC Review Conference*, October 24, 2007, http://www.sussex.ac.uk/Units/spru/hsp/Papers/421rev3.pdf [accessed May 10, 2018]; David Fidler, "The Meaning of Moscow: 'Non-Lethal' Weapons and International Law in the Early 21st Century," *International Review of the Red Cross* 87, no. 857 (2005): 525–52; Michael Gross, *Moral Dilemmas of Modern War: Torture, Assassination, and Blackmail in an Age of Asymmetric Conflict* (New York: Cambridge University Press, 2010), 77–99.

92. Nutley, *Non-Lethal Weapons*, 39–40.

93. Protocol for the Prohibition of the Use in War Asphyxiating, Poisonous or Other Gases, and of Bacteriological Methods of Warfare (adopted June 17, 1925, entered into force February 8, 1928), 94 L.N.T.S. 65.

94. Convention on the Prohibition of the Development, Production, Stockpiling and Use of Chemical Weapons and on Their Destruction (adopted January 13, 1993, entered into force April 29, 1997), 32 ILM 800, Article I (1).

95. Convention on the Prohibition of the Development, Production, Stockpiling and Use of Chemical Weapons and on Their Destruction, Article II (1,2).

96. See Fidler, "The International Legal Implications," 70.

97. Convention on the Prohibition of the Development, Production, Stockpiling and Use of Chemical Weapons and on Their Destruction, Article II (7).

98. Convention on the Prohibition of the Development, Production, Stockpiling and Use of Chemical Weapons and on Their Destruction, Article II (9).

99. Convention on the Prohibition of the Development, Production, Stockpiling and Use of Chemical Weapons and on Their Destruction, Article I (5).

100. Robinson, "Non Lethal Warfare."

101. Protocol for the Prohibition of the Use in War Asphyxiating, Poisonous or Other Gases, and of Bacteriological Methods of Warfare.

102. Convention on the Prohibition of the Development, Production, and Stockpiling of Bacteriological (Biological) and Toxin Weapons and on Their Destruction, (adopted 10 April 1972 and entered into force on 26 March 1975), 11 ILM 309, Article 1.

103. The full name is - Convention on Prohibitions or Restrictions on the Use of Certain Conventional Weapons Which May be Deemed to be Excessively Injurious or to Have Indiscriminate Effects (adopted October 10, 1980, and entered into force December 2, 1983), 19 ILM 1523.

104. Protocol on Non-Detectable Fragments (Protocol I), (adopted October 10, 1980, entered into force December 2, 1983), 19 ILM 1529.

105. Protocol on Prohibitions or Restrictions on the Use of Mines, Booby-Traps and Other Devices (Protocol II), (initially adopted October 10, 1980, entered into force December 2, 1983, review adopted in May 3, 1996, entered into force December 3, 1998), 35 ILM 1206.

106. Convention on the Prohibition of the Use, Stockpiling, Production and Transfer of Anti-Personnel Mines and on Their Destruction, (adopted September 18, 1997, entered into force March 1, 1999) 36 I.L.M 1507.

107. Protocol on Prohibitions or Restrictions on the Use of Incendiary Weapons (Protocol III), (adopted October 10, 1980, entered into force December 2, 1983), 19 ILM 1535.

108. Protocol on Blinding Laser Weapons (Protocol IV), (adopted October 13, 1995, entered into force July 30, 1998), 35 ILM 1218.

109. See Fidler, "The International Legal Implications," 68–69; Davison, *"Non-Lethal" Weapons*, 149–54.

110. Interestingly enough, already in 1994 there were calls for the development of specific new protocols covering electromagnetic weapons. See Barbara Rosenberg, "'Non-Lethal' Weapons May Violate Treaties," *The Bulletin of the Atomic Scientists* 50, no. 5 (1994): 44–45.

111. Michael Gross, "The Second Lebanon War: The Question of Proportionality and the Prospect of Non-Lethal Warfare," *Journal of Military Ethics* 7, no. 1 (2008): 1–22.

112. 1977 Additional Protocol to the Geneva Conventions of 12 August 1949, and relating to the Protection of Victims of International Armed Conflicts (Protocol I), (adopted 8 June 1977 entered into force 7 December 1978), 16 ILM 1391, Article 57(2/A/II-III).

113. Koplow, *Death by Moderation*, 190.

114. U.S. Department of Defense, *Policy for Non-Lethal Weapons*.

115. See Stephen Coleman, "Discrimination and Non-Lethal Weapons: Issues for the Future Military," in David Lovell and Igor Primoratz (eds.), *Protecting Civilians during Violent Conflict: Theoretical and Practical Issues for the 21st Century* (Surrey: Ashgate Publishing Limited, 2012), 215–29; Fidler, "The Meaning of Moscow."

116. Robin Coupland and Dominique Loye, "Legal and Health Issues: International Humanitarian Law and the Lethality or Non-Lethality of Weapons," in Dando, *Non-Lethal Weapons*, 62.

117. Fidler, "The International Legal Implications," 59.

118. Chris Jochnik and Roger Normand, "The Legitimation of Violence: A Critical History of Laws of War," *Harvard International Law Journal* 35, no. 1 (1994): 49–95.

119. Koplow, *Death by Moderation*, 202; Jez Littlewood, "NLWs and Urban Warfare: Aspects of US Thinking," in Dando, *Non-Lethal Weapons*, 21–31.

120. Charles Heal, "The Quest for the Magic Bullet," in Dando, *Non-Lethal Weapons*, 32–42.

121. Gross, "The Second Lebanon War," 15.

122. Quoted in Brad Knickerbocker, "The Fuzzy Ethics of Nonlethal Weapons," *Christian Science Monitor*, February 14, 2003, http://www.csmonitor.com/2003/0214/p02s01-usmi.html [accessed May 10, 2018].

123. See Fidler, "Non-Lethal Weapons"; Merget, "Non-Lethal Weapons."

124. Although with extensive explanatory efforts; see Robinson, "Non Lethal Warfare."

125. Metz, "Non-Lethality and the Revolution in Military Affairs," 10–20; Lewer and Schofield, *Non-Lethal Weapons: A Fatal Attraction?*, 17; Gerrard Quille, "The Revolution in Military Affairs Debate and Non-Lethal Weapons," in Nick Lewer (ed.), *The Future of Non-Lethal*

Weapons: Technologies, Operations, Ethics and Law (London: Frank Cass Publishers, 2002); Douglas Lovelace and Steven Metz, *Nonlethality and American Land Power: Strategic Context and Operational Concepts* (Carlisle: Strategic Studies Institute, 1998); Joseph Siniscalchi, "Non-Lethal Technologies: Implications for Military Strategy," *Center for Strategy and Technology Occasional Paper*, no. 3 (1998).

126. Krepinevich, Andrew, "Cavalry to Computer: The Pattern of Military Revolutions," *National Interest*, no. 37 (1994): 30–42.

127. U.S. Joint Chiefs of Staff, *Joint Publication 1-02, Department of Defense Dictionary*, 51.

128. Ofer Fridman, "Revolutions in Military Affairs that Did Not Occur: A Framework for Analysis," *Comparative Strategy* 35, no. 5 (2016): 388–406.

129. See: Swett, *Strategic Assessment: Non-Lethal Weapons*; Davison, *"Non-Lethal" Weapons*, 52.

130. Department of the Army Headquarters, United States Army, *Concept for Nonlethal Capabilities in Army Operations: TRADOC Pamphlet 525-73*, Fort Monroe VA, December 1, 1996; Siniscalchi, *Non-Lethal Technologies: Implications for Military Strategy*.

131. Department of the Army Headquarters, United States Army, *AirLand Operations: A Concept for the Evolution of AirLand Battle for the Strategic Army of the 1990s and Beyond: TRADOC Pamphlet 525-5*, Fort Monroe, August 4, 1991; See also Mark Thomas, *Non-Lethal Weaponry: A Framework for Future Integration* (Maxwell Air Force Base: Air Command and Staff College, 1998); John Barry, Michael Everett, and Allen Peck, "Nonlethal Military Means: New Leverage for a New Era," *Harvard University, John F. Kennedy School of Government, National Security Program Policy Analysis Paper*, No. 94-101, 1994; Lovelace and Metz, *Nonlethality and American Land Power*; Alexander, *Future War*, 9–56; Lewer and Schofield, *Non-Lethal Weapons: A Fatal Attraction?*, 5–23.

132. Lovelace and Metz, *Nonlethality and American Land Power*, 6.

133. Department of the Army Headquarters, United States Army, *Concept for Nonlethal Capabilities in Army Operations*, 3.

134. Lovelace and Metz, *Nonlethality and American Land Power*, 25–27; Littlewood, "NLWs and Urban Warfare"; Siniscalchi, *Non-Lethal Technologies: Implications for Military Strategy*.

135. Department of the Army Headquarters, United States Army, *Force Operating Capabilities: TRADOC Pamphlet 525-66*, Fort Monroe, March 7, 2008, 88.

136. For example, Justin Bobb, *Non-Lethal Weaponry: Applications to AC-130 Gunships*, (Maxwell Air Force Base, Alabama: Air University, 2002); Frank Siltman and John Frisbie, "Fire Support Just Got Harder: Adding Nonlethal Fires as a Core Competency," *Fires Bulletin*, No. 6, July–September 2008, 6–8; Siniscalchi, *Non-Lethal Technologies: Implications for Military Strategy*, 10.

137. The combination of simultaneously available non-lethal and lethal capabilities in assault rifles is not only technologically feasible, but has been already demonstrated by defense industries and even employed by the U.S. military. The two best examples are: (1) The M26 Modular Accessory Shotgun System (MASS)—under-barrel shotgun attachment for the M16/M4 assault rifles that offers simultaneous capability of the non-lethal 12 ga munitions and 5.56 mm lethal one (see Joint Non-Lethal Weapons Directorate, *Annual Report 2012: Non-Lethal Weapons for Complex Environments*, 2012, 9); (2) The FN 303 Mount—under-barrel compressed-air powered launcher attachment for the M16/M4 assault rifles that offers capability of the non-lethal 0.68 calibre, 8.5 gram, fin-stabilized, non-lethal projectiles and the simultaneous lethal capability of the 5.56 mm service rifle (see Joint Non-Lethal Weapons Directorate, *Non-Lethal Weapons [NLW]: Reference Book*, 13–14).

138. Citino, Robert, *The Quest for Decisive Victory: From Stalemate to Blitzkrieg in Europe 1899–1940* (Lawrence: University Press of Kansas, 2002), 204.

139. Fridman, "Revolutions in Military Affairs That Did not Occur," 402–3.

140. For example: Gil Merom, *How Democracies Lose Small Wars* (Cambridge: Cambridge University Press, 2003); Ivan Arreguin-Toft, *How the Weak Win Wars: A Theory of Asymmetric Conflict* (Cambridge: Cambridge University Press, 2005); Eric Patterson, *Just War Thinking: Morality and Pragmatism in the Struggle against Contemporary Threats* (Lanham, MD:

Lexington Books, 2007); Koplow, *Death by Moderation*; Gross, *Moral Dilemmas of Modern War*.

141. For example, Colin Gray, *Modern Strategy* (Oxford: Oxford University Press, 1999); Lawrence Sondhaus, *Strategic Culture and Ways of War* (New York: Routledge, 2006); Beatrice Heuser, *Nuclear Mentalities? Strategies and Beliefs in Britain, France and the FRG* (London: MacMillan Press, 1998); Jeremy Black, *War and the Cultural Turn* (Cambridge: Polity Press, 2012); Theo Farrell and Terry Terriff (eds.), *The Sources of Military Change: Culture, Politics, Technology* (London: Lynne Rienner, 2002).

142. Max Boot, *War Made New: Technology, Warfare and the Course of History, 1500 to Today* (New York: Gotham Books, 2006), 10.

Chapter Two

From the Frontier to the "Protector of the Free World"

The United States and Enemy Civilian Casualties

As was discussed in the previous chapter, the contemporary understanding of NLW is a fruit of American military thought of the early 1990s. During this time, the American military was mesmerized by theorizing Revolution in Military Affairs (RMA), an implementation of new military technologies combined with fundamental shifts in military doctrine and organization. Speculations about new military technologies that have revolutionary potential did not overstep NLW; for example, the Washington-based Center for Strategic and International Studies held in 1993 that: "If U.S. forces were able . . . to incapacitate or render ineffective enemy forces without destroying or killing them, the United States conduct of war would be revolutionized."[1]

In 2009, however, almost two decades after the initial rise of NLW, the U.S. Government Accountability Office (GAO) stated that "The joint non-lethal weapons program has conducted more than 50 research and development efforts and spent at least $386 million since 1997, but it has not developed any new weapon."[2] This difference between the initial enthusiasm and inability to deliver results is striking, and it is not surprising that the GAO's report points to the Department of Defense (DOD) vain management as the main reason for its failure to successfully develop and employ NLW within the U.S. military. While the employment of NLW, undoubtedly, demands a paradigm shift in military affairs, the GAO's accusation of the DOD's mismanagement seems to be a too simplistic explanation of the detrimental situation with NLW in the U.S. military. As it was discussed before, successful implementation of fundamentally new military technologies requires a particular political and cultural environment that promotes and supports the development and employment of these weapon systems. Subsequently, it seems that the roots of this astonishing gap between early expectations of NLW and their actual deployment are much deeper than a simple financial misconduct by the American military establishment.

THE STATUS OF NON-LETHAL WEAPONS IN THE UNITED STATES

As it was discussed in the previous chapter, the U.S. military had been employing NLW (mainly chemical) significantly before the concept of NLW was born in the early 1990s. However, due to the international criticism that followed the vast employment of chemical NLW in Vietnam and the Cold War–oriented military planning, which left little space to such niche technologies as NLW, the development of these technologies was mainly done by, and for, law enforcement agencies.[3]

The U.S. military started to pay more coherent attention toward NLW in 1995, during the operation "United Shield," the U.S. assistance for withdrawing UN peacekeepers from Somalia. General Anthony Zinni, who commanded the mission, became the prime advocate for the development and employment of these weapons, clearly stating that he "would never field a peace-support operation again without them."[4] The process of the institutionalization of NLW in the DOD was led by a Non-Lethal Weapons Steering Committee (NLWSC), established in 1994, and the Council on Foreign Relations. This process was finalized in 1996 with the establishment of the Joint Non-Lethal Weapons Program (JNLWP)[5] and the DOD Directive 3000.3, *Policy for Non-Lethal Weapons*, that defined NLW as "weapons that are explicitly designed and primarily employed so as to incapacitate personnel or materiel, while minimizing fatalities, permanent injury to personnel, and undesired damage to property and the environment."[6]

While, since 1996, JNLWP has been responsible for identifying and developing NLW into operationally suitable and effective solutions and for facilitating the acquisition and fielding of NLW,[7] it seems that very little innovation has been done.[8] Most of the NLW approved by JNLWP and employed within different American forces have been based on the existing solutions previously developed for law enforcement, such as small arms kinetic, electric and chemical weapons.[9] The only state-of-art technology, development of which has been led by JNLWP, is the Active Denial System (directed energy weapons that produce microwave beams).[10] Despite the fact that several systems were shipped for deployment in Afghanistan, due to operational difficulties and fear of the mischaracterization of the system as a "microwave" weapon that "fries, cooks, or sterilizes its targets," this flagship of the JNLWP's activity and investment has never been used.[11]

Interestingly enough, military theoretical interest in NLW has never faded since the late 1990s, the peak of the NLW enthusiasm. During the last decade, American officers conducted a vast amount of research considering NLW, their possible implications, and ways of deployment.[12] Nevertheless, it is very hard to say that anything from what was suggested in this purely academic research has ever been materialized. For example, on the one hand the U.S. Air Force (USAF) had expressed its initial interest in NLW already in 1999, when Major General Norton Schwartz, then the USAF Director of Strategic Planning, said: "We have yet to explore non-lethals from the air. This is the next logical step."[13] However, despite his statement, the only airborne NLW that has since been discussed by JNLWP is an Airborne Active Denial System—a system that was not taken further from the stage of concept.[14]

From very beginning, the role of NLW has been understood by the American military establishment as a tool to answer the blurring line between "law enforcement—with its restrain on the use of force—and the military"[15]—an approach that explains well the reason why JNLWP has concentrated mainly on the existing law enforcement solution. In 1991, the very beginning of the NLW debate, a research survey done by the Policy Planning Staff at the U.S. Department of State reported that NLW "will have to be carefully integrated with currently-existing technologies and systems to achieve their maximum potential."[16] More than 20 years later, it seems that very little has been done, as most of the NLW listed in the JNLWP's *Non-Lethal Weapons Reference Book* are still purely law enforcement weapons and munitions.[17] Moreover, Lieutenant General Richard Tryon, the Deputy Commander of the Marine Corps and the Chairman of the JNLWP Integrated Product Team, repeated in 2012 (20 years after the Policy Planning Staff), the very same narrative: "Non-Lethal Weapons should be compatible with existing [military] weapons/platforms."[18]

Though some integrated (lethal and non-lethal) platforms have been introduced recently, they still remain in the area of infantry weapons only. The best example is the M26 Modular Accessory Shotgun System (MASS). It is an under-barrel shotgun attachment for the M16 rifle that, while preserving the lethal capability of the main rifle, simultaneously provides a warfighter with an additional capability of 12 ga non-lethal ammunition.[19] Interestingly enough, this integration was not a fruit of JNLWP activity, but a product of the U.S. Army Program Executive Office Soldier.[20]

To summarize the current NLW situation in the United States, it is important to emphasis two main factors. The first one is the nature of the employed NLW and the level of their employment. The range of the NLW employed by the U.S. military does not much differ from the law enforcement technologies that have been available for the last four decades.[21] Excluding the Active Denial technology (that, in fact, has yet to be deployed), the JNLWP has been concentrating on existing commercially available NLW developed for law enforcement purposes, or on developing munitions with enhanced effects for these existing systems.[22] The operational use of these available NLW has been limited to specific out-of-battlefield missions, such as crowd and riot control, security patrols, checkpoints, etc.

The second factor is the technological status. Since World War II, the U.S. military has been a technological leader in military affairs, and the American military industrial complex has been able to deal with all technological challenges posed in front of it, but "without concepts for the use of Non-Lethal Weapons, developers will not be successful in focusing ideas and programs."[23] The example of the M26 MASS shows that facing a right definition of the desired capabilities, which focuses research and development efforts, American manufacturers are able to produce effective military-oriented NLW weapons. Taking NLW out of their law enforcement niche and creating technologies that will answer military requirements should not pose an enormous technological gap; it is mainly a question of political pressure put on the military establishment to make things right—i.e., minimize civilian casualties to the lowest possible number.

AMERICAN FOREIGN POLICY AND INTERNATIONAL CRITICISM

Much it has been thought and written about U.S. foreign policy. Different scholars have discussed it from different perspectives, developing different ways of analysis. Some decided to divide the history and development of American foreign policy by administrations, analyzing the approach of specific presidents,[24] while others examined different persistent narratives of U.S. foreign policy through history.[25] Some focused more on the relations between the United States and its main allies[26]; and others focused on American foreign policy and its main foes.[27] Some of these researchers generally praised American performance[28] and some others were more critical.[29] The main purpose of this chapter, however, is not to take a side in this highly intellectual debate, but rather to focus on the main inherent mechanisms of U.S. foreign policy intended to deal with international criticism, in general, and regarding civilian casualties caused by the American military, in particular.

Though the main emphasis of this chapter is on current affairs and how the American policy-makers deal with international criticism, in the second decade of the twenty-first century, it seems important to begin this analysis with different trends and narratives developed during the period of the Cold

War, as an understanding of these trends is vital to the analysis of contemporary American behavior in the international arena.[30]

At the end of World War II, while the European countries, the Soviet Union, and Japan—the great powers of World War II—were devastated, America emerged from the war relatively unscathed, finding itself as the dominant political, economic, and military power.[31] The relations between the former allies—the United States and the USSR—rapidly deteriorated. Influenced by the American foremost expert on the Soviet Union, George Kennan, President Harry Truman, on March 12, 1947, not only revealed the main American enemy: "totalitarian regimes imposed on free people [i.e., the Soviet Union] undermine the foundation of international peace and hence the security of the United States," but also articulated the narrative that will shape the American foreign policy for decades to come: "I believe that we must assist free peoples to work out their own destinies."[32] The American-Soviet rivalry of the Cold War and the American sense of manifest destiny "to support free peoples who are resisting attempted subjugation by armed minorities or by outside pressures"[33] dominated world politics for nearly half a century and was at the center of American foreign policy not only directly vis-à-vis Moscow, but all around the world.

Since then, the combination of these two narratives—America as *the protector of the free world* against *a great evil*—have shaped American foreign policy, as President Dwight D. Eisenhower put it at the dawn of the Vietnam War: "We must maintain a common worldwide defense against the menace of International Communism [i.e., *great evil*]. And we must demonstrate and spread the blessing of liberty [i.e., *protecting freedom*]."[34] As these two narratives remained the focus of American foreign policy for the entire period of the Cold War, each one of these deserves a more detailed discussion.

In his famous article "The Sources of Soviet Conduct," Kennan suggested two main directions for American foreign policy regarding the Soviet Union: (1) "the main element of any United States policy toward the Soviet Union must be that of a long-term, patient but firm and vigilant containment of Russian expansive tendencies," and (2) "American people . . . accepting the responsibilities of moral and political leadership that history plainly intended them to bear."[35] These two were highly emphasized by Truman and Eisenhower and continued to dominate American foreign policy for the rest of the Cold War.

President Kennedy was defending "the entire Western Hemisphere" from the "Soviet threat to world peace"[36] and was ready to "pay any price, bear any burden, meet any hardship, support any friend, oppose any foe to assure the survival and the success of liberty."[37] President Johnson was "concerned with threats to the peace and security of the peoples of southeast Asia" coming from "the Communist government of North Viet-Nam"[38] as "there is no one else who can do the job. Our [American] power alone in the final test can stand between expanding communism and independent Asian nations."[39] President Nixon believed that although "the defense and progress of other countries must be first their responsibility" facing "the emerging polycentrism of the Communist," the role of the United States "remains indispensable,"[40] and President Ford promised that "the United States will fulfil its responsibilities as a leader among nations."[41] While President Carter famously stated that he has "no fear of communism and no inordinate concern about communism," he saw the "inherent strength of the United States . . . [as] superior to that of the Soviet Union" and was ready "to act promptly and decisively to help countries whose security is threatened by external forces [i.e., communism]."[42] President Reagan was concerned about "becoming powerless to deter or counter Soviet aggression around the world" and had been unable to "assist friendly governments facing threats from the Soviet Union, its surrogates, and from other radical regimes."[43]

And then the Cold War came to an end. Already at the end of 1989, in his final message, President Reagan was "confident that relations between our two countries [the United States and the USSR] will continue on the positive" and that "we have been able to find some common ground."[44] His successor, President Bush, observing the collapse of the Soviet Empire cautiously stated that the "whole policy for the United States . . . will be to watch for those changes and try to facilitate them and work with those who are willing to move towards freedom and democracy."[45] Bush, whose political views were shaped by the strategy of containment, felt very uncomfortable trying to design a new strategy for the rapidly changing world. His main problem was best formulated by an American commentator, Norman Ornstein: "What does a superpower do in a world no longer dominated by superpower conflict?"[46] With the disappearance of the communist *great evil*, the Bush team of closest advisers, who by themselves were a product of the Cold War and, in fact, were more prepared to contain the non-existent Soviet Union rather than reshape America's international agenda in the new world,[47] tried to stick to the Cold War dogma. Dealing with the Persian Gulf Crisis in 1991, Iraqi President Saddam Hussein was described as the "worldwide threat to democracy" against whom America was "ready to use force to defend a new order emerging among the nations of the world—a world of sovereign nations living in peace."[48]

An attempt to define a new *great evil*, however, was not an easy task. While the case of Iraq eventually corresponded with both of the traditional narratives of American foreign policy, in other cases the Bush administration had to compensate for the lack of *an evil enemy*, by amplifying the second narrative of the United States as *the protector* of freedom and peace: "in a world where we are the only remaining superpower, it is the role of the United States to marshal its moral and material resources to promote a democratic peace. It is our responsibility, it is our opportunity to lead. There is no one else." This amplification is best demonstrated by the American involvement in the 1989 Philippine coup attempt, which was described as a move "to safeguard democracy in the Philippines;" the 1989 invasion of Panama that was advocated as the protection of "the security of the Canal;" and the 1992 U.S. involvement in Somalia that was an attempt to "stem this human tragedy of Somalia."[49]

The outcome of the 1992 elections symbolized George Bush's failure to read the American voters' hearts in a world without a Soviet threat. Americans decided to give a chance to Bill Clinton, not because of his (mostly lacking) experience in foreign affairs, but because it seemed that he, unlike Bush, knew how to deal with American problems at home.[50] And, though it was interpreted as if the United States had lost its "missionary zeal," the narratives of Clinton's foreign policy did not differ much from his predecessor. Clinton, indeed, put much effort to make the U.S. economic agenda a centerpiece of his foreign policy, successfully promoting the North American Free Trade Agreement (NAFTA), the Asia-Pacific Economic Cooperation (APEC), and the General Agreement on Tariffs and Trade (GATT).[51] American military involvement during the Clinton presidency, however, proved that America had no intentions of abandoning the world. Clinton was the first American president in a truly post–Cold War world and the narrative of *the great evil* definitely did not suit any more. In the absence of other alternatives, the Clinton administration had to persist with maintaining the remaining Cold War narrative of America as *the protector of free people and democracy*.

Clinton inherited the conflict in Somalia from George Bush and could not immediately withdraw American forces from this troubled intervention; therefore, he continued to explain that "we went to Somalia because without us a million people would have died. We, uniquely, were in a position to save them."[52] The following American reaction to the conflict in the Balkans in 1999 persisted to promote this narrative: "We act to protect thousands of innocent people in Kosovo from a mounting military offensive. We act to prevent a wider war, to diffuse a powder keg at the heart of Europe that has exploded twice before in this century with catastrophic results."[53] Unlike Bush, who rushed to title Saddam as "an unusual and extraordinary threat,"[54] Clinton was very cautious in playing the narrative of *a great evil*—describing Milošević's activity in Kosovo, he hesitated to use the term "genocide."[55] It did not stop him, however, from emphasizing the narrative of America as *the protector of free people and democracy*, as "ending this tragedy is a moral imperative" and "America has a responsibility . . . to save innocent lives and preserve peace, freedom, and stability in Europe."[56]

The arrival of the new millennium did not forecast any significant changes for American foreign policy. Bush Junior, like Clinton before him, struggled to find a proper *great evil* at a time when Eurasia's greatest powers had a potential to challenge "the peace" that America had sought—China was only a "competitor" and the United States, in fact, was "Russia's ally in self-reform."[57] Consequently, it is not surprising that from the beginning of his presidential campaign, his vision of American foreign policy was shaped mainly by America's success as *the protector* (and promoter) *of free people and democracy*:

> Our world, shaped by American courage, power and wisdom, now echoes with American ideals. We won a victory, not just for a nation, but for a vision. A vision of freedom and individual dignity—defended by democracy, nurtured by free markets, spread by information technology, carried to the world by free trade. The advance of freedom—from Asia to Latin America to East and Central Europe—is creating the conditions for peace.[58]

Then, on the morning of September 11, 2001, the presidency of George W. Bush, as well as American foreign policy, were turned upside down by two highjacked airplanes that brought down the World Trade Center in Manhattan and a third that seriously damaged the Pentagon. Later that day, in his address to the nation, Bush stressed that "today our Nation saw evil, the very worst of human nature" but "we go forward to defend freedom and all that is good and just in our world."[59] In this speech he emphasized that America is not any more *the protector of free people and democracy* against no enemy, as it was in the previous decade; but there is *a great evil*—"terrorists who committed these acts and those who harbor them."[60] After 9/11, the Bush Administration rediscovered the two main foreign policy narratives of the Cold War—America is *the protector of the free world* against *a great evil*, with the only difference was Soviet communism replaced by international terrorism. Both narratives of the Cold War foreign policy were, undoubtedly, brought back by Bush after 9/11 as the declared Global War on Terror was to "lift a dark threat of violence . . . [and] rally the world . . . by our efforts, by our courage."[61]

Moreover, after the devastating attacks on American soil, Bush enjoyed the same unlimited freedom of action as Franklin D. Roosevelt did after the Pearl Harbor attack 60 years earlier. But, while Roosevelt's *evils* were just "arrogant rulers whose selfish purpose is to destroy free institutions"[62] (i.e., Japan, Nazi Germany, and Italy), Bush's *great evils* were infused by the Cold War heritage:[63] on a general level there was the Islamic radicalism that "like the ideology of communism . . . is dismissive of free peoples,"[64] and on a personal level there was Saddam Hussein who was "a student of Stalin, using murder as a tool of terror and control."[65]

The following American interventions, in Afghanistan in 2001 and Iraq in 2003, were presented not only as a "monumental struggle of good versus evil,"[66] but, maximizing the narrative of *protector of freedom and democra-*

cy, Afghan people were promised help "in this time of confusion and crisis in their country,"[67] and the Iraqi people, once "freed from the weight of oppression," were assured of "help [to] rebuild their economy and create the institutions of liberty in a unified Iraq at peace with its neighbors."[68] Consequently, it is not surprising that the intervention in Afghanistan was quickly renamed from the initial "Operation Infinite Justice" to one more suitable with foreign policy goals, "Operation Enduring Freedom" (though the given reason for the change was to sidestep the objections in the Muslim world to the first title),[69] and the intervention in Iraq from the very beginning was dubbed "Operation Iraqi Freedom." Despite the rising opposition to the lengthening interventions in Iraq and Afghanistan, Bush was re-elected for the second term, persisting with the narratives of the American fight against a *great evil* to *protect the free world*: "Our military has brought justice to the enemy and honor to America. Our Nation has defended itself and served the freedom of all mankind."[70]

Undoubtedly, the maintenance of these two narratives played an important role in fending off criticism of American policies. One of the best examples is the case of the Guantanamo Bay Detention Camp—"a safe and secure location to detain and interrogate enemy combatants"[71] established after the beginning of the Iraqi and Afghani campaigns. Almost immediately after its establishment, the Guantanamo Camp fell under extensive criticism from the international community. For example, German Chancellor Angela Merkel said in 2006 that "an institution like Guantanamo in its present form cannot and must not exist . . . We must find different ways of dealing with prisoners,"[72] and Irene Khan, the Secretary General of Amnesty International, stated in the foreword to the 2005 Annual Report that "the detention facility at Guantanamo Bay has become the gulag of our times, entrenching the practice of arbitrary and indefinite detention in violation of international law."[73] To fend off the rising international criticism that protested the legal status and physical condition of detainees at Guantanamo Camp, American leadership firmly persisted with the narratives of fighting the *great evil* to *protect the free world*. Defending the existence of the detention camp, Bush stated in 2006 that the people who were held at Guantanamo "aren't common criminals or bystanders accidentally swept up on the battlefield, [but] suspected bomb-makers, terrorist trainers, recruiters and facilitators, and potential suicide bombers . . . and individuals involved in other attacks that have taken the lives of innocent civilians across the world" and that keeping them there and questioning them "has given us information that has saved innocent lives by helping us stop new attacks, here in the United States and across the world."[74]

President Barack Obama, who replaced Bush in 2008, had already during his election campaign described the Guantanamo Camp as a "sad chapter in American history" and promised to close it.[75] The real reason behind this attempt, however, probably was an outcome of internal political sparring, rather than the result of international pressure. Newly elected, Obama was looking for an opportunity to criticize Bush and advocated the closure of the camp because "our [previous] government made a series of hasty decisions . . . based upon fear rather than foresight, and all too often trimmed facts and evidence to fit ideological predispositions."[76] Though Obama's vision of foreign policy was seen by many as a U-turn away from the core principles of Bush's worldview, he was not a "non-interventionist calling for a retreat to Fortress America."[77] On the one hand, Obama, unlike Bush, demonstrated high reluctance to embrace physical humanitarian intervention facing different crises (e.g., the Arab Spring events in 2011 and the subsequent lengthening Syrian Civil War; the rise of the Islamic State of Iraq and the Levant [ISIL] on the territories of Iraq and Syria; the Ukrainian Revolution in 2014 and the subsequent Russian annexation of Crimea). On the other hand, in many ways the relationship between Bush's and Obama's foreign policy rhetoric is marked more by continuity, rather than change. After all, according to Obama, the *great evil* was still there:

> Russian aggression in Europe recalls the days when large nations trampled small ones in pursuit of territorial ambition. The brutality of terrorists in Syria and Iraq forces us to look into the heart of darkness.[78]

As well as the American duty to *protect the free world*:

> I can promise you that the United States of America will not be distracted or deterred from what must be done. We are heirs to a proud legacy of freedom, and we're prepared to do what is necessary to secure that legacy for generations to come.[79]

Since the end of World War II, and especially since the end of the Cold War, the United States has been enjoying a position of unprecedented economic, technological, political, and above all, military power. While many scholars accuse the Americans of imperial ambitions, the establishment of a unipolar world and becoming a "New Rome,"[80] others argue that the United States enjoys the position of leadership, rather than autocracy, and it has to work hard to preserve it.[81] Since the mid-twentieth century, the United States has considered itself "the leader of European-American civilization against upstart challengers to that civilization,"[82] first Nazi Germany and Imperial Japan, then Soviet communism and, finally, global terrorism. Consequently, it is not surprising that American leaders continuously have addressed their country as *the protector of the free world* against *the great evil.* While this demonstration of good will, undoubtedly, helped Americans to fend off some of the international criticism, especially those originating outside the "free world," it was not enough to create a politically acceptable environment for

American military operations. The additional mechanism that helped American leadership to facilitate hostile international opinion has always been their ability to create alliances and coalitions as a part of their strategy, thus minimizing the pool of possible criticizers and spreading the remaining criticism among as many players as possible.[83]

In the American view of international relations, alliances have a special place as they not only "presumably transcend and subordinate separate national interests," but also "represent indigenous harmony and initiative,"[84] allowing the United States to be the first among several "equal" partners. While in both World Wars the United States was a latecomer and naturally joined the existing alliances, since the beginning of the Cold War, American leadership has made great efforts to attract as many potential partners as possible to support American actions in any possible way: militarily, materially, or just verbally.

An interesting point is that since Korea and Vietnam, through to the Gulf War, and up to the wars in Afghanistan and Iraq (with probably one exception, Kosovo),[85] the United States overwhelmingly dominated its allies not only in terms of military power, but also in the decision-making process regarding the direction of the war.[86] It is not the military forces that the United States wanted most from their allies, but their politically visible support; as President Johnson told British Prime Minister Wilson in 1966, after the UK declined to send troops to Vietnam: "a platoon of bagpipers would be sufficient; it was the British flag that was wanted."[87] From the American viewpoint, the major role of the "coalition of willing" is sharing the political burden of the military action, rather than the physical one. American leadership has always desired to preserve its freedom of action, simultaneously enjoying the political benefits of collective actions while dispersing the political responsibility and potential consequential liability. For that reason, the American first choice of action has never been working through formal institutions and alliances (e.g., NATO) that consists of players who do not necessarily share the American desire and tend to bargain for their political support, but establishment of ad-hoc multinational cooperation with international players, who were ready, due to their own interests, to comply with the American subduing leadership.[88] The American actions during the interventions in Afghanistan in 2001, and in Iraq in 2003, were very good examples (the first was very successful, the second almost a failure) of this mechanism to minimize international political pressure.

In the aftermath of the 9/11 attacks, the international community expressed a strong solidarity with the United States. While immediately after the attacks NATO invoked Article 5 and offered assistance, the American reaction was lukewarm. The Bush administration, enjoying overwhelming international support, desired immediate action and did not want to complicate the decision-making process with NATO bureaucratic procedures that could slow the pace of operations. The American determination to act outside the NATO framework, taking aboard only partners that were ready to abide to American leadership, was best demonstrated by Donald Rumsfeld, the U.S. Secretary of Defense, who stated in September 2001: "The mission determines the coalition. And the coalition must not be permitted to determine the mission."[89] Only in 2003, when U.S. leadership started to attract too much criticism over its uncompromising decision to invade Iraq, did Washington begin to consider a larger NATO role in Afghanistan, officially transferring the command to the alliance in August 2003. The main purpose of the coalition's operational involvement in Afghanistan was not only to lessen the burden of war on the United States, but also to provide the required legitimacy for military activity.[90] In this way, despite the fact that the United States remained the main contributor to the International Security Assistance Force (ISAF) in Afghanistan, the political responsibility for the consequences of the military activity was formally shifted to NATO.

The attack on Iraq was a completely different story. Unlike Afghanistan, there was no direct linkage to the 9/11 attacks and Saddam Hussein did nothing (at least in the terms of an immediate threat) to provoke international intervention in his country. Moreover, while the American attack on Afghanistan materialized within four weeks after 9/11 (the bombing of Taliban positions and al-Qaeda bases had already begun on October 6, 2001),[91] the idea of attacking Iraq was treading water for a year and half. It is not surprising that many NATO members, such as France, Germany, and Turkey, as well as other countries such as Russia and China, which had backed the U.S. actions in Afghanistan, were very critical of the Iraqi campaign. Despite the hollow claim of the United States that it had a coalition of some 44 members, in reality, the Washington "coalition of the willing" consisted of the United States, Britain, and Spain, on the political front, and of the United States, Britain, and Australia, who contributed their forces for the action.[92] Compared with Afghanistan, the American attempt to share the burden of political responsibility in the Iraqi War was a complete fiasco. Despite the fact that the British went all the way with Americans come hell or high water, due to their traditional belief in the "special relationship" with the United States,[93] it was mainly Washington, which had to deal with the political consequences on the international arena.[94]

One of the main assumptions of this research is that international political criticism can be one of the main drivers that would force political leadership to press their military to significantly decrease civilian casualties, i.e., employ NLW. The analysis of the different mechanisms of American foreign policy presented above, and the ways in which Washington fends off international criticism, clearly shows the independence of American military activity from international pressure.

American political-military decision-makers are not meaningfully preoccupied by international public opinion for several major reasons. First, there are structural conditions that improve the U.S. ability to act without the political approval of the rest of the world: the size and vitality of the American economy and its military might; the lack of political unity within Europe; and Russia, which still struggles to recover its positions in the global arena.[95] Referring to the Bush decision to go to war in Iraq without explicit UN backing, UN Secretary General Kofi Annan accused Washington of "proliferation of the unilateral and lawless use of force, with or without justification."[96] Even this unambiguous criticism, however, did not stop the United States from invading Iraq.

The second reason is the fact that after World War II the United States defined itself as the leader of the democratic free world. As it was discussed above, the narrative of the United States as *the protector of the free world* against *the great evil* has dominated American foreign policy from the Cold War until today. While at the end of the Cold War, Gorbachev's adviser Georgiy Arbatov said that: "We are doing something really terrible to you—we are depriving you of an enemy,"[97] new great evils have been continuously reinvented—Soviet communism was replaced by global terrorism, and global terrorism seems to have been replaced by ISIL and, probably, Putin's Russia. The success of American leadership in maintaining the principle that "we are good"[98] significantly eased international pressure on American political decision-makers. From 2002 to 2006 alone, approximately 200,000 people were killed as a result of the violence of the war that began with the U.S. invasion of Iraq[99]—it seems that the United States has emerged relatively clear from any significant political consequences.[100]

The third and final reason is the success of Washington in recruiting other countries, even if only by flying their flags, sharing the political responsibility, and minimizing international pressure. Avoiding the slow negotiation within existing bureaucratic alliances and utilizing the national interests of independent nations, which try to cash in the coin of their own political advantages from supporting the United States, Washington succeeded in defusing international criticism and continuing with the course of its military actions.

The combination of these three narratives of American foreign policy has allowed American decision-makers to be relatively relaxed about their military performance regarding enemy civilian casualties, and they feel no need for fundamental changes to make their military less lethal. With this absence of significant external political need, the only factor that has the potential to influence American leadership to change its mind is political pressure from home, in other words, domestic opposition that values the lives of the enemy's population. To understand the prospects (or lack of them) of such internal political pressure, it is important to analyze the American cultural perspective on war and the extent to which American people appreciate human life—their own, as well as the lives of their enemies.

AMERICAN CULTURE AND ENEMY CIVILIAN CASUALTIES

Defining the American mind, Henry Steele Commager famously stated that it is "the product of an interplay of [the American] inheritance and environment, both varied and complex."[101] In other words, the American character is a brew of a cramped Europe with its values rooted in Ancient Greece and Rome and the American environment—"the sense of spaciousness, the invitation to mobility, the atmosphere of independence, the encouragement to enterprise and to optimism."[102] An analysis of the American cultural context demands an understanding of the fundamental influence of the so-called "free land" phenomenon on the minds of European immigrants and how modified inherited institutions and values shaped the "American Way," "American Dream," and "American Mind."

The significance of the "free land" concept in shaping American character was highly advanced by renowned American historian Frederic Jackson Turner in his famous paper titled "The Significance of the Frontier in American History" that he read before the Chicago meeting of the American Historical Association in 1893. Since then his thesis that "the existence of an area of free land, its continuous recession, and the advance of American settlement westwards, explain American development"[103] has been at the center of a vast discourse among American historians, attracting many supporters, as well as detractors.[104] And while it is not the place of this work to take a side in this discourse, Turner's thesis seems to make a vital contribution to the understanding of the American culture, as the frontier was "the line of most rapid and effective Americanization."[105]

The following analysis does not intend to produce the whole picture of the American cultural context; this mission is beyond this work, its narrow purpose is to understand how American history and cultural predispositions have shaped what Americans think about war, in general, and enemy civilian casualties, in particular. Those who are interested in a broader discussion of the distinctive American way of thought and its influence on the American conduct of war, are advised to focus more on the principal works of American historians that rest at the core of this analysis, such as Henry Steele Commager, Stow Persons, Daniel Boorstin, and Russell Weigley.[106] In addition to these classic titles, the following analysis is also based on more contemporary scholars focusing on the cultural-socio-political context of American history, such as John Grenier, John Mueller, Thomas Mahnken, and John Tirman.[107]

Thus, American character is an outcome of the impact of a wilderness environment upon an old established European culture. Although America was settled by immigrants who transplanted cultures centuries old, the New World was genuinely new and different from the Old. And, yet, an understanding of this newness requires a comprehensive knowledge of the cultural baggage brought by first settlers and then waves of immigrants across the Atlantic. Unfortunately, given constraints of space, this research will merely focus on the final product of this "most ambitious experiment ever undertaken in the intermingling of peoples"[108]—the American character.

American Character and the Value of Human Life

The story of America is very much a story of a conquest. In contrast to the European understanding of conquest, when it is a successful outcome of a violent struggle between two counterparts, the conquest that shaped American character was a subjugation of ferity and the darkness of wilderness on behalf of progress and enlightenment. And the most essential and influential part of this expansion was the frontier—"the meeting point between savagery and civilization"[109] that was "one of the forces which did most to shape American life from the beginning."[110]

With their arrival in North America, Europeans found a continent with a very healthy climate that demanded neither acclimatization, nor a painful process of adjustment. Moreover, the local inhabitants were too few and too primitive to imperil the colonization. With the establishment of the first significant foothold it was quickly acknowledged that life in the New World might be not only prosperous, but can be obtained by anyone, as it offered a boundless, almost ungraspable for the European mind, "free to take" land. Since the mid-seventeenth century America acquired the image of a New World, "a potential New Eden in the West,"[111] with unlimited wilderness that was just waiting to be conquered. And while this New World offered opportunities, new hopes, and potential success for the European immigrants, its main appeal was its limitless resources. In contrast to the Old World that was full of political, social, and economic restrictions, built and maintained through centuries, America offered free land, and "so long as free land exists, the opportunity for a competency exists, and economic power secures political power."[112]

The most important factor in the American conquest of the wilderness was the fact that even with the establishment and the development of the East coast, the promised opportunity of the New World did not stop there—the prospect of free land areas continually existed on the western border of the settled parts of the United States until the end of the nineteenth century, when "the year 1890 is usually taken to mark the date when there was no more frontier available."[113] Undoubtedly, the slow conquest of the West brought with it social and economic limitations to the settled areas, however "whenever capital tended to press upon labor or political restrains to impede the freedom of mass, there was this gate of escape to the free conditions of the frontier."[114] The possibility of escaping this burden, the chance of restarting from the beginning—whether from the Old World to the New, or from the established New to the wilderness—exemplifies better than anything else one of the most important characteristics of the American character shaped by the frontier phenomenon—indefinite optimism. One's past, mistakes, and errors do not bind one irretrievably—"fresh starts could be made, tomorrow promised to be better than today, and progress always seemed to be possible."[115]

The main source of this optimism was the fact that this progress was a part of the frontier experience. A frontiersman witnessed it daily he was a part of the transformation of the wilderness into farms, villages into towns and cities. As the past was left behind, Americans were dreaming of the future and they were used to seeing their visionary plans become reality. The conquest of the wilderness was not an easy labor, but being an American meant to have "dreams and behold visions . . . to have faith in man, [and] an unbounded confidence in his ability to make his dreams come true."[116] This leads to the next characteristic that the frontier sculpted within the American mind, as it transformed the European immigrant into an American—with boundless self-confidence and belief in America's destiny. Hence, an essential part of the American character is an engraved presumption that an American can solve any problem. A strong belief in the "can do it" and a predisposition to "get on with it" have characterized the American psychological mind-set since the conquest of the frontier.[117] In fact, the famous 2008 victory speech of the first African American president, "Yes, We Can!"[118] demonstrates, better than anything, the actuality of this narrative in the beginning of the twenty-first century. Moreover, based on the Puritan heritage of its first settlers, Americans deeply believe in American exceptionalism and that the United States is inherently different from all other nations. Since the times of Puritan John Winthrop, who, in 1630, still aboard his ship *Arbella,* admonished the future Massachusetts Bay colonists that their new community would be "as a City upon a Hill,"[119] the American mind-set has been predisposed to believe that, as President Reagan put it, "Americans [are] awed by what has gone before, proud of what for them is still . . . a shining city on a hill."[120]

Little by little the wilderness transformed the European germs brought from the Old World into a new product that was purely American. Spacious and boundless land did not only beget spacious ideas, it magnified two very European ideals that were restricted by the European realities—individualism and personal freedom. Limitation of resources, especially land, engraved within the European mind-set the necessity to search for more efficient ways of exploiting natural resources, ultimately leading to the strong narratives of individualism, on the one hand, and personal freedom, on the other. Both of these characteristics found flourishing ground on American soil that was free from the Old World's boundaries—"out of his [frontiersman's] wilderness experience, out of the freedom of his opportunities, he fashioned a formula to social regeneration—the freedom of the individual to seek its own."[121]

Free land promoted individualism. The conquest of the wilderness was driven by individuals seeking their own benefits, rather than a well-planned activity led by the authorities or great enterprises. The first was the individual pioneer, then the settler, and only then the men of capital and enterprise. Each one was driven by his own dreams and ambitions, stimulated by his individualistic aim to build a better tomorrow. This social flexibility of American life, so different from over-crowded Europe, predisposed the American character toward a very strong sense of individualism. It does not mean that the frontiersman preferred social isolation, as pioneers and, obviously, settlers could definitely appreciate the benefits of community. However, the membership in the community (*company*) was secondary to the individual's aims—"it was not unusual at all for people to leave one company to join a second or third, abandoning each when it no longer served their purpose."[122] The European social boundaries of a complex society did not withstand the limitlessness of opportunities offered by free land; the frontier's influence was to some degree anti-social, admiring individualism and producing "antipathy to control, and particularly to any direct control."[123]

This strong sense of individualism ultimately led to very keen relations between the American character and personal freedom. The concept of free land and endless opportunities magnified the inherent traditions of the Magna Carta, brought by the first waves of British immigrants. Americans, as nowhere else, even in Britain, were able to exercise personal liberty and dared to contest any attempt by the authorities to limit it. The frontier, "where individual liberty was something confused with the absence of all effective government,"[124] and new generations raised into this reality played a significant role in shaping the cultural predispositions that explain the relations between Americans and their state.

Due to inherent individualism, independence, and personal freedom, "American's attitude toward authority, rules, and regulations was the despair of bureaucrats and disciplinarians."[125] The initial communities, and then villages and towns, were organized by people who not only personally chose to join them, but essentially helped to build them. The conquest of the Western wilderness was a business of Americans rather than the state, and frontiersmen, facing an absence of legal authority, had to developed ways to preserve the order. Consequently, it is not surprising that in reality, when the individual was not ready to limit his freedom by complex legal regulation, "a crime was more an offence against the victim, than a violation of the law of the land."[126]

Raised and bred in the wilderness, "the American was always taking a short cut to freedom, a short cut to fortune, a short cut to learning, and a short cut to heaven"[127] and he was always eager to protect his freedom to make his short cuts. This avidity of the American character to safeguard his freedom ultimately led to a kind of built-in resistance to any type of governmental invasion of privacy and general assumption that an excessive conformity with authority is "akin to cancer."[128] While the strong senses of individualism and personal freedom, mixed with the narrative of strong resistance to any attempts to restrict these qualities of the American character, explain the widely discussed American predisposition toward lawlessness,[129] they are also one of the main reasons behind the American tendency toward violence. The conquest of the American wilderness was a violent affair, and this experience that lasted for nearly three centuries undoubtedly played a role in shaping the American character. It is important to note, however, that the violent origins of the frontier stem from the people, rather than the rough environment. From the very beginning of the colonies, the Anglo-American settlers were violent. Free from the European socio-cultural restrictions that had limited violence in the Old World, on top of being continuously involved in violent struggles against the indigenous Indians, as well as against other colonies, "the Anglo-American colonial people were prone to violence."[130]

Consequently, it seems that an American attitude toward violence is an outcome not only of American freedom created by the frontier, but an inherent right, even a duty of an American, to protect his freedom at any cost. The fact that American individualism led to a general distrust of authorities played an essential role in shaping the American violent nature. In the American mind, one's freedom and privacy are the most precious values that have to be protected and, as one does not trust the authorities providing this protection, or wants them to do so (as such protection will limit his freedom), one would preserve this task for oneself. In other words, in an American's eyes, the protection of his life and his property is first and foremost his own duty and right, rather than the responsibility of the state. And American people, raised on the traditions of the frontier, value the use of violence as an effective method not only to protect, but also to solve any disagreement or conflict.[131]

This fundamental belief in the legitimacy of each citizen to apply violence explains best not only the fact that "Americans, more than any other people in the world, are able to own destructive weapons for private use,"[132] but also the reason why Americans interpret the Second Amendment to the Constitution of the United States as a constitutional right to own these weapons. In fact, the amendment states that: "A well-regulated militia being necessary to the security of a free State, the right of the people to keep and bear arms shall not be infringed"—and the interpretation of this amendment as a right to possess firearms in contemporary America proves that the violence of the frontier, and the amendment that represents it, is still deep in the heart of American society.[133]

Other evidence that the frontier's violent experience is still influential in the American mind is the prevalence of capital punishment in law and practice. The straightforward violence of the frontier created lynching as "a punishment for anyone accused of an offence which outraged the community."[134] Effectively, over the years, with the empowerment of the state and the rule of law, the tradition of lynching transformed into legalized capital punishment, which is still practiced by most of the states in the United States. When capital punishment was largely abolished by the rest of the modern Western World, in the United States, in 2015, it still enjoys a vast public support, with 60 percent of the population considering it morally acceptable.[135]

While these narratives generally describe the inherent motifs of violence engraved in the American character, they also point to the general attitude toward value of life. Americans appreciate first and foremost American lives (as individual and as whole) and anyone who endangers their lives, property, and freedoms deserves a violent and decisive response. The experience of the American frontier brought European individualism into extremes, and, therefore, Americans value human life in a more individualistic and selfish manner than any other nation. This very self-centered value of life, based on American traditional self-reliance, has shaped American social life for generations and anyone (whether a fellow American, American authority, or a reckless and uncouth non-American) who dared to stand between an American and his "dream"—"that dream of land in which life should be better and richer and fuller for every man, with opportunity for each"[136]—deserved to experience American violence to its full capacity.

War and the American Character

The experience of the frontier, and the previously discussed narratives, implanted into the American character have an exceptional power to explain the American attitudes toward wars in general and toward others' suffering in particular.[137] While the constructive perspective (strategy, tactics, etc.) of what is called the "American Way of War" has been widely discussed in the literature,[138] the main focus of this work is the cultural dimensions of this phenomenon. It is, however, important to understand the historical process that has designed contemporary American public attitudes about war.

The historiography of the "American Way of War" generally divides this phenomenon into three main periods: "The First American Way of War" (from the beginning of the colonies until the War of 1812), "The American Way of War" (from the War of 1812 until the end of the Cold War), and "The New American Way of War" (since the end of the Cold War until today).[139] And while all three are highly interconnected and the characteristics of the previous periods ultimately shaped the following ones, each one of these periods played its own cultural, social, and military role designing the unique American understanding of the war phenomenon and, therefore, deserves a short introduction.

As discussed, the impact of the frontier on the European germs shaped the American character, creating its own specific cultural narratives. It also played a vital role in shaping what is characterized as "The First American Way of War." As many historians already pointed to the fact that the encounter between European military culture and the wilderness with its inhabitants (i.e., native Americans) and their military culture created something very distinctive from the European practice of war—"the interaction of military cultures in New England . . . produced a compound more toxic than either of its elements."[140] While both the European colonists and the native Americans used to have unwritten codes of warfare that regulated and restricted violence predominantly to battlefields, their encounters produced a startlingly wanton outcome. The frontier freed the European mind not only from the social restrictions discussed above, it unleashed European-style warfare from the formalized European practice of war, where "armed conflict remained a ritualized activity, regulated by a code of honor and fought between armies, not entire populations."[141] From the very beginning the military experience of the New World was characterized by a violent style of what the eighteenth-century writers called *la petite guerre*—early Americans understood war not as a clash between military organizations, but rather as "disrupting enemy troops, supply, and support networks . . . ambushing and destroying enemy detachments . . . and, most important, destroying enemy villages and fields and killing and intimidating non-combatant population."[142]

The American Revolution was a culminating point of the "First American Way of War" on both strategic and tactical levels. While the struggle against British forces were managed more or less according to the old European traditions, the military operations on the frontier, against the indigenous Indians, between 1774 and 1783 emphasized the extravagant violence of the American *petite guerre* tactics.[143] In other words, in terms of fighting against native Americans, the goal was their subjugation through violence; in terms of fighting Britain, it was a mild revolution "staged by reluctant rebels for limited goals, which were well within their reach."[144] From this perspective, the American Revolution is the best example of what today's American military calls *asymmetric irregular warfare*[145]—"to win, the British army had to destroy the Continental Army. The Continental Army had only to survive."[146]

It is important to note, however, that the war on the frontier, rather than the civilized battles with the British, were those which shaped the "First American Way of War" and its brutality. Moreover, the conquest of the wilderness from its natives engraved in the American mind the two main ways to justify war. The European *jus ad bellum*, based on Christian traditions, generally justified wars when "military actions [were] undertaken by lawful sovereigns for legitimate causes, such as self-defense or avenging the wrong."[147] The frontier, lacking "lawful sovereigns" by its definition, simplified this concept to pure vengeance and a strong belief that through facing the primitive and heathen population with violence, Americans do God's work. While the first justification of war as vengeance derives from human nature that is not unique to Americans, the second one deserves a more detailed discussion.

The origins of the beliefs that war against Native Americans, those "savage beasts or evil instruments of the devil,"[148] is God's work are based on the Christian traditions of the Puritans, who were the first to wage these wars on the frontier and held a "strong hostility towards cultures unlike their own."[149] During over four centuries of American history, the sense of promotion of God's Will, as a justification of war, has been transformed into a narrative of the rightness of war fought against a savage enemy. The "First American Way of War" engraved in the American mind that war is about "the righteous cause, . . . the brutish enemy, . . . and the cold indifference to the suffering of the natives."[150]

The successful end of the Revolution and the establishment of the U.S. Army in 1784 opened a new chapter in American military history. The nineteenth century, and especially the Civil War, shaped what Russell Weigley considered as "The American Way of War." Based on the traditions of the *petite guerre* against Native Americans, rather than on restricted war against the British, Americans encouraged the belief that "the object of war is nothing less than the enemy's destruction."[151] This belief helped in shaping one of the most distinctive American qualities of conducting wars—extreme aggressiveness "at all levels of conflict, from tactical (the conduct of individual fights) to operational (the orchestration of battles in time and space) to the strategic (the planning out of campaigns to achieve the objectives of war)."[152] The "American Way of War," molded by the Civil War and tuned through World War I, triumphed in World War II.[153] Describing the four main characteristics of this "American Way of War," Eliot Cohen emphasized excessive aggressiveness, decisiveness, lack of constraints (political or material), and "bright" lines between politicians and military;[154] it is important, nonetheless, to understand the socio-cultural roots of these narratives embedded in the American character.

Constructed by the frontier philosophy and decisive, as well as destructive, wars with the Native Americans, waged with the purpose of a pure military defeat rather than political resolution or compromise, American military minds tended to give "little regard to the non-military consequences of what they were doing."[155] Unlike the classic Clausewitzian view that "war is a mere continuation of policy by other means,"[156] the "American Way of War" is built on the assumption that "[an] inter-state war [is] not a continuation of political intercourse, but a symptom of its failure."[157] The U.S. Army's textbook on strategy, published in 1936, simply stated that "politics and strategy are radically and fundamentally things apart. Strategy begins where politics end. All that soldiers ask is that once policy settled, strategy and command shall be regarded as being in a sphere apart from politics."[158]

From that perspective, in the American mind, a war is an unfortunate obstacle that stands between them and their "dream," and, as discussed above, it has to be overcome as soon as possible, or, according to Colin Gray, "Americans have approached warfare as regrettable occasional evil that has to be concluded as decisively and rapidly as possible."[159] This predisposition, quite an apolitical one, toward war offers a possible explanation of an interesting and, to some degree, contradictory aspect of "The American Way of War." While the American character is built on the sense of opportunity and exploiting available prospects for one's own benefits, through their history Americans generally failed to utilize the political opportunities created by wars, and too often the U.S. military efforts "have not been suitably cashed in the coin of political advantage."[160]

While some scholars would argue that this "American Way of War" is still prevalent in the American way of conducting wars in the beginning of the twenty-first century, others argue that since the end of the Cold War, a "New American Way of War" has been developing.[161] The main argument in favor of this new approach is that in the era of a proactive exercise in the policing of the so-called *Pax Americana*, the American way of war has been adapting itself to the use of warfare as a political instrument.[162] Regardless of this theoretical discourse, the realities of American armed confrontations in the last decades clearly show that American cultural predispositions, which originated from the previous two periods, are still at the heart of American military activity.

The American mind-set still considers war as a temporary unfortunate disruption and, fueled by historical inherent violence, it believes that the most moral approach to wage war is "at the highest possible intensity in order to finish it as soon as possible."[163] Moreover, based on the narrative of the "First American Way of War" that Americanized the wilderness, Americans still justify war as a right act of vengeance and as a rightful act intended to "free" or "protect" local populations, promoting (American) democracy, (American) freedom and (American) rule of law. On the one hand, the extreme sense of vengeance can be clearly seen in the request of a mourning father, who had lost his son in the 9/11 attack, "to put his [son's] name on some piece of armament in the Iraq War."[164] On the other, the intense sense of the right cause is emphasized by another father, who, meeting his son coming back from almost five years in the Taliban's captivity, said "he was proud how far his son was willing to go to help the Afghan people."[165] Addressing these two vital narratives of American cultural ways to justify war, President Bush told Congress and the public just days after the 9/11 attack that "tonight, we are a country awakened to danger and called to defend freedom. Our grief has turned to anger, and anger to resolution. Whether we bring our enemies to justice, or bring justice to our enemies, justice will be done."[166]

This chapter explores the reasons behind the fact that the American military has failed to employ NLW. American political decision-makers are, undoubtedly, very sensitive to public attitudes regarding their (and American military as well) performance. Facing domestic political pressure to decrease enemy civilian casualties during American military operations, the American leadership would definitely look for a way to solve this problem. The main question, however, is whether American public cultural attitudes, based on their inherent violence, aggressiveness, justification of revenge, and an excessive belief in the rightness of everything American, are predisposed towards valuing the lives of the enemy's population.

The analysis of the "First American Way of War" clearly shows that the roots of the American approach to conducting warfare are everything but an appreciation of enemy population; "the European settlement of America was a violent affair, arguably genocide . . . [and] that experience, lasting nearly three centuries, shaped Americans' attitudes toward war, particularly the depiction of the enemy."[167] The different ways (against native Americans and against the British) in which Americans fought during the Revolution engraved in the American mind that rules of war apply differently to somebody who does not share the same race, religion, and culture.[168] The Civil War was definitely a crucial turning point in designing American cultural attitudes toward war, however, it seems that the then established "American Way of War" changed very little, if at all, the American attitude toward the enemy's population—"of all the enemies the United States has fought in its history, the enemy population most intensely hated was probably the Japanese in World War II"[169]—the culminating point of the "American Way of War." Analyzing American public attitudes from World War II, through Korea, Vietnam, and the rest of American military operations up to Kosovo, John Mueller concluded that "[Americans] seem to be remarkably insensitive to casualties suffered by foreigners including essentially uninvolved—that is, innocent—civilians."[170] Although an analysis of American views about the use of nuclear weapons at Hiroshima and Nagasaki in 1945 that targeted the civilian population shows that the approval of the bombing has been diminishing over time—from 85 percent in 1945 to 55 percent in 1994—it still remains very high.[171] More contemporary surveys show that this tendency has not changed much, as in 2015 "the share of Americans who believe the use of nuclear weapons [against civilian populations] was justified is 56%."[172]

If something has changed in the American attitude toward war that can be related to the so-called "New American Way of War" it is sensitivity to friendly casualties, especially when war is at discord with American inherent cultural perceptions of war—decisive aggressiveness with a fast resolution. The experiences of the Korean War and, especially the war in Vietnam, proved that "Americans don't like to see their fathers and sons dying, especially in long wars fought over unclear limited objectives."[173] Moreover, since the evolution of a high-technology way of warfare, this understandable sensitivity to friendly casualties in wars has been fueled by an expectation for an exceedingly low casualty rate, due to the fact that new technologies allow minimizing mortal danger to which American soldiers are exposed.[174]

It is important to note that even this increasing value on Americans' lives is relative. When a war is congruent with American cultural predispositions, they are ready to tolerate casualties. As polls showed, in October 2001, directly after the 9/11 attacks, "anywhere from about 60 to 75 percent expressed a tolerance for 'substantial' casualties . . . and said they still would consider the war to be worth its costs."[175] Moreover, the following wars in Afghanistan and Iraq proved again that the American public shows "relatively little concern about civilian casualties incurred in the war."[176] Even after high profile incidents with a high number of enemy civilian casualties, such as the Wedding Party Incident in Afghanistan in 2002,[177] or the Marketplace Incident in Iraq in 2003,[178] the American public showed an astonishing obtuseness, continuing to support "brave American soldiers" fighting against a "brutish enemy." Despite vast and mainly negative coverage of these incidents in the domestic and foreign media outlets, after the first incident, official polls showed that about 83 percent of Americans still approved U.S. military action in Afghanistan[179]; and after the second incident, the approval even rose from 75 percent to 77 percent.[180]

An analysis of the American public reaction to civilian deaths caused by the U.S. military in the last two decades (Iraq, 1991; Kosovo, 1999; Afghanistan 2001; Iraq, 2003) conducted by the RAND Corporation stated that "very large majorities [of Americans] consistently stated their belief that civilian casualties in these wars were unavoidable accidents of war."[181] Even during the peak of American anti-war protests during the Vietnam War, the argument that American military caused the deaths of innocent civilians was extremely marginal—in 1967, only 6 percent of responses to the question "What two or three things about the war in Vietnam most trouble you personally?" referred to civilian casualties.[182] American public opinion, based on the inherent American attitudes toward war, is far from pressuring political and defense decision-makers to minimize enemy civilian casualties in wars. In the light of this American cultural context, it is difficult to disagree with

Ralph Peters, who stated in his very controversial book, *Fighting for the Future: Will America Triumph?*, that:

> The citizenry of the United States, in fact, will tolerate the killing of enormous numbers of foreigners, so long as that killing does not take too long, victory is clear-cut, friendly casualties are comparatively low, and the enemy dead do not have names, faces, and families. [183]

AMERICAN MILITARY CULTURE AND NLW

The existing research clearly shows that many of the previously discussed features of the American character have also had a great influence on American military culture, as Colin Gray concluded: "American strategic culture . . . [is] the product of the significantly unique American historical experience."[184] For example, researchers often interpret the U.S. interventions in Iraq as a continuation of the American frontier,[185] or point to the fact that American traditional apolitical strategy (i.e., the separation between politics and military activity)[186] has led the United States "to fight the wars as if they were battles . . . [confusing] the winning of campaigns . . . with the winning of wars."[187]

This chapter, however, focuses on the failure of the American military to successfully employ NLW, i.e., combine military technology with accordingly formulated concepts of operations and military organization. Consequently, it seems important to analyze the role of technology in the American military culture in general, and what American military thought has to say about NLW in particular.

American Military Culture and Technology

One of the most prominent features of American military culture is a great emphasis on technology. Reflecting the growth of American industrialism and consequent capitalism in the nineteenth century, and the American pioneering of mass production processes in the early twentieth century, American military culture has been shaped around the concept of overwhelming the enemy by logistic superiority and state-of-the-art technology. While some argue that the American way of war can be more often described as logistic, rather than strategic,[188] the majority agree that "reliance on advance technology has been a central pillar of the American way of war . . . [and] no nation in recent history has placed greater emphasis upon the role of technology in planning and waging war than the United States."[189]

The military fascination with technology is deeply rooted in American culture itself. In the nineteenth century, America's productivity powered by European immigrants suffered from a shortage of skilled craftsmen, who preferred to stay in Europe, and Americans were forced to "invent and use machines as substitutes for human skill and muscle."[190] This technological enthusiasm and general faith in technological solutions have been successfully adopted by the American military culture that started to seek refuge from difficult problems of strategy in technology. Moreover, America's geographical position in the world created a drive for technological means to project power. While European armies (except for the British) could just march into battle, Americans historically developed and emphasized technologically based abilities: first of all, the navy, then the air force, and then missile-based nuclear power.[191]

In the American mind, as already mentioned, war is an unfortunate, but temporary, trouble that has to be solved as quickly and decisively as possible. This approach to war combined with the traditional pro-technological orientation created almost an instinctive reliance of American strategists on technology. Despite the fact that the experience of Vietnam created certain skepticism of technological omnipotence, the general techno-craze and the trust of technology—a "silver bullet" promising a decisive victory—have never been abandoned. The American military has always been preoccupied by the application of technological solutions to strategic problems, thus driven (and usually capable of) to acquire exceedingly expensive state-of-the-art technology, often unable to comprehend that technical sophistication does not guarantee success in conflicts.[192]

The experience of the First Gulf War reinforced American technological romanticism in military affairs, putting an end to the skepticism of the Vietnam generation. In the wake of the overwhelming victory, the American trust in technology has never been stronger, creating vast debate whether the world had witnessed a Revolution in Military Affairs.[193] Mesmerized by their technological triumph, American military signified the revolutionary technology itself as a driving change, refusing to understand that operational concepts and organizational adaptations are "more important than either new technology or getting it fielded in a significant number of systems."[194]

The extended irregular warfare in Iraq and Afghanistan, so similar to Vietnam, shook the American reliance on advanced technology as a key solution to any military puzzle. It seems, however, that despite the desperate battle of the American strategists to argue for doctrinal and organizational changes, American culture struggles to accept that war is not simply a matter of applying technology correctly. In the American military mind, the best (if not the only) way to improve the performance of the military is by an amelioration of technology, rather than the way it is used, or misused, through overuse. In other words, to put it simply: if the performance is poor, it is the fault of the gadget, rather than the way it is used.

Consequently, this techno-utopia in military affairs traditionally has led American military to embrace new technologies first, speculating about their revolutionary promises, and think about their conceptual and organizational requirements later. The best example is the early-1990s information-technology-led transformation that integrated precision-guided munitions, novel command and control technologies, and advance reconnaissance capabilities.[195] Fascinated by the overwhelming success of American technology in the Gulf War, Admiral William A. Owens stated in 1995, that "we have decided to build what some of us call the system of systems; namely, interactions that will give us dominant battlespace knowledge and the ability to take full military advantage of it."[196] Eight years later, the United States, which was "the first nation to emerge in the post-revolutionary era" and was "equipped with proficiencies" that promised "change the character of warfare as it has been known for centuries,"[197] found itself in Iraq, realizing that many military problems just cannot be solved purely in technical terms.[198] It seems, however, that this disappointment has not changed the American way of thinking about technology. For example, a document published in 2013, *Capstone Concept for Joint Operations: Joint Force 2020*, intended to provide general guides to future force development and transformation, was awash with prudent words on the future security environment; at its core, however, rested carefully crafted requirements for the next generation of technological capabilities.[199]

American minds consider technology as the lodestone of military success, and American military history, undoubtedly, is full of technological marvels. Americans are easily enthralled by novel technologies, but facing operational disappointment, they, similarly, easily accept their delusion. Yet, due to their cultural predispositions, when they hit their fingers, Americans will always seek to improve the hammer, rather than the way they hold the nail.

American Military Thought and NLW

As Americans have approached warfare as the regrettable occasional evil, and a war as a temporary affair, it is not surprising that scholars report a strong preference of American military tradition to value practice, rather than theory, and to neglect the importance of professional military education required for the constant preparation for a future conflict.[200] Americans traditionally believe in their technological and logistical superiority, and take for granted that it will enable them to concentrate enough forces and firepower, achieving what they consider as the main objective of war—"the complete overthrow of the enemy, [and] the destruction of his military."[201] Consequently, it is not surprising that existing research suggests that historically, the United States did not develop excellence in strategy and comprehensive military thought—it simply did not have to.[202]

The presence and the quality of professional military education have been considered by scholars an important factor in the analysis of the strategic thinking and conceptual military thought.[203] While the American military has an extensive network of professional education institutions intended to tutor officers' corps at different levels, the view prevails that "education is a luxury for the American military rather than a necessity."[204] Professional education has never been considered by the American military establishment as a career-enhancing experience. In other words, as Williamson Murray concludes: "[the military] failed to enunciate a clear vision of why it believes professional military education to be important."[205]

The American military mind, trapped in the industrial age with its techno-utopia, tends to reduce strategic problems to technical equations and undervalue (or even scorn) the ability of systematic conceptualization and theorizing. Moreover, significant cultural differences between different services (Air Force, Army, Navy, and Marines),[206] combined with a strong sense of inter-rivalry and the fact that traditionally the Joint Chiefs of Staff (JCS) has only advisory power[207]—all have led to a situation where no single institution in the American military possesses an integral grasp of the strategic environment enabling the development of all-inclusive theoretical concepts. For example, in the 1990s, when different services had their own conceptual understanding of the information-technology-led RMA, and the JCS produced concepts that showed a "lack of any significant intellectual content" with "interminable laundry lists of bureaucratic concerns" that were "best suited to insomniacs,"[208] the American military was just unable to realize the revolutionary implications of the novel technology—i.e., technological seeds preceded the maturation of the conceptual ones.[209]

Consequently, it is not surprising that an analysis of American official documents, conceptualizing development and employment of NLW, shows a very similar situation—a convoluted nexus of advisory, consultative, and coordinating policies unable to produce one clear comprehensive concept that will define the required technological capabilities of NLW and the conceptually novel ways of their employment.

For example, according to the Department of Defense NLW Policy, NLW have the potential to enhance the commander's ability to:

> (1) Deter, discourage, delay, or prevent hostile and threatening actions; (2) Deny access to and move, disable, and suppress individuals; (3) Stop, disable, divert, and deny access to vehicles and vessels; (4) Adapt and tailor escalation of force options to the operational environment; (5) Employ capabilities that temporarily incapacitate personnel and materiel while minimizing the likelihood of casualties and damage to critical infrastructure; (6) De-escalate situations to preclude lethal force; (7) Precisely engage targets; (8) Enhance the effectiveness and efficiency of lethal weapons; (9) Capture or incapacitate high value targets; (10) Protect the force.[210]

But, according to the Joint Non-Lethal Weapons Program (JNLWP), a directorate that "stimulates and coordinates" NLW requirements for all U.S. Armed Services, yet is, in fact, located under the Commandant of the Marine Corps,[211] NLW are supposed to be able to:

> (1) Deny access into/out of an area to individuals (open/confined) (single/few/many); (2) Disable individuals (open/confined) (single/few/many);(3) Move individuals through an area (open/confined) (single/few/many); (4) Suppress individuals (open/confined) (single/few/many); (5) Stop small vehicle; (6) Stop medium vehicle; (7) Stop large vehicle; (8) Disable vehicle/many vehicles; (9) Stop small vessel; (10) Stop large vessel; (11) Disable vessel/many vessels; (12) Stop fixed-wing aircraft on the ground; (13) Disable aircraft on the ground; (14) Divert aircraft in the air; (15) Deny access to facility.[212]

While, facing these differences, one can argue that DOD defines more conceptual aspects and JNLWP translates them into more technical requirements, the U.S. Army has produced its own list, entirely different from these two, according to which NLW should enhance the capability of a Joint Force in the accomplishment of the following objectives:

> (1) Discourage, delay, or prevent hostile actions; (2) Limit escalation; (3) Take military action in situations where the use of lethal force is either not the preferred option, or is not permitted under the established Rules of Engagement (ROE); (4) Better protect our forces; (5) Disable equipment, facilities, and enemy personnel; (6) Engage and control people through civil affairs operations and Psychological Operations (PSYOP); (7) Dislodge enemy from positions without causing extensive collateral damage; (8) Separate combatants from noncombatants; (9) Deny terrain to the enemy.[213]

It seems that the story of NLW in the U.S. military detrimentally follows the traditional narrative of the American military culture—the inability to produce and coordinate comprehensive concepts for conceptually new technologies. While JSC argue that "minor lapses in conduct or application of fires could seriously damage the international reputation of the United States"[214] and the DOD Defense Science Board claims that "with respect to the human toll on innocent civilians, the U.S. strategy is to reduce collateral damage,"[215] only the U.S. Army successfully addresses this true demand from NLW: to "dislodge enemy from positions without causing extensive collateral damage" and "separate combatants from noncombatants."[216] Moreover, the Army goes further in its conceptualization, trying to suggest a required organizational change that will accommodate NLW:

> The future Modular Force, specifically, must be provided with organic nonlethal capabilities to disrupt, dislocate, disorganize, disintegrate, fix, isolate, suppress, and destroy enemy functions. Joint force commanders (JFCs), fur-

> thermore, must be provided with multifunctional/multirole lethality options in integrated multipurpose system configurations. . . . The future Modular Force Soldier must have the ability to employ a wide array of lethal and nonlethal munitions based upon mission need and force protection.[217]

It seems that the U.S. Army is the one oasis in the desert that is the theoretical conceptualization of NLW. Nevertheless, it is important to state that similar to the case of the information RMA of the 1990s, where academic strategists such as Andrew Marshall, Williamson Murray, Andrew Krepinevich, and many others tried to put the U.S. military on the right track, NLW are also not parentless. On the academic forefront of the NLW debate, a special place is reserved for Douglas Lovelace Jr. and Steven Metz,[218] director and the director of research at the U.S. Army War College Strategic Studies Institute, as well as David Koplow,[219] a professor of law at Georgetown University, who has held a variety of senior defense-related positions in the U.S. government. Since the late 1990s, these scholars theorized and anticipated changes that NLW might bring to the way the U.S. military fights its war; however, it seems that the military has not been able to develop an appropriate philosophy, nor to find leaders willing to bridge the gap between these theories and practice.

In the very beginning, following its traditional techno-utopia and tendency to seek a "silver bullet" solution, American military thought embraced NLW as the next big thing. In 1998, Lieutenant Colonel James C. Duncan reported:

> General Zinni's decision to equip the Marines of I Marine Expeditionary Force with non-lethal weapons was revolutionary. Without question, his decision has had a tremendous impact on all U.S. military forces by providing the stimulus to change the perception of non-lethal weapons. In addition, Operation United Shield prepared the groundwork for the development of a new concept for the employment of nonlethal weapons and moved the U.S. military toward a new age of warfare.[220]

But the first users and main NLW advocates, General Zinni, Colonel Lorenz, and Lieutenant Colonel Stanton, after their return from Somalia generally labelled NLW as tools of "peace operations and armed interventions short of war"[221] that "broaden the scope of possible responses to riots and other low-intensity tactical situations."[222] Despite some enthusiasts who tried to promote NLW as systems that "have broad application across the entire spectrum of conflict" including armed conflicts, as "battlefields may be shaped through operations which employ non-lethal weapons"[223]; this vision has never been materialized. Two decades after their first employment in Somalia, NLW remain to be considered by American military thought as the means for law enforcement activity only—the *2012 Non-Lethal Weapons*

Training and Readiness Manual, published by the Marine Corps, speaks about performing riot control formations and deploying civil disturbance units, rather than an employment of NLW on the battlefield.[224]

The case of NLW in the U.S. military perfectly demonstrates that, as per usual with the American military culture, American military thought struggles to produce and implement a comprehensive theoretical and conceptual basis for new technologies. Despite the warning of the early NLW users that "Tools are not the problem. The solution . . . is not to produce more nonlethal technology; it is to set clear policy goals and objectives,"[225] due to its cultural characteristics, the American military has failed to produce these goals and policies and take NLW out of their niche of strictly law enforcement purposes.

CONCLUSIONS: THE FAILURE OF THE RMA OF NLW IN THE UNITED STATES

Analyzing the history of NLW in the United States, it is important to highlight that the peak of American interest in NLW coincided with the employment of other revolutionary technologies related to information-led RMA, in general, and precision-guided munitions, in particular. During that time, the mainstream discussion was that the major purpose of these technologies was to provide the U.S. military with higher precision, faster reaction and, more importantly, higher lethality. Thus, it is not surprising that the role of NLW remained at the periphery of American thinking at that time.[226]

Moreover, since then, the American military establishment has never felt stressed by its political leadership to reduce civilian casualties from the politically acceptable level to the lowest possible—the war in Iraq proved that American political tolerance of enemy civilian casualties is very high. While in Iraq there were no World War II–style city bombings like in Hamburg, Dresden, or Tokyo, or the carpet bombing of the 1960s in Vietnam,[227] after four years of war, American public opinion expressed very little concern about 200,000 Iraqi civilian deaths (as direct and indirect outcomes of the U.S. intervention).

While the frontier officially disappeared at the end of the nineteenth century, it has never disappeared from the American military culture that has continued to maintain the myth of the "First American Way of War." Analyzing the linguistic and practical discourses of the U.S. military in Iraq, anthropologists highlight that, on the metaphorical level, the "Wild West" is very much alive and kicking within the American military.[228] During the Iraqi War, statements like "Anbar [a province of Iraq] has the savagery, lawlessness, and violence of America's Wild West" or "We refer to our base as 'Fort Apache' because its right in the middle of Indian Country" had become the norm, representing not only an attempt to "recapitulate the presumed success of the U.S. military against Native Americans,"[229] but also the relevance of the violent approach of the "First American Way of War" toward the local "savage and uncivilized" population.

On the international arena, Washington has succeeded in maintaining the "good" intentions of the U.S. military activity. Shaped by the inherent belief in American destiny as *a City upon a Hill*, American politicians have successfully projected this image onto the international arena, as President Bush put it in 2004:

> Like generations before us, we have a calling from beyond the stars to stand for freedom. This is the everlasting dream of America . . . we go forward, grateful for our freedom, faithful to our cause, and confident in the future of the greatest nation on Earth.[230]

The "we are good" image of America—as *the protector of the free world* against *a great evil*, two main narratives of the Cold War–era foreign policy—has been successfully utilized to minimize the influence of international criticism, as well as to improve the already high domestic support. In addition, the strategy of "collations of the willing" (when it was implemented) successfully eliminated American unilateral political responsibility for the consequences of U.S. military activity, helping to spread the criticism among as many players as possible.

One of the main reasons for the lack of progress with NLW is the fact that in the context of American culture, the idea of non-lethality sounds like an oxymoron. As discussed, the American mind, bred on the violence and the sense of vengeance of the frontier, perceives war as a temporary unfortunate phenomenon that has to be resolved as quickly and aggressively as possible—this approach leaves very little space for the appreciation of enemy populations. Furthermore, American military culture, predisposed toward projection of excessive firepower, combined with a view that the military activity is temporary, leaves even less room for weapons that are not intended to improve the range, accuracy, speed, or lethality of the forces. Despite the fact that the wars in Iraq and Afghanistan were anything but short, due to the inability of American military culture to deal with prolonged conflicts, these interventions were, in fact, "truly long wars fought one year at a time."[231]

While the American military cultural context paid very little respect to the enemy's population, it will be incorrect to claim that the U.S. military did not try to reduce civilian casualties in Iraq and Afghanistan. U.S. Army Lieutenant General Curtis Scapaaotti, the former commander of ISAF Joint Command in Afghanistan, stated in his foreword to the handbook titled *Afghanistan Civilian Casualties Prevention*, that:

> Lethal force is part of war, and we must ensure our soldiers can protect themselves at all times. However, we must take measures to mitigate the impact on the civilian populace we are protecting. Through deliberate planning, training, tactical patience, and effective mission execution, the number of civilian casualty incidents can be significantly decreased and these negative effects minimized. [232]

Interestingly enough, neither his address, nor the handbook itself, paid attention to the technological way to reduce civilian casualties. While the handbook has a very small chapter dedicated to NLW that, in fact, is no more than just a list of available weapons, the main emphasis of the handbook is on the use of lethal force in a way that will try to reduce civilian casualties, by force training, planning considerations, and consequence management.

On the one hand, as was discussed above, in the American military mindset technology is a lodestar of military success. On the other, facing the problem of civilian casualties, the U.S. military has decided to adapt the application of lethal forces, rather than utilize NLW—the most promising technological solution. The explanation for this contradictory behavior is simple. A full implementation of any RMA demands significant political pressure, in the case of NLW—political pressure to minimize enemy civilian casualties. Collateral damage, civilian casualties, and even "small massacres" are considered by the American mind as unfortunate, accidental, but inevitable outcomes of war, and therefore politically defendable. It is important to note that after a decade of American involvement in Iraq and Afghanistan, the U.S. military began to understand that "the impact of CIVCAS [civil casualties] has increased to the point that single tactical actions can have strategic consequences and limit overall freedom of action."[233] Consequently, it will be wrong to claim that there was no political pressure at all; it was definitely there, just not powerful enough to demand radical changes that are required for the implementation of revolutionary technologies.

This lack of political pressure, however, was not the only reason for the failure of the RMA of NLW in the United States. Due to traditional military style and culture, the American military was unable to produce a clear and comprehensive vision of what is required from NLW. Based on the early assumption that NLW have a merely law enforcement role, due to its cultural, structural and historical contexts, the American military thought was incapable of developing the conceptual basis required for the RMA of NLW. Without a clear military concept regarding NLW, and without extensive political pressure that will require reducing enemy civilian casualties to the lowest possible number, NLW are condemned to remain a marginal and oxymoronic phenomenon in the history of the technological adventures of the U.S. military.

NOTES

1. Michael Mazarr, *The Military Technical Revolution: A Structural Framework. Final Report of the CSIS Study Group on the MTR* (Washington, DC: Center for Strategic and International Studies, 1993), 43.

2. U.S. Government Accountability Office (GAO), *DOD Needs to Improve Program Management, Policy, and Testing to Enhance Ability to Field Operationally Useful Non-Lethal Weapons: Report to Congressional Requesters*, Washington, DC, April 2009, 18.

3. John Alexander, "Shoot, But Not to Kill—Non-Lethal Weapons Have Yet to Establish a Military Niche," *Jane's International Defense Review*, June 1, 1996; David Boyd, "The Search for Low Hanging Fruit: Recent Developments in Non-Lethal Technologies," in Malcolm Dando (ed.), *Non-Lethal Weapons: Technological and Operational Prospects*, Jane's Special Report, November 2000.

4. Quoted in Alexander, "Shoot, But Not to Kill," 219.

5. Neil Davison, *Non-Lethal Weapons* (Basingstoke: Palgrave MacMillan, 2009), 52.

6. U.S. Department of Defense, *Policy for Non-Lethal Weapons, Directive 3000.3*, Washington DC, July 9, 1996.

7. Joint Non-Lethal Weapons Program, *Purpose*, https://jnlwp.defense.gov/About/Planning-Guidance-Vision-Mission-Definition/ [accessed May 10, 2018].

8. U.S. Government Accountability Office (GAO), *DOD Needs to Improve.*

9. Joint Non-Lethal Weapons Directorate, *Non-Lethal Weapons Reference Book*, June 30, 2011.

10. David Koplow, *Death by Moderation: The U.S. Military's Quest for Usable Weapons* (New York: Cambridge University Press, 2010), 196.

11. Tracy Tafolla, David Trachtenberg, and John Aho, "From Niche to Necessity: Integrating Non-Lethal Weapons into Essential Enabling Capabilities," *Joint Force Quarterly* 66 (2012): 71–79; U.S. Government Accountability Office (GAO), *DOD Needs to Improve*, 25.

12. For example, see James Ogawa, *Evaluating the U.S. Military's Development of Strategic and Operational Doctrine for Non-Lethal Weapons in a Complex Security Environment* (Fort Leavenworth: Command and General Staff College, 2003); Richard Scott, *Conflict Without Casualties: Non-Lethal Weapons in Irregular Warfare* (Monterey: Naval Postgraduate School, 2007); Jeffrey Voetberg, *Non-Lethal Weapons: Considerations for the Joint Force Commander* (Newport: Naval War College, 2007); Ryan Whittemore, *Air-delivered Non-Lethal Weapons in Counterinsurgency Operations* (Maxwell Air Force Base: Air Command and Staff College, 2008); Timothy Cullen, *Lethality, Legality, and Reality: Non-Lethal Weapons for Offensive Air Support* (Maxwell Air Force Base: Air Command and Staff College, 2008); Dewey Granger, *Integration of Lethal and Nonlethal Fires: The Future of the Joint Fires Cell* (Fort Leavenworth: Command and General Staff College, 2009); Team Bravo/Cohort 19, *Viable Short-Term Directed Energy Weapon Naval Solutions: A Systems Analysis of Current Prototypes* (Monterey: Naval Postgraduate School, 2013).

13. Quoted in Bryan Bender, "USAF Seeks Air-Dropped Non-Lethal Weapons," *Jane's Defence Weekly*, March 24, 1999.

14. Joint Non-Lethal Weapons Directorate, *Non-Lethal Weapons Reference Book*, 60.

15. Steven Metz, "Non-Lethal Weapons: A Progress Report," *Joint Force Quarterly* 28 (2001): 18–22; Douglas Lovelace and Steven Metz, *Nonlethality and American Land Power: Strategic Context and Operational Concept*s (Carlisle: Strategic Studies Institute, 1998); Joseph Siniscalchi, "Non-Lethal Technologies: Implications for Military Strategy," *Center for Strategy and Technology Occasional Paper*, no. 3 (1998); Gerald Norbut, *Non-Lethal Weapons: Force Enabler for the Operational Commander Conducting Peace Operations* (Newport: Naval War College, 2001).

16. U.S. Department of State, Policy Planning Staff, *Non-Lethal and Discriminate Weapons and Technologies*, Washington, 1991.

17. Joint Non-Lethal Weapons Directorate, *Non-Lethal Weapons Reference Book.*

18. "Interview with Lieutenant General Tryon," published in Joint Non-Lethal Weapons Program, *Annual Report*, 2012, 7.

19. Dan Parsons, "Army, Marine Corps Succeed in Rapidly Fielding Specialized Individual Weapons (UPDATED)," *National Defence*, January 2013.

20. The U.S. Army Program Executive Office Soldier, *Equipment Portfolio*, 2014; Joint Non-Lethal Weapons Program, *Annual Report*, 2012, 9.

21. See Security Planning Corporation, *Non-Lethal Weapons for Law Enforcement: Research Needs and Priorities: A Report to the National Science Foundation*, Washington, DC, 1972.

22. U.S. Government Accountability Office (GAO), *DOD Needs to Improve* ; Joint Non-Lethal Weapons Directorate, *Current Non-Lethal Weapons*, http://jnlwp.defense.gov/CurrentNonLethalWeapons.aspx [accessed May 10, 2018]; Joint Non-Lethal Weapons Directorate, *Developing Non-Lethal Weapons*, http://jnlwp.defense.gov/DevelopingNonLethalWeapons.aspx [accessed May 10, 2018].

23. National Security Council, *An Assessment of Non-Lethal Weapons Science and Technology* (Washington, DC: National Academies Press, 2003), 99.

24. For example, James Giglio, *The Presidency of John F. Kennedy* (Lawrence: University Press of Kansas, 1991); Laurence Barrett, *Gambling with History: Ronald Reagan in the White House* (Garden City, NJ: Doubleday, 1983); Stephen Cimbala (ed.), *The George W. Bush Defense Program: Policy, Strategy & War* (Washington, DC: Potomac Books, 2010).

25. For example, Michael Hunt, *Ideology and the U.S. Foreign Policy* (New Haven, CT: Yale University Press, 1987); James Meernik, *The Political Use of Military Force in US Foreign Policy* (Aldershot: Ashgate Publishing Limited, 2004).

26. For example, Philip Gordon and Jeremy Shapiro, *Allies at War: America, Europe and the Crisis over Iraq* (New York: McGraw-Hill, 2004); John Peterson and Mark Pollack (ed.), *Europe, America, Bush: Transatlantic Relations in the Twenty-First Century* (London: Routledge, 2003); Robert Kagan, *Paradise & Power: America and Europe in the New World Order*, Revised and Expanded Edition (London: Atlantic Books, 2004).

27. For example, Gordon Chang, *Friends and Enemies: The United States, China, and the Soviet Union, 1948-1972* (Stanford: Stanford University Press, 1990); Norman Graebner, Richard Burns, and Joseph Siracusa, *Reagan, Bush, Gorbachev: Revisiting the End of the Cold War* (Westport: Greenwood Publishing Group, 2008); Donette Murray, *US Foreign Policy and Iran: American-Iranian Relations since the Islamic Revolution* (New York: Routledge, 2009).

28. For example, Andrew Bacevich (ed.), *The Imperial Tense: Prospects and Problems of American Empire* (Chicago: Ivan R. Dee, 2003).

29. For example, Robert Kagan and William Kristol (eds.), *Present Dangers: Crisis and Opportunity in American Foreign and Defense Policy* (San Francisco: Encounter Books, 2000).

30. Richard Betts, *American Force: Dangers, Delusions, and Dilemmas in National Security* (New York: Columbia University Press, 2013), chapters 1–4.

31. John Spanier and Steven Hook, *American Foreign Policy since World War II*, 13th edition (Washington, DC: Congressional Quarterly Inc, 1995), 21.

32. Harry Truman, *Special Message to the Congress on Greece and Turkey*, Washington, DC, March 12, 1947, http://www.presidency.ucsb.edu/ws/index.php?pid=12846&st=truman&st1=special+message [accessed May 10, 2015].

33. Truman, *Special Message to the Congress*.

34. Dwight Eisenhower, *Radio and Television Address to the American People on the Need for Mutual Security in Waging the Peace*, Washington, DC, May 21, 1957, http://www.presidency.ucsb.edu/ws/index.php?pid=11042&st=&st1 [accessed May 10, 2018].

35. "X" [George Kennan], "The Sources of Soviet Conduct," *Foreign Affairs* 25, no. 4 (1947): 566–82.

36. John Kennedy, *Radio and Television Report to the American People on the Soviet Arms Buildup in Cuba*, Washington, DC, October 22, 1962, http://www.jfklibrary.org/Research/Research-Aids/JFK-Speeches/Cuba-Radio-and-Television-Report_19621022.aspx [accessed May 10, 2018].

37. John Kennedy, *Inaugural Address*, Washington, DC, January 20, 1961, http://www.presidency.ucsb.edu/ws/?pid=8032 [accessed May 10, 2018].

38. Lyndon Johnson, *Remarks at Syracuse University on the Communist Challenge in Southeast Asia*, New York, August 5, 1964, http://www.presidency.ucsb.edu/ws/?pid=26419 [accessed May 10, 2018].

39. Lyndon Johnson, *Remarks to Committee Members on the Need for Additional Appropriations for Military Purposes in Viet-Nam and the Dominican Republic*, Washington, DC, May 4, 1965, http://www.presidency.ucsb.edu/ws/?pid=26938 [accessed May 10, 2018].

40. Richard Nixon, *Second Annual Report to the Congress on United States Foreign Policy*, Washington, February 25, 1971, http://www.presidency.ucsb.edu/ws/?pid=3324 [accessed May 10, 2018].

41. Gerald Ford, *Address Before a Joint Session of the Congress Reporting on the State of the Union*, Washington, January 15, 1975, http://www.presidency.ucsb.edu/ws/?pid=4938 [accessed May 10, 2018].

42. Jimmy Carter, *The President's News Conference*, Chicago, May 25, 1978, http://www.presidency.ucsb.edu/ws/?pid=30852 [accessed May 10, 2018].

43. Ronald Reagan, *Message to the Congress Transmitting the Fiscal Year 1984 Budget*, Washington, DC, January 31, 1983, http://www.presidency.ucsb.edu/ws/?pid=40539 [accessed May 10, 2018].

44. Ronald Reagan, *New Year's Messages of President Reagan and President Mikhail Gorbachev of the Soviet Union*, Washington, DC, January 1, 1989, http://www.presidency.ucsb.edu/ws/?pid=35313 [accessed May 10, 2018].

45. George Bush, *The President's News Conference Following the North Atlantic Treaty Organization Summit Meeting in Brussels*, Brussels, May 30, 1989, http://www.presidency.ucsb.edu/ws/?pid=17077 [accessed May 10, 2018].

46. Norman Ornstein, "Foreign Policy and the 1992 Elections," *Foreign Affairs* 72, no. 3 (1992): 1–16.

47. Melanson, *American Foreign Policy since the Vietnam War*, 200–222.

48. George Bush, *Radio Address to the Nation on the Persian Gulf Crisis*, Washington, DC, January 5, 1991, http://www.presidency.ucsb.edu/ws/?pid=19193 [accessed May 10, 2018].

49. George Bush, *Remarks at the United States Military Academy in West Point*, West Point, New York, January 5, 1993, http://www.presidency.ucsb.edu/ws/?pid=20414 [accessed May 10, 2018].

50. Michael Cox, *US Foreign Policy after the Cold War: Superpower Without a Mission?* (London: Royal Institute of Foreign Affairs, 1995), 12.

51. Cox, *US Foreign Policy after the Cold War*, 29–36; Spanier and Hook, *American Foreign Policy*, 265–68.

52. William Clinton, *Message to the Congress Transmitting a Report on Somalia*, Washington, DC, October 13, 1993, http://www.presidency.ucsb.edu/ws/?pid=47197 [accessed May 10, 2018].

53. William Clinton, *Address to the Nation on Airstrikes Against Serbian Targets in the Federal Republic of Yugoslavia (Serbia and Montenegro)*, Washington, DC, March 24, 1999, http://www.presidency.ucsb.edu/ws/?pid=57305 [accessed May 10, 2018].

54. George Bush, *Letter to Congressional Leaders Reporting on the National Emergency With Respect to Iraq*, Washington, DC, February 11, 1991, http://www.presidency.ucsb.edu/ws/?pid=19295 [accessed May 10, 2018].

55. William Clinton, *Interview with Dan Rather of CBS News*, Washington, DC, March 31, 1999, http://www.presidency.ucsb.edu/ws/?pid=57340 [accessed May 10, 2018].

56. Clinton, *Address to the Nation on Airstrikes Against Serbian Targets*.

57. George Bush, *A Distinctly American Internationalism*, Simi Valley, November 19, 1999, https://www.cnsnews.com/news/article/distinctly-american-internationalism [accessed May 10, 2018].

58. George Bush, *A Period of Consequences*, Charleston, September 23, 1999, http://fas.org/spp/starwars/program/news99/92399_defense.htm [accessed May 10, 2018].

59. George Bush, *Address to the Nation on the Terrorist Attacks*, Washington, DC, September 11, 2001, http://www.presidency.ucsb.edu/ws/?pid=58057 [accessed May 10, 2018].

60. Bush, *Address to the Nation on the Terrorist Attacks*.

61. George Bush, *Address Before a Joint Session of the Congress on the United States Response to the Terrorist Attacks of September 11*, Washington, DC, September 20, 2001, http://www.presidency.ucsb.edu/ws/?pid=64731 [accessed May 10, 2018].

62. Franklin Roosevelt, *Christmas Eve Message to the Nation*, Washington, DC, December 24, 1941, http://www.presidency.ucsb.edu/ws/?pid=16073 [accessed May 10, 2018].

63. Martin Shaw, *The New Western Way of War* (Cambridge: Polity Press, 2005), 24–25.

64. George Bush, *Remarks to the National Endowment for Democracy*, New York, October 6, 2005, http://www.presidency.ucsb.edu/ws/?pid=73821 [accessed May 10, 2018].

65. George Bush, *Address to the Nation on Iraq From Cincinnati, Ohio*, Cincinnati, October 7, 2002, http://www.presidency.ucsb.edu/ws/?pid=73139 [accessed May 10, 2018].

66. George Bush, *Remarks Following a Meeting with the National Security Team*, Washington, DC, September 12, 2001, http://www.presidency.ucsb.edu/ws/?pid=58058 [accessed May 10, 2018].

67. George Bush, *The President's Radio Address*, Washington, October 6, 2001, http://www.presidency.ucsb.edu/ws/?pid=24998 [accessed May 10, 2018].

68. Bush, *Address to the Nation on Iraq From Cincinnati.*

69. "Infinite Justice, out - Enduring Freedom, in," *BBCNews*, September 25, 2001, http://news.bbc.co.uk/1/hi/world/americas/1563722.stm [accessed May 10, 2018].

70. George Bush, *Remarks in a Victory Celebration*, Washington, DC, November 4, 2004, http://www.presidency.ucsb.edu/ws/index.php?pid=62664 [accessed May 10, 2018].

71. Donald Rumsfeld, *DoD News Briefing with Secretary Rumsfeld and General Pace*, Washington, DC, June 14, 2005, http://archive.defense.gov/Transcripts/Transcript.aspx?TranscriptID=3854 [accessed May 10, 2018].

72. "SPIEGEL Interview: Merkel: Guantanamo Mustn't Exist in Long Term," *Spiegel International Online*, January 9, 2006, http://www.spiegel.de/international/spiegel-interview-merkel-guantanamo-mustn-t-exist-in-long-term-a-394180.html [accessed May 10, 2018].

73. Amnesty International, *Amnesty International Annual Report 2005*, ii, https://www.amnesty.org/en/documents/pol10/0001/2005/en/ [accessed May 10, 2018].

74. George Bush, *Remarks on the War on Terror*, Washington, DC, September 6, 2006, http://www.presidency.ucsb.edu/ws/index.php?pid=779 [accessed May 10, 2018].

75. Alex Spillius, "Barack Obama 'proposes to move terrorist suspects from Guantanamo Bay,'" *The Telegraph*, November 10, 2008, http://www.telegraph.co.uk/news/worldnews/northamerica/usa/3417913/Barack-Obama-proposes-to-move-terrorists-suspects-from-Guantanamo-Bay.html [accessed May 10, 2018].

76. Barack Obama, *Remarks at the National Archives and Records Administration*, Washington, DC, May 21, 2009, http://www.presidency.ucsb.edu/ws/?pid=86166 [accessed May 10, 2018].

77. James Lindsay, "George W. Bush, Barack Obama and the Future of US Global Leadership," *International Affairs* 87, no. 4 (2011): 765–79.

78. Barack Obama, *Remarks to the United Nations General Assembly in New York City*, New York, September 24, 2014, http://www.presidency.ucsb.edu/ws/index.php?pid=107615 [accessed May 10, 2018].

79. Obama, *Remarks to the United Nations General Assembly.*

80. For example, John Ikenberry, "America's Imperial Ambitions," *Foreign Affairs* 81, no. 5 (2002): 44–60; Bacevich, *The Imperial Tense*.

81. Melvyn Leffler and Jeoffrey Lergo (eds.), *To Lead the World: American Strategy after the Bush Doctrine* (Oxford: Oxford University Press, 2008).

82. Samuel Huntington, "The Erosion of American National Interests," *Foreign Affairs* 76, no. 5 (1997): 30.

83. Eliot Cohen, "Kosovo and the New American Way of War," in Andrew Bacevich and Eliot Cohen (eds.), *War over Kosovo* (New York: Columbia University Press, 2001), 51.

84. Robert Osgood, *Alliances and American Foreign Policy* (Baltimore: The John Hopkins Press, 1968), 2.

85. Bacevich and Cohen, *War over Kosovo*.

86. Cohen, "Kosovo and the New American Way of War," 51; Stanley Sloan, *Permanent Alliance?—NATO and the Transatlantic Bargain from Truman to Obama* (London: Bloomsbury Publishing, 2010), chapters 9–11.

87. Quoted in Rhiannon Vickers, "Harold Wilson, the British Labour Party, and the War in Vietnam," *Journal of Cold War Studies* 10, no. 2 (2008): 56.

88. Bruce Jentleson, *American Foreign Policy: The Dynamics of Choice in the 21st Century*, 3rd edition (New York: W.W. Norton & Company, 2007), 368–69; Gordon and Shapiro, *Allies at War*.

89. Quoted in Sloan, *Permanent Alliance?*, 187.

90. G. K. Herring, "The War in Afghanistan: A Strategic Analysis," in Williamson Murray (ed.), *National Security Challenges for the 21st Century* (Carlisle: US Army War College, 2003), 178.

91. Shaw, *The New Western Way of War*, 26.

92. See David Malone, *The International Struggle Over Iraq: Politics and the UN Security Council 1980–2005* (Oxford: Oxford University Press, 2006), chapter 7; Michael Clarke, "The Diplomacy that Led to War in Iraq," in Paul Cornish (ed.), *The Conflict in Iraq, 2003* (New York: Palgrave Macmillan, 2004).

93. Jane Sharp, "The US-UK 'Special Relationship' after Iraq," in Cornish, *The Conflict in Iraq*.

94. Samantha Power, "Legitimacy and Competence," in Leffler and Lergo, *To Lead the World*.

95. Robert Jervis, *American Foreign Policy in a New Era* (New York: Routledge, 2005).

96. Quoted in Dana Milbank, "At UN, Bush Is Criticized over Iraq," *The Washington Post*, September 24, 2003, http://www.washingtonpost.com/archive/politics/2003/09/24/at-un-bush-is-criticized-over-iraq/a8b7d7eb-15be-4b47-a197-6275867a26e7/ [accessed May 10, 2018].

97. Quoted in Huntington, "The Erosion of American National Interests."

98. Noam Chomsky, *Interventions* (London: Penguin Group, 2007), 101–3.

99. This number is based on a "safe" estimation by the author, as is difficult to estimate the exact number of civilian casualties and various scientific surveys offer different numbers; between 150,000 to over one million Iraqis died as a result of conflict during this time. Consequently, the estimation of 200,000 seems to be "safe" enough. See Gilbert Burnham et al., *The Human Cost of the War in Iraq: A Mortality Study, 2002–2006* (Baltimore: Johns Hopkins University, 2006); Amir Alkhuzai et al., "Violence-Related Mortality in Iraq from 2002 to 2006," *New England Journal of Medicine* 358, no. 2 (2008): 484–93; "More than 1,000,000 Iraqis Murdered," *Opinion Research Business*, September 2007. For the methodological problems related to calculating civilian casualties during military operations see Taylor Seybolt, Jay Aronson, and Baruch Fischhoff (ed.), *Counting Civilian Casualties: An Introduction to Recording and Estimating Nonmilitary Deaths in Conflict* (Oxford: Oxford University Press, 2013).

100. Carl Boggs, "Outlaw Nation: The Legacy of U.S. War Crimes," in Carl Boggs (ed.), *Masters of War: Militarism and Blowback in the Era of American Empire* (New York: Routledge, 2003).

101. Henry Commager, *The American Mind: An Interpretation of American Thought and Character Since the 1880's* (New York: Yale University Press, 1950), 3.

102. Commager, *The American Mind*, 4–5.

103. Frederic Turner, "The Significance of the Frontier in American History," in George Taylor (ed.), *The Turner Thesis: Concerning the Role of the Frontier in American History* (Lexington: D.C. Heath and Company, 1972), 3.

104. For example, Ray Billington (ed.), *The Frontier Thesis: Valid Interpretation of American History?* (New York: Holt, Rinehart and Winston, 1966); Taylor, *The Turner Thesis*.

105. Turner, "The Significance of the Frontier," 5.

106. Henry Commager, *The Empire of Reason: How Europe Imagined and America Realized the Enlightenment* (Garden City: Anchor Press, 1977); Stow Person, *American Minds. A History of Ideas* (New York: Henry Holt and Company, 1958); Daniel Boorstin, *The Americans: The Colonial Experience* (New York: Vintage Books, 1958); Daniel Boorstin, *The Americans: The National Experience* (New York: Vintage Books, 1965); Daniel Boorstin, *The Americans: The*

Democratic Experience (New York: Vintage Books, 1973); Russell Weigley, *The American Way of War* (Bloomington: Indiana University Press, 1973).

107. John Grenier, *The First Way of War: American War Making on the Frontier, 1607–1814* (Cambridge: Cambridge University Press, 2005); John Mueller, "Public Opinion as a Constraint on U.S. Foreign Policy: Assessing the Perceived Value of American and Foreign Lives," unpublished text prepared for presentation at National Convention of International Studies Association at Los-Angeles, California, March 14–18, 2000; Thomas Mahnken, *Technology and the American Way of War Since 1945* (New York: Columbia University Press, 2010); Thomas Mahnken, *United States Strategic Culture* (Washington: Science Applications International Corp, 2006); Thomas Mahnken, "The American Way of War in the Twenty-first Century," *The Review of International Affairs* 2, no. 3 (2003): 73–84; John Tirman, *The Deaths of Others: The Fate of Civilians in America's Wars* (Oxford: Oxford University Press, 2011).

108. Allan Nevins and Henry Commager, *America: The Story of Free People* (Oxford: Oxford University Press, 1977), vi.

109. Turner, "The Significance of the Frontier," 4.

110. Nevins and Commager, *America: The Story of Free People*, 184.

111. Robert Fossum and John Roth, *The American Dream* (South Shields: Peterson Printers, 1981), 7.

112. Frederic Turner, "Later Explanations and Developments," in Billington, *The Frontier Thesis*, 24.

113. Walter Webb, "The Frontier and the 400 Year Boom," in Taylor, *The Turner Thesis*, 131.

114. Turner, "Later Explanations," 24.

115. Fossum and Roth, *The American Dream*, 6.

116. Turner, "Later Explanations," 24.

117. Fossum and Roth, *The American Dream*, 21.

118. "Full text: Obama's victory speech," *BBC News*, November 5, 2008, http://news.bbc.co.uk/1/hi/world/americas/us_elections_2008/7710038.stm [accessed May 10, 2018].

119. Boorstin, *The Americans: The Colonial Experience*, 3–4.

120. Ronald Reagan, *Election Eve Address: "A Vision for America,"* Washington, DC, November 3, 1980, http://www.presidency.ucsb.edu/ws/?pid=85199 [accessed May 10, 2018].

121. Turner, "Later Explanations," 23.

122. Boorstin, *The Americans: The National Experience*, 69.

123. Turner, "The Significance of the Frontier," 22.

124. Ibid., 23.

125. Commager, *The American Mind*, 19.

126. Turner, "Later Explanations," 23.

127. Commager, *The American Mind*, 21.

128. Fossum and Roth, *The American Dream*, 25.

129. For example, Peter Lupsha, "American Values and Organized Crime: Suckers and Wiseguys," in Sam Girgus (ed.), *The American Self: Myth, Ideology, and Popular Culture* (Albuquerque: University of New Mexico Press, 1981); Boorstin, *The Americans: The Democratic Experience,* chapter 8; Commager, *The American Mind.*

130. Marshall Smelser, "An Understanding of the American Revolution," in Walter Nicgorski and Ronald Weber (ed.), *An Almost Chosen People: The Moral Aspirations of Americans* (Notre Dame, IN: The University of Notre Dame Press, 1976).

131. See Alphonso Pinkney, *The American Way of Violence* (New York: Random House, 1972), chapter 6; Tirman, *The Deaths of Others*, chapter 2.

132. Pinkney, *The American Way of Violence*, 169.

133. Ibid., 168–71.

134. Ibid., 79.

135. Poll completed by Gallup, http://www.gallup.com/poll/1606/death-penalty.aspx [accessed May 10, 2018].

136. James Adams, *The Epic of America* (London: George Routledge & Sons Ltd., 1940), 415.

137. Tirman, *The Deaths of Others*, chapter 2.

138. For example, Antulio Echevarria, *Reconsidering the American Way of War: US Military Practice from the Revolution to Afghanistan* (Washington, DC: Georgetown University Press, 2014); Adrian Lewis, *The American Culture of War: The History of U.S. Military Force from World War II to Operation Iraqi Freedom* (New York: Routledge, 2007); Benjamin Buley, *The New American Way of War: Military Culture and the Political Utility of Force* (New York: Routledge, 2008); Colin Gray, *Irregular Enemies and the Essence of Strategy: Can the American Way of War Adapt?* (Carlisle: Strategic Studies Institute, 2006); Weigley, *The American Way of War*; Grenier, *The First Way of War*.

139. "The First American Way of War" generally corresponds with the phenomenon defined in Grenier, *The First Way of War*; "The American Way of War" generally corresponds with the phenomenon defined in Weigley, *The American Way of War*; and "The New American Way of War" generally corresponds with the phenomenon defined in Buley, *The New American Way of War*.

140. Adam Hirsch, "The Collision of Military Cultures in Seventeenth-Century New England," *Journal of American History* 74, no. 4 (1988), 1204; See also Ian Steele, *Warpaths: Invasions of North America* (New York: Oxford University Press, 1994); John Ferling, *A Wilderness of Miseries: War and Warriors in Early America* (Westport: Greenwood Press, 1980); Ronald Karr, "'Why Should You Be So Furious?': The Violence of the Pequot War," *Journal of American History* 85, no. 2 (1998): 876–909; Grenier, *The First Way of War*.

141. Hirsch, "The Collision of Military Cultures in Seventeenth-Century New England," 1188.

142. Grenier, *The First Way of War*, 1.

143. Ibid., chapter 5.

144. Smelser, "An Understanding of the American Revolution," 17.

145. Asymmetric is—"In military operations the application of dissimilar strategies, tactics, capabilities, and methods to circumvent or negate an opponent's strengths while exploiting his weaknesses"; Irregular Warfare is "a violent struggle among state and non-state actors for legitimacy and influence over the relevant population(s)." U.S. Department of Defense, *The Joint Publication 1-02, Department of Defense Dictionary of Military and Associated Terms*, Washington, DC, 8 November 2010 (as amended through 15 February 2016).

146. Smelser, "An Understanding of the American Revolution," 13.

147. Karr, "'Why Should You Be So Furious?,'" 880.

148. Hirsch, "The Collision of Military Cultures," 1206.

149. Karr, "'Why Should You Be So Furious?,'" 888.

150. Tirman, *The Deaths of Others*, 18.

151. Weigley, *The American Way of War*, xxi.

152. Cohen, "Kosovo and the New American Way of War," 40–41.

153. Weigley, *The American Way of War*.

154. Cohen, "Kosovo and the New American Way of War," 40–46.

155. Weigley, *The American Way of War*, xviii.

156. Carl Von Clausewitz, *On War*, edited and translated by Michael Howard and Peter Paret (Princeton: Princeton University Press, 1989), 87.

157. Buley, *The New American Way of War*, 2.

158. The Command and General Staff School, *Principles of Strategy for an Independent Corps or Army in a Theater of Operations* (Fort Leavenworth: The Command and General Staff School Press, 1936), 19.

159. Gray, *Irregular Enemies and the Essence of Strategy*, 44.

160. Ibid., 30.

161. For example, Max Boot, "The New American Way of War," *Foreign Affairs* 82, no. 4 (2003): 41–58; Buley, *The New American Way of War*; Cohen, "Kosovo and the New American Way of War."

162. Buley, *The New American Way of War*, 6–15.

163. Ibid., 20.

164. Eugene Jarecki, *Why We Fight* (Axiom Films, 2005).

165. "Bowe Bergdahl release: US parties clash on Afghan deal," *BBC New*, June 2, 2014, http://www.bbc.co.uk/news/world-asia-27658821 [accessed May 10, 2018].

166. "Text of George Bush's speech," *The Guardian*, http://www.theguardian.com/world/2001/sep/21/september11.usa13 [accessed May 10, 2018].
167. Tirman, *The Deaths of Others*, 22.
168. Boggs, "Owtlaw Nation," 191–94.
169. Mueller, "Public Opinion," 1.
170. Ibid.
171. Sadao Asada, "The Mushroom Cloud and National Psyches," in Laura Hein and Mark Selden (ed.), *Living with the Bomb: American and Japanese Cultural Conflicts in the Nuclear Age* (Armonk: M. E. Sharpe, 1997), 177–79; see also Beatrice Heuser, *The Bomb: Nuclear Weapons in their Historical, Strategic and Ethical Context* (London: Longman, 2000), chapter 4.
172. Bruce Stokes, "70 Years after Hiroshima, Opinions Have Shifted on Use of Atomic Bomb," *Pew Research Center*, August 4, 2015, http://www.pewresearch.org/fact-tank/2015/08/04/70-years-after-hiroshima-opinions-have-shifted-on-use-of-atomic-bomb/ [accessed May 10, 2018].
173. Mark Lorell and Charles Kelly, *Casualties, Public Opinion, and Presidential Policy during Vietnam War* (Santa Monica: The RAND Corporation, 1985), 28.
174. Gray, *Irregular Enemies and the Essence of Strategy*, 47–49.
175. Eric Larson and Bogdan Savych, *American Public Support for U.S. Military Operations from Mogadishu to Baghdad* (Santa Monica: The RAND Corporation, 2005), 121.
176. Eric Larson and Bogdan Savych, *Misfortunes of War: Press and Public Reaction to Civilian Deaths in Wartime* (Santa Monica: The RAND Corporation, 2006), 157.
177. On July 1, 2002, U.S. aircraft attacked six sites in an Afghan village where a wedding was being celebrated, reportedly killing 48 people and wounding another 117. See Alissa Rubin, "The Conflict in Iraq: 40 Killed in U.S. Attack on Iraqi Border Village," *Los Angeles Times*, May 20, 2004, http://articles.latimes.com/2004/may/20/world/fg-iraq20 [accessed May 10, 2018].
178. On March 26, 2003, Iraqi officials claimed that a coalition missile fell on a Baghdad marketplace killing 14 and injuring more than 30 civilians (later the toll of dead was raised to more than 36). See "'36 killed' in raids on Baghdad," *The Guardian*, March 27, 2003, http://www.theguardian.com/world/2003/mar/27/iraq.sallybolton [accessed May 10, 2018].
179. Larson and Savych, *Misfortunes of War*, 155.
180. Ibid., 195.
181. Ibid., xx.
182. Lorell and Kelly, *Casualties, Public Opinion and Presidential Policy*, 25.
183. Ralph Peters, *Fighting for the Future: Will America Triumph?* (Mechanicsburg: Stackpole Books, 2001), 105.
184. Colin Gray, "National Style in Strategy: The American Experience," *International Security* 6, no. 2 (1981): 44.
185. Tirman, *The Deaths of Others*.
186. Jeffrey Record, *Beating Goliath* (Washington, DC: Potomac Books, 2009), chapter 5; Gray, *Irregular Enemies*, 30–32.
187. Antulio Echevarria, *Towards an American Way of War* (Carlisle: Strategic Studies Institute, 2004), 10.
188. Thomas Kane, *Military Logistics and Strategic Performance* (London: Frank Cass, 2001); Gray, *Irregular Enemies*, 45–47.
189. Mahnken, *Technology and the American Way*, 5; see also Richard Lock-Pullan, "U.S. Military Strategy, Strategic Culture, and the 'War on Terror,'" in John Owens and John Dumbrell (ed.), *America's "War on Terrorism": New Dimensions in U.S. Government and National Security* (Lanham, MD: Lexington Books, 2008), 161.
190. Gray, *Irregular Enemies*, 35–36.
191. Williamson Murray, "The Future of American Military Culture: Does Military Culture Matter?," *Orbis* 43, Issue 1 (1999): 27–42; Lock-Pullan, "U.S. Military Strategy," 161.
192. Gray, *Irregular Enemies*, 35–36; see also Dima Adamsky, *The Culture of Military Innovation: The Impact of Cultural Factors on the Revolution in Military Affairs in Russia, the US, and Israel* (Stanford: Stanford University Press, 2010), 85–87.

193. Theodor Galdi, *Revolution in Military Affairs?— Competing Concepts, Organizational Responses, Outstanding Issues*, Congressional Research Service Report to the Congress (CRS 951170F, 11 December 1995).

194. Barry Watts, *Six Decades of Guided Munitions and Battle Networks: Progress and Prospects* (Washington: Center for Strategic and Budgetary Assessments, 2007), 77.

195. Adamsky, *The Culture of Military Innovation*, 2.

196. William Owens, "The American Revolution in Military Affairs," *Joint Force Quarterly* 10 (1996): 37.

197. Ibid., 38.

198. Lock-Pullan, "U.S. Military Strategy," 177.

199. U.S. Joint Chiefs of Staff, *Capstone Concept for Joint Operations: Joint Force 2020*, Washington, September 10, 2012.

200. Murray, "The Future of American Military Culture."

201. Weigley, *The American Way of War*, xxi; Mahnken, *United States Strategic Culture*, 10–12.

202. Richard Lock-Pullan, *US Intervention Policy and Army Innovation: From Vietnam to Iraq* (London: Routledge, 2006), 13; Adamsky, *The Culture of Military Innovation*, 80.

203. Williamson Murray, "Transformation and Professional Military Education: Past as Prologue to the Future," in Williamson Murray (ed.), *National Security Challenges for the 21st Century* (Carlisle: US Army War College, 2003); Murray, "The Future of American Military Culture"; Adamsky, *The Culture of Military Innovation*.

204. Murray, "Transformation and Professional Military Education," 10–11.

205. Murray, "Transformation and Professional Military Education," 11; see also Williamson Murray, "Introduction," in Williamson Murray (ed.), *The Emerging Strategic Environment: Challenges of the Twenty-first Century* (Westport: Greenwood Publishing Group, 1999), xxvi–xxvii.

206. Carl Builder, *The Masks of War: American Military Styles in Strategy and Analysis* (Baltimore: John Hopkins University Press, 1989).

207. Adamsky, *The Culture of Military Innovation*, 83–85.

208. Murray, "The Future of American Military Culture," 40.

209. Adamsky, *The Culture of Military Innovation*, 88.

210. U.S. Department of Defense, *Policy for Non-Lethal Weapons*.

211. Joint Non-Lethal Weapons Program, *Organization*, http://jnlwp.defense.gov/About/Organization.aspx [accessed June 24, 2015].

212. Joint Non-Lethal Weapons Program, *Non-Lethal Weapons Requirement Fact Sheet*, http://jnlwp.defense.gov/pdf/pressroom/NLW%20Requirements%20Fact%20Sheet%20Oct%202011.pdf [accessed June 24, 2015].

213. Department of the Army Headquarters, United States Army, *Force Operating Capabilities: TRADOC Pamphlet 525-66*, Fort Monroe, March 7, 2008, 88.

214. U.S. Joint Chiefs of Staff, *Capstone Concept for Joint Operations*.

215. Defense Science Board, Department of Defense, *Challenges to Military Operations in Support of U.S. Interests*, Volume II, Main Report, Washington DC, December 2008, 9.

216. Department of the Army Headquarters, United States Army, *Force Operating Capabilities*, 88.

217. Ibid.

218. Lovelace and Metz, *Nonlethality and American Land Power*; Steven Metz, "Non-Lethality and the Revolution in Military Affairs," in Dando, *Non-Lethal Weapons: Technological and Operational Prospects*; Metz, "Non-Lethal Weapons: A Progress Report."

219. David Koplow, *Non-Lethal Weapons: The Law and Policy of Revolutionary Technologies for the Military and Law Enforcement* (New York: Cambridge Press, 2006); David Koplow, "Red-Teaming NLW: A Top Ten List of Criticisms about Non-Lethal Weapons," *Case Western Reserve Journal of International Law* 47, Issue 1 (2015): 229–38; Koplow, *Death by Moderation*.

220. James Duncan, "A Primer on the Employment of Non-Lethal Weapons," *Naval Law Review* 45 (1998): 1.

221. F. M. Lorenz, "Non-Lethal Force: The Slippery Slope to War?," *Parameters* XXVI, no. 3 (1996): 52–62.

222. Martin Stanton, "What Price Sticky Foam?," *Parameters* XXVI, no. 3 (1996): 63–68.

223. Duncan, "A Primer on the Employment," 33; Also see Timothy Lamb, *Emerging Non-Lethal Weapons Technology and Strategic Policy Implications for 21st Century Warfare* (Carlisle Barracks: U.S. Army War College, 1998); Lovelace and Metz, *Nonlethality and American Land Power*.

224. United Stated Marine Corps, *Non-Lethal Weapons (NLW) Training and Readiness (T&R) Manual*, Washington, 2012.

225. Stanton, "What Price Sticky Foam?"

226. Metz, "Non-Lethality and the Revolution in Military Affairs."

227. Howard Zinn, *Vietnam: The Logic of Withdrawal* (Boston: Beacon Press, 1967), Chapter 6.

228. For example, Daniel Bolger, *Savage Peace: Americans at War in the 1990s* (New York, Presidio, 1995); Robert Kaplan, *Imperial Grunts: The American Military on the Ground* (New York: Random House, 2005); Stephen Silliman, "The 'Old West' in the Middle East: U.S. Military Metaphors in Real and Imagined Indian Country," *American Anthropologist* 110, Issue 2 (2008): 237–47.

229. Silliman, "The 'Old West' in the Middle East."

230. George Bush, *Remarks Accepting the Presidential Nomination at the Republican National Convention in New York City*, New York, September 2, 2004, http://www.presidency.ucsb.edu/ws/index.php?pid=72727& [accessed May 10, 2018].

231. Command Sergeant Major Robert Rush quoted in Thomas Ricks, "Rotation Policy: Iraq & Afghanistan Were Really the Wars We Fought 1 Year at a Time," *Foreign Policy*, May 27, 2015, http://foreignpolicy.com/2015/05/27/rotation-policy-iraq-afghanistan-were-really-the-wars-we-fought-1-year-at-a-time/ [accessed May 10, 2018].

232. Center for Army Lessons Learned (CALL), *Afghanistan Civilian Casualties Prevention,* Handbook 12–16, June 2012, p. IV.

233. Ibid., 1.

Chapter Three

"Besieged Fortress" and the Politics of "Double Standards"

Russia and Enemy Civilian Casualties

The development and use of non-lethal weapons for law enforcement purposes in the former USSR was more or less similar to the developments in the West. Different chemical, kinetic, and acoustical special munitions have been employed by both the Ministry of Interior (MVD) Forces and the Russian Ministry of Defense. The MVD has employed NLW at least since the 1970s.[1] However, it is important to note that until today it uses the term *special'nie sredstva* (special means) that has no generic definition, but represents a list of different means, tools, and other equipment, including NLW, which police officers have the right to use in specific situations for specific purposes. The list of equipment, as well as when and how it can be used, is regulated by Russian criminal law.[2] The Russian Ministry of Defense, interestingly enough, employs a different term—*orug'ie neleta'nogo de'stviya* (weapon with non-lethal action),[3] which is closer to the Western approach.

The difference between these two terms is not only semantic. While the first is a part of federal legislation and represents a very concrete list of means and weapons approved for police use, the second one presents a more abstract concept, mainly based on Western approaches.[4] Despite the fact that the Russian Ministry of Defense made an attempt to formalize the status of NLW in the Russian military forces, by publishing in 2009 the *Conception to the Development and Employment of Weapons with Non-Lethal Action in the Armed Forces of the Russian Federation*, it seems that not only has this attempt failed,[5] but also the reason for this failure was more of political nature rather than a simple military malfunction.

THE STATUS OF NON-LETHAL WEAPONS IN RUSSIA

While, as discussed above, the development and employment of NLW by Soviet and later Russian law enforcement agencies were independent but followed similar lines to those of the West, the growth of interest toward these weapons in the Russian military has been directly influenced by Western (especially American) military attitudes that emerged in the mid-1990s.[6] The analysis of Western policies and views on the future of NLW brought the Russian decision-makers to the understanding of the important potential of these technologies for future military operations. This understanding culminated in the second part of the 2000s in the publication of two pivotal documents: the *Conception to the Development and Employment of Weapons with Non-Lethal Action* (*Kontseptsiya sozdaniya i primeneniya oruzhiya neletal'nogo deystviya*) published by the Military-Industrial Committee of the Government of the Russian Federation in 2005, and the *Conception to the Development and Employment of Weapons with Non-Lethal Action in the Armed Forces of the Russian Federation* (*Kontseptsiya sozdaniya i primeneniya oruzhiya neletal'nogo deystviya v Vooruzhennykh Silakh Rossiyskoy Federatsii*) approved by the Russian Minister of Defense in 2009.[7]

This apparent enthusiasm emanating from the highest level of political-military decision-makers was swiftly accepted by different defense industries. In an attempt to answer this new fashion for NLW, the main weapon developers and manufacturers directed their efforts in two different directions. The first one was the adjustment of the existing lethal ordnance with non-lethal (mainly chemical) effects. This led to the development and demonstration of a spectrum of non-lethal munition filled with chemical irritant-action pyrotechnic composition, such as: 105 mm munition for RPG grenade launcher, 120 mm mortal projectile, Helicopter Container Dispenser, and even 500 kg Air Cluster Bomb.[8] The second direction was the development of new weapons specifically designed as NLW. Although none of the manufacturers has demonstrated functioning prototypes, records in open literature indicate at least two main fields of research and development activities: directed energy weapons[9] and electrical weapons for significant distances.[10]

The military force, chosen to be the test-bed for the employment of NLW, was the Collective Rapid Reaction Force (CRRF) from the Russian-led Collective Security Treaty Organization (CSTO). This unit, rather than the ordinary contingent from the Russian military, was selected due to "its primary tasks, as well as political-military, economic and social-demographic conditions, in which these tasks have to be accomplished."[11] In other words, as CRRF is intended to operate exclusively in the territory of the CSTO countries,[12] within friendly population, the reduction of collateral damage and civilian casualties seemed to be important. The introduction of NLW to CRRF occurred in 2010 during the large training operation "Vzaimodeystviye-2010" (Cooperation-2010) in Russia. During the training, a company from the Russian 98th Guards Airborne Division and a unit from the Kyrgyz Special Forces "Ilbirs" Brigade, with the assistance of NLW, performed three missions: (1) counteracting an enemy ambush, (2) securing the free movement of military convoys through the roadblocks organized by a violent civil population, and (3) liberating a village captured by irregular forces.[13] While this could be a good start for the military usage of NLW in Russia, the reality was completely the opposite. Firstly, this training did not involve any NLW developed specifically for military use, but was based on existing weapons currently employed by the Russian MVD forces, such as irritant and flash-bang hand grenades, 40 mm non-lethal munition, multipurpose non-lethal pistols, etc.[14] Secondly, despite the positive demonstration of the role of NLW in future military operations, their further employment has not been initiated.

The initial enthusiasm has been successfully buried by the Russian bureaucracy machine. Without any integrating systematic mechanism that organizes, regulates, and focuses the development and employment of NLW, each main governmental service (i.e., MVD, Federal Security Service [FSB], and Armed Forces) has developed its own approach, and any attempts to coordinate inter-departmental committees were initially doomed to fail due to traditional rivalry between these organizations.[15] In light of this bureaucratic uncertainty, it is not surprising that NLW projects have suffered from insufficient governmental funding,[16] and without visible potential contracts, the industries have reduced their investments to the minimum. Moreover, it is important to state that despite the fact that CRRF is built mainly from pure military units, it is designated to operate inside the country—a task that is the traditional responsibility of MVD internal armed forces. This contradiction has played an additional role in the process of the general stagnation regarding NLW after the optimistic training in 2010. On the one hand, MVD forces have already had a satisfying arsenal of NLW (for their own purposes),[17] hence this military employment was a mere adaptation of the existing law enforcement weapons and tactics. On the other hand, due to the domestic nature of the CRRF's operational scenarios, the Ministry of Defense and Armed Forces saw no reason to expand the experience achieved by CRRF to their regular forces, intended to fight on conventional battlefields.

To summarize the current situation with NLW in Russia, it is important to focus on two main factors. The first one is the technological status. As noted before, Russian defense industries have the required technological potential to supply efficient NLW to the military forces. While some of these weapons, apparently, will be based on the existing solutions for law enforcement agencies,[18] Russian manufacturers have the necessary capabilities to produce NLW intended specifically for military use (and have already demonstrated that).[19] The second factor is the actual employment of NLW within Russian

Armed Forces. It seems that the CRRF experience with NLW was a very limited one-time experiment, rather than a test-bed for the possible expansion of NLW future employment. In an addition to CRRF, the only other military unit that employs NLW is the Russian Military Police,[20] which uses these weapons to fulfil its security missions, rather than on the battlefield.

While there are no visible substantial plans to expand this situation to the regular Armed Forces, it seems important to understand the nature of the main obstacles that prevent the further development of NLW. The following two parts discuss the reasons behind the general impotence of political decision-makers to coordinate activity in the field of NLW and provide substantial funding.

RUSSIAN FOREIGN POLICY AND INTERNATIONAL CRITICISM

The foundations of Russian foreign policy in the second decade of the twenty-first century lie in Russia's political and economic transformation processes since the dissolution of the Soviet Union. Russian scholars divide this 25-year period into three main parts, waves, or generations: "Blind" Westernization (first part of 1990s), Selective Partnership (from mid-1990s to mid-2000s), and Revitalization of Russia (since mid-2000s).[21] While this research is mainly interested in the nature of the last generation of Russian foreign policy, a brief discussion of the previous two seems vital as they provide a key to the understanding of the current "revitalization" concept and its role in the Russian response to international criticism.

According to Professor Alexei Bogaturov, with the dissolution of the USSR, Russian political decision-makers sincerely believed that the Western community was looking for an opportunity for political and economic integration with Russia, similar to the processes started with other post-Soviet countries in Eastern Europe.[22] The end of the Cold War symbolized the unconditional victory of democracy, and the foreign policy of the newborn democratic Russia was based on two main concepts: (1) securing favorable international conditions for the building of democracy in Russia, and (2) that the main and universal political value is the democratization of the world. The political notion, in the early 1990s, was that Russian domestic and international success had to be based, not only on a Russian attempt to be more similar to the West, but, in fact, on the Russian effort to be a part of the West. The best example of Russian attempts to be integrated with the West was the 1992 Charter for American-Russian Partnership and Friendship. It was the first time in Russian history that an agreement, which was signed with another state, defined Russian internal politics, where, from the Russian perspective, Moscow accepted the United States as an external referee of Russian internal reforms. The Russian leadership of the early 1990s avoided any clear indication of their national interests in foreign policy, yet identified these interests with the interests of the democratic countries, the so-called "global community of democracies." Moscow strongly urged foreign partners to recognize that the support of the Western initiatives was, in fact, the main goal of Russian foreign policy. The best example of this logic of universal "democratic solidarity" was Russian political behavior in the initial period of the Yugoslavian Crisis, when Moscow unconditionally supported the creation of new independent states in the territory of the former federation. It declared its diplomatic recognition of the new governments of the former Yugoslavia at the same time as the European Union, and before the United States. The Kremlin's refusal to support the central government in Belgrade against separatists in Croatia and Slovenia was a surprise to many Western diplomats and many Russian citizens, as well.[23]

Russian political analysts and scholars have claimed that not only this policy of "Blind" Westernization had no success, domestically or internationally, but in fact, was a complete failure. Domestically, years of economic crisis and the process of political decentralization almost tore Russia apart, delegitimizing the Kremlin in the eyes of regional leaders and the general population. Internationally, the first part of 1990s was a period of rapid contraction of Russian influence. Economic inability to maintain relations established during the Soviet era, and absolute adaptation of Western policies at the expense of the Kremlin's interests, not only stopped Russia from preserving its interests in former socialist European countries, but also led the Kremlin to abandon its partners in the Middle East, South Asia, Africa, and Latin America. The conclusion of Russian political observers was that in less than five years, Russian political status was reduced from a self-confident global power to a politically irrelevant appendage of the West.[24] This self-effacing policy started to change during the mid-1990s due to two main reasons. The first one was domestic. Internal pro-Western decentralization policies, formulated in an attempt to be admired by the West, created economic chaos, notions of separatism, and danger of additional secession. The second reason was the Western reaction to the Russian attempt to westernize itself. The West showed little appreciation for Russian solidarity and did not show any desire for integration with an economically struggling, politically unstable state that until recently was its number one enemy. As a reaction to these developments, the Kremlin started to shift its foreign policy doctrine from unconditional compliance to the so-called "Selective Partnership."[25]

While the course of Russian foreign policy, in the second half of the 1990s, generally preserved its pro-Western narratives, the main emphasis of Moscow's new doctrine changed from an unequivocal agreement to political bargaining. Utilizing the political advantages of being the only successor to the USSR, i.e., the biggest nuclear power and a permanent member of the UN Security Council, Moscow started to replace the romanticism of the early

1990s with a more pragmatic approach.[26] According to Bogaturov, the Kremlin carefully began to demonstrate and defend its interests. The best example of this transformation was Moscow's reaction to the Kosovo crisis. The Kremlin's gingerly support for NATO-led peacekeeping operations in 1995 and 1996 (mainly due to the fact that Russia was a part of the solution) was replaced by strong opposition to the NATO intervention in 1999.[27]

The Kremlin was very disappointed by the inability of the Western countries to accept the idea that Russian historical ambitions to be a leading actor in the geopolitical arena did not disappear with the collapse of the USSR. In 2015, in a famous documentary, *President*, Putin recalled:

> I served in the KGB, in foreign intelligence, for almost 20 years, and even I thought that with the collapse of this ideological obstacle, in the shape of the Communist Party's monopoly on power, everything will cardinally change. No . . . it did not change in essence, because, as it turned out, . . . there are geopolitical interests that are completely disconnected from any ideology. This is where our partners [the West] were required to understand that Russia is a country that has, and it cannot be without, its own geopolitical interests. And it is required to respect each other, while searching for balances and mutually accepted resolutions.[28]

In the beginning of the twenty-first century, the Russian position in the international arena was very selective and non-partisan. On the one hand, the Kremlin finally accepted NATO's expansion into Eastern Europe, and, following the 9/11 events, unconditionally supported the NATO-led anti-terror coalition, created in 2002.[29] On the other hand, Russia signed the 2001 Sino-Russian Treaty of Friendship and refused to provide support for the 2003 U.S.-led invasion of Iraq.[30]

Meanwhile, the prices for fossil fuels had remained high, and heavily based on their export, the Russian economy started to recover. Building on rising oil and gas revenues, the Russian economic growth of the first part of 2000s was not only substantial; it was unprecedentedly high and fast.[31] This success had an immediate influence on domestic public opinion that considered Putin as "savior of the nation," "restorer of order," and "distributor of the wealth."[32] Consequently, it was not surprising that on the wave of economic success and domestic support, the Kremlin started to revalue its position in global affairs. The culminating point of this revaluation was President Putin's 2006 speech at the Meeting with the Ambassadors and Permanent Representatives of the Russian Federation, where, for the first time, Russia clearly formulated its political ambitions. This speech symbolized the beginning of a new approach in Russian foreign policy, an approach that stated that the Russian role and its influence on global affairs have to be proportional to and based on its geopolitical history and growing economic capabilities.[33] While this speech was mainly aimed at a Russian internal audience, its "international twin" was the 2007 President Putin's speech at the Munich Security Conference: "Russia—[is] a country with more than a thousand years of history, and it has almost always exercised the privilege to conduct an independent foreign policy. We are not going to change this tradition today."[34]

Building on its economic success, the Kremlin finally stopped knocking on the firmly closed Western door, and sent a clear message to its Western partners that they did not need to pretend any more that they liked the "new-style democratic Russia" and believed in "its bright future."[35] The combination of economic growth, the Soviet legacy in foreign policy, and the international, mainly non-Western, demand for alternative solutions to world crises, transformed Russia back into one of the most significant international players. It was the beginning of a new period in Russian foreign affairs—a period of Russian Revitalization based on segregation from the West (especially the United States), allegations of double standards, and the claim that Western politicized decisions are based on outdated Cold War prejudices.[36]

The "blind" compliance with the West in the 1990s and walking a fine line between Western and Russian interests since the mid-2000s had been replaced by an uncompromising defense of national interests and an open criticism of the West,[37] as President Medvedev put it in 2009:

> Russia, remaining one of the leading economies, a nuclear power and a permanent member of the UN Security Council, must speak openly and directly about its position, defending it at all places. Do not play up and adapt. And in the case of threat to its own interests—decisively defend them.[38]

The best representation of this transformation can be seen in a comparative analysis of Russian self-definition in the subsequent *Foreign Policy Concepts*.[39] While the 2000 Concept gingerly stated that "The Russian Federation has a real potential for ensuring itself a worthy place in the world,"[40] the 2008 Concept proclaimed Russia as "the largest Euro-Asian power . . . one of the leading States of the world and a permanent member of the UN Security Council,"[41] and the 2013 Concept concluded this transformation, clarifying that "Russia's foreign policy . . . reflects the unique role our country has been playing over centuries as a counterbalance in international affairs and the development of global civilization."[42]

Since his comeback to the presidential chair in 2012, Vladimir Putin, surrounded by a team of his most loyal politicians, has paid special attention to the implementation of the Russian Revitalization concepts in foreign policy that were first formulated before his leave.[43] It was not an aggressive change of course, but an amplification of several major narratives that have underlain Russian foreign policy since the mid-2000s. The first one was the argument that "the independence of Russian foreign policy is based on [Russia's] geographical sizes, unique geopolitical position, centuries-old historical tradition, culture and the self-consciousness of our nation."[44] After the desperate attempts of the 1990s to define Russia as part of European Civilization (a view that was shared neither by European, nor by most of the Russians themselves),[45] at the beginning of the twenty-first century the Kremlin started to present Russia as an independent (Slavic/Orthodox) civilization, highlighting its historical and cultural richness. The best examples of this new narrative can be found among the highest-level politicians, quoting Orthodox-patriotic ideas of Aleksandr Solzhenitsyn[46]; reinforcing their views on foreign policy by citing historical figures, such as Alexander Nevsky[47] or Alexander III[48]; and backing the ideas of Ivan Ilyin—the mid-twentieth-century political philosopher, Slavophile, conservative monarchist, and defender of Orthodox-Slavic civilization.[49]

This self-identification based on a narrative of independence was a direct answer to the 1990s' illusions of Westernization, which, according to the Russian politicians, was destroyed by Western antagonism that was driven by Cold War perceptions. While Putin says "Russian foreign policy was, is and will be substantive and independent,"[50] Western attempts to intervene in Russian internal affairs, in the early 1990s, were not forgotten even 20 years later: "we cannot consider, even hypothetically, an option of a 'warped' Russia as a player manipulated by some other superior player on the international arena."[51] This determination to keep the West out of Russian internal affairs, on the one hand, and the desire, based on Soviet legacy, to restore the Russian traditional role as a "counterbalancing factor in international affairs,"[52] on the other, emphasized the second narrative of modern Russian foreign policy—multi-polarity at the expense of Western hegemony.

On the one hand, the policy of multi-polarity has come to enforce Russian positions by supporting different new international organizations (e.g., the Group of Twenty, the Shanghai Cooperation Organization, BRICS, and others) and therefore counterbalancing the Western traditional union between the United States and Europe.[53] On the other hand, the Russian promotion of multi-polarity in international affairs is intended to cool down American ambitions as "ruler-of-the-world," as Sergey Lavrov, the Russian Minister of Foreign Affairs, put it:

> Now it is possible to speak, not of preserving [Western] domination, which is impossible under the conditions of the rapidly increasing influence of cultural-civilizational diversity on the World's destiny, but of confirming European civilization's claim for [just another] place among the team of leaders.[54]

This opposition to the West dictating the line in global affairs has led to the next Russian foreign policy narrative—criticism of the West (mainly the United States) of double standards.[55] Building on the tractable nature of the international legal system, the Kremlin has continuously promoted the idea that its "Western partners, headed by the United States of America, in their practical policy, prefer to be guided not by the rule of international law, but by the rule of the strongest."[56] This idea is supported by the Russian claim of Western illegal behavior in Kosovo in 1999, in Iraq in 2003, in Libya in 2011, and other occasions.[57]

This leads to the final major narrative of Russian foreign policy—the intensification of the UN positions, in general, and the role of the UN Security Council, in particular. While this narrative generally supports the previously discussed concept of multi-polarity, as, according to Sergey Lavrov, the "Charter of the United Nations is 'programmed' as a regulator of the multi-polar world," it emphasizes the importance of the international legal system in the eyes of the Russian leaders. Built on the amendable nature of international law, the message from the Kremlin is that "We can guarantee that we will never violate our international obligations."[58]

There is no doubt that all these narratives come first and foremost to defend Russian national interests and reinforce its position in the international arena.[59] Their understanding, however, allows an identification of two major ways in which the Kremlin resists international pressure (mainly from the Western community) that criticizes Russian political decisions. While the first one is framing its activity within the context of international law, the second one is the repudiation of Western criticism because "[they] themselves are not perfect in many senses."[60] In other words, facing international criticism, Russian leaders powerfully dismiss it, arguing that their activities comply with international law, at least to the same degree that Western actions do. The best examples are two cases in which Russian leadership had to deal with intensive Western criticism: the recognition of Abkhazia and South Ossetia in 2008, and the annexation/accession of Crimea by Russia in 2014.

On August 26, 2008, after a short military conflict with Georgia, Russia recognized the Republic of Abkhazia and the Republic of South Ossetia as independent states. Despite the immediate wave of criticism from Western governments and accusations of the illegality of this move, the Kremlin presented firm legal ground to its decision, claiming that it was "guided by the provisions of the UN Charter, the 1970 Declaration on the Principles of International Law Governing Friendly Relations Between States, the CSCE Helsinki Final Act of 1975 and other fundamental international instruments."[61] Furthermore, Moscow's response also included other typical narratives of the Russian modern foreign policy, such as the accusation of double standards:

> [while] Western countries rushed to recognise Kosovo's illegal declaration of independence from Serbia. We argued consistently that it would be impos-

sible, after that, to tell the Abkhazians and Ossetians . . . that what was good for Kosovo-Albanians was not good for them. In international relations, you cannot have one rule for some and another rule for others.[62]

The second example is the annexation/accession of Crimea by Russia following the political crisis in Ukraine, in the winter of 2014. While most of the international community did not recognize this move, labelling it an illegal annexation, according to the Russian view, every detail of the process was in complete compliance with international law.[63] Despite the fact that Putin acknowledged the deployment of Russian Special Forces to Crimea, according to him, everything was legal:

> I did not even have to use the permission of the [Russian] Federal Council to deploy Russian troops in Ukraine. And I did not even bend the truth, as under the related international treaty, we had the right to have 20,000 men in our base in Crimea, even slightly more. But even with the troops that we had added, as I have mentioned, we were still under 20,000. As we did not exceed the number of our troops in Crimea, strictly speaking, we did nothing wrong.[64]

In addition, the election of a new pro-Russian Prime Minister of Crimea, according to Putin, was an entirely lawful outcome of the events:

> The Crimea's Parliament was an absolutely legitimate, fully fledged representative body in Crimea that was formed long before any of these difficult and tragic events. So these representatives convened a meeting, held a vote and elected a new Prime Minister of Crimea, Sergey Aksyonov. And de jure President Yanukovich approved his appointment. In terms of the Ukrainian law, everything was honored and respected. Of course, it is possible to throw words around and interpret these events in a variety of ways. But, if you look from the legal perspective, you cannot pick a hole in it.[65]

The Kremlin's figurative response to Western accusations of incompliance with international law was best expressed by Sergey Lavrov: "It is good that they [the Western countries] at least have recalled the existence of international law. Thanks for nothing! But it is better late than never." Again, Russian reaction to Western pressure was the accurately explained (although still unaccepted by the West) foundation of Russian actions in the international legal system, and the already discussed comparison to past and present Western actions, which are considered by Moscow as illegal, or, at least immoral, in protecting pure Western political interests.[66]

While the narratives of Russian foreign policy outlined above offer a general explanation of Russian reaction to international pressure and criticism, Russia clearly has a very powerful "political will" in the international arena. Building on its status as a nuclear power and permanent member of the UN Security Council, its growing economy and broad domestic political support, the Kremlin, without any doubt, is able to withstand international criticism and accusations of the violations of international law, in general, as well as during military operations, in particular. An analysis of the previously discussed case of the Russian 2008 Operation to Force Georgia to Peace,[67] clearly shows how the Kremlin successfully fended off the allegations of disproportional use of force.[68] Moscow's reaction followed the general pattern of Russian foreign policy: a demonstration of a powerful and independent Russia in the multi-polar international environment—a "Russia to be reckoned with"[69]; utilization of the tractable nature of the international legal system—posing the question: "Who did define the criteria of the proportionality of the use of force?"[70]; and a comparison between Russian actions and similar Western activities in favor of the first.[71]

The main purpose of this chapter is to understand why NLW are not employed by the Russian military. As it was discussed before, international political pressure can be one of the main drivers behind the increasing employment of NLW in an attempt to decrease civilian casualties and thus international criticism. The analysis of Russian foreign policy presented above, and the ways in which the Kremlin fends criticism off clearly show Russia's forceful (even brawny) reaction to international (mainly Western) pressure. Answering the question of whether the readiness of Russian nuclear forces during the Crimea crisis was raised to the highest alert, Putin replied:

> We were ready to do so. But I was already speaking with my colleagues [Western leaders], and I was saying it directly . . . and openly that this is our historical territory, and Russian people are living there, they came under a threat and we cannot abandon them. It was not us who staged the coup [in Ukraine], it was the nationalists and people with extreme views. You supported them. But where are you? Thousands of kilometers from here! And we are here and this is our land. What do you want to fight for? You don't know? But we know and we are ready for it![72]

In light of this effective defensive foreign policy, there is no doubt that Russian decision-makers feel no need to demand fundamental changes in their military forces in an attempt to make them less lethal than their Western counterparts. The lack of this political need is based on three main conditions: Russian economic strength; Western military performance, which provides comparable cases for Russian criticism (i.e., the lethality of Western military forces); and the lack of a domestic opposition that cares about the enemy's population.

Interestingly enough, the first condition creates a sort of paradox as economic strength is necessary not only for a strong foreign policy, but also for military transformation (i.e., economic decline will weaken Russian foreign policy, but it will make any fundamental military transformation—including the RMA of NLW—economically impossible). The level of lethality of the

Western (American—the most important in the Russian eyes) military and its attitudes toward NLW were discussed in the previous chapter, and it does not seem that NLW are going to be employed widely in the near future. Consequently, the only remaining influential factor on the political desire to decrease lethality on the battlefield (i.e., employ NLW) is domestic public attitudes toward enemy civilian casualties.

RUSSIAN CULTURE AND ENEMY CIVILIAN CASUALTIES

In an attempt to understand Russian historical and cultural background, it is important to remember that "in the national body of Russia, there are islands and oases of Europe and Asia as well."[73] In other words, due to Russia's intermediate geopolitical and cultural position between the West and the East, "Russian people had to solve problems that demanded a special strategy."[74] Russia's historical-cultural place between European civilization in the West and Islamic, Confucian, and Indus civilizations in the East combined with its geographical and climate conditions have played an instrumental role in designing the Russian character throughout the whole of Russian history.[75]

In 1944, Ivan Ilyin, one of the famous Russian philosophers who studied Russian political, historical, social, and spiritual narratives, wrote: "Russia, as a nation and culture, still appears to Western Europe as a hidden world, as a problem that cannot be understood, a kind of Sphinx."[76] Consequently, when analyzing Russian cultural context, it seems right to focus, first and foremost, on Russian authentic sources and materials, rather than on their existing Western interpretation.[77] In other words, in an attempt to understand "what Russian people think about war and civilian casualties," this research will focus on the writing of traditional Russian political philosophers and anthropologists, such as Ivan Ilyin, Nikolay Lossky, Ivan Solonevich, and Nikolai Berdyaev.[78] Most of these thinkers, who wrote about the Russian cultural mind-set in the first part of the twentieth century, had to escape or were expelled from the USSR due to their anti-Bolshevik views, and their publications were banned by Soviet censorship. Although the restrictions on their work were annulled with the dissolution of the Soviet Union and their books were finally published in Russia, their real success came only in the 2000s, when they were widely embraced by Russian political leaders. It is not surprising that national conservative views of Ivan Ilyin, Ivan Solonevich, and others have been heavily used by Vladimir Putin, Dmitry Medvedev,[79] and other leading politicians to enforce their policy of Russian revitalization:

> The problems of Russia, its history, and the ways for the social and spiritual revival of the Motherland have a central place in Ilyin's thoughts. The search for national self-identification and the protection of moral and spiritual val-

> ues—all these goals are actual today no less than in the first part of the previous century.[80]

To avoid possible criticism that the validity of these thinkers is outdated and their reputation is artificially recreated due to political reasons of the Kremlin's leaders, this research will be complemented by the work of various Russian contemporary scholars, who specialize in the identification of Russian mentality and its roots, such as Alexander Panarin, Vladimir Pantin, Vladimir Lapkin, and Vyacheslav Gubanov.[81]

Russian Culture and the Value of Human Life

Ivan Ilyin famously claimed that "the history of Russia is akin to the history of a besieged fortress."[82] While it is hard to imagine a way in which enormous Russia can be besieged, its people have been beleaguered during the entirety of Russian history, surviving in extreme climatic conditions, braving repetitive invasions from the East and the West, and an uncompromising, almost tyrannical, treatment by their own rulers. It, indeed, seems accurate to agree with the claim of the Russian philosophers that the history of the Russian people reads like the history of a besieged people.

The extreme Russian climate and boundless territories are seen by Russians as the main reasons behind many major aspects of the Russian character, such as: collectivism, absolute conformity with political centralism, and strong emphasis on endurance and patience.[83] Owing to its agricultural heritage, Russian philosophers define the Russian character in terms of communality. Subsequently, and like any other agricultural society, the rural commune in Russia was created to allow labor cooperation in the conditions of high-risk farming. While definitely not the only country with extreme seasons and barren land that made cooperation vital for survival, Russia was the only one with an unlimited amount of this land. And while in other (Western) countries the limitation of sources (land, water, labor, etc.) led to the quest for more efficient ways to exploit natural resources, Russia's way to increase productivity was always based on the conquest of additional lands, involving additional resources and additional labor. This Western race for better efficiency, which was the core reason for the development of individualism and private property, was never a part of Russian culture, where the one's contribution to the collective interest was always prevalent over one's private life.[84]

The Russian character and culture, similar to the American experience of the frontier, is an outcome of the conquest of boundless wilderness, though the process of the Russian conquest was entirely different from the American one, as was its influence on the Russian mind. In both cases, it was about "civilizing" the wilderness; however, while the conquest of America was led by individuals seeking their own destiny and benefits, the conquest of Russia was led by the state. In the American frontier experience, the development of villages into towns and into cities was set in the hands of the community—in Russian history, it was always the responsibility of the authorities. While American cities were built by businessmen who, in the pursuit of their own benefits, established themselves as community leaders and propelled small towns into large cities, such as William Ogden in Chicago or William Larimer in Denver,[85] the development and prosperity of Russian Ekaterinburg,[86] Samara,[87] or Novosibirsk[88] were the outcome of governmental policy. When the frontier concept of "free land" was responsible for the American individualism, in the Russian case, the state-led expansions, based on the state's ownership of the land, concomitantly shaped in the Russian mind the importance of the collective.

The collectivistic characteristics of Russian culture were significantly fueled by the enduring serfdom of the seventeenth and eighteenth centuries (the period when in Europe it was already abandoned). The history of serfdom is seen as having obvious and major influences on the cultural character of the Russian people. An increasing oppression of individualism led to the lack of feeling of responsibility for land, family, and life, which, in turn, led to the absolute conformity with the representative of the authorities, as a famous Russian proverb says: "[the] Master will come—[and the] master will judge us" (*Vot priyedet barin—barin nas rassudit*).[89] While the Emancipation Reform of 1861, under the reign of Alexander II of Russia (who was also known as Alexander the Liberator), officially liquidated the serfdom, it had a small impact on the established cultural environment because, in fact, it gave former serfs neither independence, nor full responsibility. After the reform, the peasants remained under the supervision of the rural commune that had the power to redistribute the land among its members.[90]

This undeveloped sense of personal self-realization led to one of the most paradoxical phenomena in Russian culture: anarchical behavior on the one hand, and a full political submissiveness on the other.[91] Interestingly enough, these two characteristics succeeded in coexisting, being, in fact, interdependent—understanding their own anarchical predispositions, Russian people demanded strong and even ruthless treatment. The best example of this characteristic is provided by Nikolay Lossky:

> In St. Petersburg in the spring, when the ice on the Niva River started to melt, the ice-crossing of the river became unsafe. The mayor ordered to place policemen on the bank of the river, who will stop the ice-crossers. A peasant, despite warning shouts from a policeman, went on the ice, fell and began to drown. The policeman saved him from death, but the peasant, instead of being grateful, started to criticize him: "Where have you been?" The policeman answered him: "But I shouted out a warning"—"Shouted?! You had to sock me in the face!"[92]

On the one hand, lack of the responsibility for one's land and life generated a feeling of unlimited personal freedom in the Russian cultural environment, creating a strong narrative that probably (*avos'*), supposedly (*nebos'*) or somehow (*kak-nibud'*) problems would be resolved. On the other hand, it created a strong belief in, even demand for, a powerful authority that would guide the people and protect them from their own carelessness.[93] While American individualism and the cultural predisposition that an American can solve any problem led the American mind to lock down upon bureaucrats and the authorities that they represented, the Russian concept of personal freedom from responsibility led to a higher dependency on the state.

Flourishing throughout the Soviet era, this under-developed sense of personal responsibility and self-realization remains relevant in the beginning of the twenty-first century. In 2009, President Medvedev described one of the main social weaknesses of the Russian people as their "confidence that all problems should be resolved by the state. By somebody else, but not by each one in his place. The desire to 'build yourself,' to achieve personal success, step by step, is not our national habit."[94]

When Russian political philosophers emphasized a natural freedom inherent within the "Russian Soul,"[95] they did not mean the Western understanding of freedom (i.e., human rights based on cultural individualism), but an entirely Russian interpretation of freedom: "freedom from earthly worries."[96] This understanding of freedom explains two very important cultural narratives in the Russian character: political submissiveness, and a rejection of the idea of formal human rights and freedoms. The Russian character underlines the prevalence of spirituality over materialism and pragmatism, and therefore the idea of spiritual freedom, rather than a physical-material one, was always at the core of the Russian character—"Our people are not materialistic . . . they are contented spiritually."[97] The way to achieve this freedom was by an utter transfer of political responsibility to an authority (master/monarch/government), ultimately accepting powerful, even totalitarian rule as a required precondition for organizing a stable society. While an American values the material achievements of a person, a Russian appreciates more the spiritual: "a person achieves his social status according to his communion with the prevalent system of ideological values, the Slavic-orthodox values in pre-revolutionary Russia, or communism in the USSR."[98]

Interestingly enough, this narrative of an overwhelming desire for spiritual freedom and unwillingness to deal with material problems (because dealing with the material fetters one's spirit), on the one hand; and a long history of ruthless governance, which fully exploited this unmaterialistic nature of their people, on the other, led to this general disbelief in the idea of formal human rights and freedoms bestowed upon a people by their political authority. According to Ivan Ilyin, "a Russian is characterized by internal freedom [and] there are no artificially invented bans for him"[99]—and therefore there is no need for formal (artificial) freedoms. While for Americans, their material rights and freedoms (e.g., speech, movement, security, etc.) are integral parts of their self-determination, for the Russian mind these rights are something outlandish and unwarranted by spiritual happiness or earthly existence, something unnecessary "that had been awarded by the generous fatherland or a kind master and, at any given moment, can be taken away without reasonable explanations."[100]

There is no doubt that political and ideological limitations of the Soviet era did not contribute to the development of individuality and a more Western understanding of freedom and human rights. In fact, more than 20 years after the fall of the Soviet Union and hasty attempts to democratize (or Westernize) Russia, the Russian masses, in general, continue to uphold their historical cultural preferences for strong authority, at the expense of formal human rights. For example, several surveys conducted in 2014 show that between 18 different merits, Russian people placed "human rights" only at the fifth place, preferring "order," "security," "peace," and "rule of law." Placing "private property" (ninth) and "democracy" (twelfth) after such merits as "stability" (seventh), and "patriotism" (eighth)—the Russian people had articulated their true mentality based on their unique history and experience.[101]

This reliance on authority, which secures peace and the rule of law, combined with the traditional sense of collectivism, shaped the Russian cultural appreciation of human life. While American culture values human life due to very individualistic, even selfish, reasons, the Russian appreciation of the community comes before the individual and, therefore, his life should be dedicated to the collective.[102] The Russian character, shaped by the idea of collective responsibility and punishment (*krugovaya poruka*), which was a part of the domestic law until the twentieth century,[103] promotes the responsibility of an individual for the whole collective, and vice versa. While this Russian cultural naive readiness to serve the common good was vastly exploited by the Russian totalitarian governments,[104] it also projected on the Russian cultural perspective on war and the readiness of the Russian people to pay unprecedented costs to protect their land and state.[105]

War and the Russian Character

One of the most prevalent Russian historical narratives is the traditional assumption that Russia has always enjoyed the access to a massive territory and, as Ivan Ilyin puts it, the lack of natural geographical borders, such as seas or mountain massifs, left this enormous flatland unprotected and exposed to invasions.[106] This geopolitical "podium" of Russian history had one of the most significant influences on the composition of the Russian character. Geographically unprotected and placed between the centers of two different civilizations—the Western and the Asian (Mongols and Chinese)—Russia suffered repetitive massive invasions through its history: in 1236, the Mongol Empire (Golden Horde) conquered most of Kievan Rus, establishing more than 200 years of Tatar-Mongol Yoke; in 1571, the Crimean Khan Devlet I Giray raided the Tsardom of Russia, setting capital Moscow on fire; in 1610, the Polish King Sigismund III occupied most of the western part of the Tsardom of Russia, putting his son Wladyslaw IV Vasa on the Russian throne (although the son never actually ruled); in 1812, Napoleon invaded Russia, defeating the Russian army and conquering Moscow; and finally, in 1942, German forces occupied most of the European part of the USSR, besieging Leningrad (St. Petersburg), Moscow, and Stalingrad (Volgograd).[107]

According to Vyacheslav Gubanov, the Russian mind-set perceives Russia's people as the defenders of Europe, as an outpost of Western civilization that protects it from the barbaric East. In return, instead of giving the Russians gratitude and support, European countries took advantage of Russian backwardness, the price that the Russian nation had to pay to stop the Mongols and allow the Renaissance.[108] This ruthless military experience predetermined the Russian mind "to defense without support, to rely only on your own forces, to live on an open flatland, to create and wait for the danger of a new invasion or attack that can come at any time."[109]

Without any doubt, not all Russian wars were defensive or even were justified as such. Since the late eighteenth century and until the Bolshevik Revolution in 1917, the Russian Empire was "The Gendarme of Europe," saving Austria in the War of the Second Coalition in 1799–1800, helping to defeat Napoleon in 1813–1814 by invading Prussia, and going all the way to Paris, in the War of the Sixth Coalition, and rescuing the Austrian Empire from the Hungarian Revolution in 1849. But these were not the wars that influenced Russian mentality, as above all, it had been marked by the tragic experience of the Tatar-Mongol Yoke, the Patriotic War against the French invasion in 1812, and the Great Patriotic War against the Nazi German invasion in 1941 (the title of "patriotic" was awarded to these wars for a reason).

While in English the last two are titled as 'patriotic,' their original Russian name is *otechestvennyye* (Fatherland's). Russian language has two different words for the English term "motherland"—*rodina* (motherland) and *otechestvo* (fatherland). These two symbolize the female and male characteristics of the Russian character:

> Motherland [and] motherhood are associated with the language, with the folklore, with the national character and the undefined, but powerful domain of unconsciousness. Fatherland [and] fatherhood—[are associated] with duty and citizenship, with the social-political, conscious life.[110]

While *rodina* symbolizes a more spiritual connection with *zemlya-mat'* (earth-mother), *otechestvo* (the land of fathers) is associated with the harsh reality of the real world as "father leads the son to the economic and political world: makes him a worker, a citizen, a soldier."[111] Consequently, a Father-land's war is not an ordinary war, it is a war in which every son sees his duty and right to participate, defending his Fatherland at any cost.

In the Russian mind, Russian history is "the history of defense, struggle and sacrifice: from the first attacks of nomads on Kiev in 1037 until today. Accessible from all flanks, completely unprotected, Russia was a kind of 'sweet booty' for the nomadic East, as well as for the settled West."[112] One of the main outcomes of this reality is the fact that, historically, Russian warfare had always been a national affair. Since the first regular units (*strel'tsy*) established by Ivan the Terrible, through more systemized recruiting systems created by Peter the Great, and up to the truly national armies (first the Imperial and then the Soviet) of the twentieth century, the Russian military had always a very national character built on the "obligation of every Russian person to protect its Fatherland."[113] Russian military history, unlike the Western one, did not know the phenomenon of mercenaries. And despite the fact that during different periods, the Russian army recruited foreign officers, the process of their enlistment was similar to that of Russian subjects and included taking an oath to protect the Russian state. While the Western term "soldier" originated from the Old French *souled*—soldier's pay that was paid by *solidus* (Latin for the golden coin issued in the Late Roman Empire),[114] the Russian *vóin* has completely different origins.[115] The roots of the word *vóin* are in old Russian *vói*—people's militiaman (*narodnyy opolchenets*)—the major defender of the Fatherland.[116] Therefore, *voyn á* (war in Russian) is an activity done by *vóinami* (people's soldiers) to protect their people.

All major wars that shaped the Russian character (the Mongol Yoke, the 1812 Patriotic War, and the Great Patriotic War) occurred on Russian territory, extensively involving the Russian population and encoding in the Russian mind that war is a choice between fighting or being destroyed: "to wage a war or be turned into slaves and disappear."[117] Obviously, the general Russian reaction to war involves more generic and universal merits, such as protection of the nation, religion, state, etc.; but as a result of the historically established belief that war is a people's affair (rather than a state's), the most fundamental reason for war in Russian mentality is the protection of one's life and one's family. While a Westerner fights to make world better, a Russian fights to survive.

While this state of a constant fight for survival has been described by famous Russian philosopher Ivan Solonevich as the major reason behind the Russian cultural predisposition toward strong and decisive authority, rather than democracy,[118] an additional outcome of this state has been the increased importance of war in Russian mentality:

> The whole history of the Russian state is an extreme struggle for its security against external and internal threats, a struggle that has left a recognizable trace on the consciousness of the Russian people.[119]

Consequently, it is not surprising that this permanent state of war (in addition to its extreme climate and ruthless governance) created in the Russian character a general worship for suffering; as Fyodor Dostoyevsky put it, "it seems that the Russian people enjoy their suffering," as "the most important, the most fundamental spiritual need of the Russian people is a need for suffering."[120]

It seems that the narratives of war, permanent struggle, and suffering have played constructing roles in the Russian cultural context throughout all of its history. Interestingly enough, the history of the twentieth century contributed to this phenomenon not less, but, if anything, more than any other period in Russian history. The first reason for this was the ideology enforced by Soviet leadership that glorified the idea of class conflict.[121] It was, after all, the political-ideological environment of the Soviet Union that was responsible for the absence of conceptual-philosophical and practical development of the ideas of non-violence and peacemaking.[122]

The second reason was the Great Patriotic War. Without any doubt, this war was one of the most noteworthy events in Russian history; its contribution to the Russian cultural context, however, was significantly increased by the fact that the victory in this war has been, in fact, the only positive point of reference in the national self-identification of the late-Soviet and then post-Soviet Russian society. Even at the beginning of the twenty-first century, the importance of the victory in this war has remained overwhelmingly high for Russian society. For example, in 2003, surveys showed that 87 percent of the Russian people, thinking about the whole of Russian history, believe that the victory in the Great Patriotic War is an event that makes them the most proud.[123] The notion of war is deep within the Russian character, and its modern perspective on war is not far away from the Russian traditional explanation of war described above. The Russian historical narrative of war—protecting one's life and one's family, fighting to survive rather than bettering the world—seems to be as concrete as ever at the beginning of the twenty-first century. The 2005 surveys about the justification of war showed that 68 percent of the population call a war just when people protect their houses, relatives and family, and their country; and only 6 percent call a war just when it is waged for exalted goals and ideals, religious principles, or for national honor and glory.[124] Most interestingly, these surveys also show that only one-fifth of the population actually use ethical or humanitarian aspects in their justification for war, and for all the rest—territorial and population protection are justifiable factors by themselves.[125]

While the American mind perceives war as a temporary and unfortunate phenomenon and, therefore, approaches war as apolitical problem that has to be solved as quickly as possible, Russians see it as a part of their *esse*, and waging wars is always a matter of internal politics. Moreover, due to the discussed Russian cultural predisposition toward strong leadership, Russian people justify war only when they are led by powerful leaders. Russians were always ready to defend their Fatherland facing a foreign enemy. From the thirteenth century's Tatar-Mongol Yoke, and through to Napoleon's and Hitler's campaigns, Russian people have always demonstrated exceptionally high levels of patriotism and readiness to fight the adversary and suffer the consequences of protecting their country. Russian history, however, suggests that people are ready to suffer war not only because they think it is just, but mostly, because they expect their leaders to stand firm and lead them to victory. The tsarist government during World War I, or the Soviet leaders during the Afghanistan War, did not prove themselves as strong leadership that deserve to be followed, and as a proof of that, Nikolas II was the last tsar and Mikhail Gorbachev—the last Soviet leader.[126]

Moreover, the readiness of the Russian people to fight depends more on leadership than on the nature of the conflict itself. The best example is the Russian involvement in Chechnya in the 1990s. During the first Chechen campaign from 1994 to 1996, Yeltsin's government was associated with a high level of corruption, economic failure, and an inability to project central power on to the regions.[127] In light of this weak leadership, Russians saw this war as Boris Yeltsin's attempt to prove his power, rather than defend Russian land, and the Russian public wanted "to get rid of both its hapless president and his Caucasian adventure."[128] Lacking strong leadership, Russians were not ready to fight: in 1995, the surveys coined out that 40 percent of the population preferred a peaceful solution, with an additional 24 percent who rejected any military involvement, asking to withdraw forces from Chechnya. Moreover, in June 1995, another survey showed that in answering the question "What, in your opinion, is necessary to do in relation to Chechnya?" about 38 percent of the population approved its exit from the Russian Federation.[129]

This approach to the very same conflict changed entirely in 1999 with the arrival of Putin, who was seen as an energetic, decisive, and strong leader on his way to power. Putin not only took the full responsibility for the war's prosecution, but also produced an image of a determined and capable leader. In 1999, directly after his appointment as Prime Minister, during a briefing with the Russian generals in Chechnya, Putin paid tribute to the fallen soldiers, which had a very ordinary beginning:

> I would like, according to Russian tradition, . . . to raise this glass and drink it for the memory of those who have fallen . . . We have no right to allow ourselves any moment of weakness, because if we will, then those, who have fallen, died with no reason.

But, then again, had a completely unexpected end:

> This is why, I suggest, to put this glass away today. We definitely will drink for them [fallen soldiers], but we will drink later, when the immediate goals of principal character will be solved. This is why, I suggest to have a bite and start working.[130]

This act, of firmly putting the glass back on the table, symbolizes in Russian cultural interpretation the decisiveness and power of a leader who deserves to be followed. It is not surprising then that under the Putin leadership, the readiness of the Russian people to fight cardinally changed: in 1999, 70 percent of the respondents approved the second campaign in Chechnya that began that year, with 62.5 percent wanting the war to go on to "final victory."[131]

Russian people, led by a strong leader, demonstrate levels of endurance and sacrifice that cannot be grasped by the Western mind. But, without powerful leadership, the same people, who had successfully defended their Fatherland against the all-powerful Napoleon and Hitler, managed to bring down their own state twice within a single century: in 1917 and again in 1991.[132]

This chapter explores the reasons behind the fact that the Russian military does not employ NLW. Public opinion can influence decision-makers in favor of these weapons by applying domestic political pressure to decrease civilian casualties during military operations. The main question, however, is whether the cultural context that lies in the roots of public opinion is predisposed toward valuing the lives of the enemy's population.

As discussed above, due to their historical and cultural background, Russians see human rights quite differently from their American counterparts. While an American demands and protects his rights from his government, a Russian sees no need for that. The nature of the Russian people is characterized by a predisposition toward internal spiritual freedom and, consequently, they do not value formal (artificially created) rights and freedoms given by strong (even totalitarian) authority, to which they zealously transfer their materialistic troubles. On the level of the unconscious cultural context, Russian people are ready to suffer and make sacrifices to sustain their internal spiritual freedom. While in the West, the struggle for human rights has been translated into an increasing valuing of human life, Russians always have preferred to leave these materialistic worries to the authorities. This situation has not differed much from the reigns of Ivan the Terrible, Peter the Great, or

Alexander the Liberator through the rules of Stalin, Khrushchev, or Brezhnev, up to Putin's regime in the beginning of the twenty-first century.

The narratives of constant war and eternal struggle are an integral part of the Russian mentality. Major General (ret.) Alexander Vladimirov, a Russian contemporary military theoretician and author of the imperative monograph titled *Fundamentals of General Theory of War*, argues that "an eternal war—this is what the history of humanity offers instead of eternal peace, which philosophers and moralists have dreamt about."[133] Russian people are ready to suffer and sacrifice fighting a just (according to their interpretation) war, a war that is intended to protect their lands and families, rather than promote an ideology or a high moral standard. Moreover, as already discussed, Russians traditionally support wars when they are led by strong and decisive leaders who, historically, were not only ruthless with Russia's enemies, but also with the Russians themselves. The paradox of the Russian character is that when it comes to fighting just wars (i.e., just causes and strong authority), the Russian people are ready to pay any price. The best example of their readiness for sacrifice is shown in surveys done after the 2002 Nord-Ost siege in Moscow, in an attempt to analyze people's tolerance of victims in counterterror operations involving hostages. Interestingly enough, the surveys showed that more than 40 percent of the population define an operation's outcome as a successful one, when 10 to 20 hostages out of 100 are killed.[134] Furthermore, despite the 125 hostages who had died during the law enforcement operation in the Nord-Ost crisis, a month after the siege, 23 percent of respondents thought that the operation was successful and 55 percent approved it as satisfying.[135]

This example perfectly demonstrates the readiness of the Russian people, due to their cultural predispositions described above, to suffer and sacrifice in the fight for a just war. And if the readiness for such self-victimization is so high, it subsequently leaves no place for concern for enemy civilian casualties. In other words, the suffering (and casualties) of the enemy's population during military operations is generally approved in the Russian cultural context, as Russians are ready to suffer themselves. Moreover, the Russian cultural collectivism and the idea of collective responsibility is relevant not only for the Russian community; it is also projected on the enemy. To the Russian mind, the community is responsible for its members' actions, and therefore the enemy's population is frequently considered as responsible for the actions of its military. For example, in 2002, surveys of Russian public opinion showed that 36 percent of respondents believed that the Chechen population was responsible for its own troubles and Chechens got what they deserved, 67 percent believed that it was impossible to negotiate with Chechens as equals as "they understand only the language of force," and 47 percent of respondents denied the right to be equal citizens of the Russian Federation to the Chechen population.[136]

Consequently, in light of this cultural context and mentality, it seems right to argue that Russian public opinion is far from pressuring politicians and defense decision-makers to consider the minimization of civilian casualties as a decisive factor in their strategies.

RUSSIAN MILITARY THOUGHT AND NLW

Previous research on Russian military thought undoubtedly shows deep interconnections between the Russian mentality, national cultural context, and Russian military culture.[137] Consequently, many of the arguments presented and discussed in the following section will be based on the Russian cultural factors analyzed previously.

According to Russian tradition, a war is an integral part of the human conditions (*esse*), and, therefore, it is not surprising that in analyzing Russian military culture, scholars state that Soviet and then Russian military thinkers "interpreted each event as a sign signaling a change in the whole and . . . saw problems as interrelated and requiring a generalized frame of reference to determine their meaning." This inherent Russian cultural ability to see the big picture has been, undoubtedly, reflected in Russian military theory conceptualization. In Russian tradition, "military science is an all-encompassing discipline . . . that regulates professional practice from strategy and doctrine, on the national level, to tactics on the battlefield."[138] While in American tradition, military experience is the predominant ingredient of military professionalism, in Russian military culture, it is the ability of systematic military thinking that enjoys the status of an essential and highly valued skill.

Unlike their American counterparts, Russian military thinkers never considered advanced technologies as means to fight better with fewer numbers. Russians have never denigrated the relevance and importance of technology; when faced with a problem, however, they do not instinctively seek a technological solution.[139] This might explain how the tendency to concentrate on big abstract ideas often led Russian military thinkers to develop state-of-the-art doctrines and concepts that were incompatible with the material capacity of the Russian military establishment.

This chapter, however, follows the development of NLW and their conceptualization by Russian military thought. Consequently, as NLW is a military technology, it focuses on the relation between Russian military culture and technological developments in general, and NLW in particular.

Russian Military Thought and Technology

The Industrial Revolution—which essentially occurred in Russia during the twentieth century—with its stressful and strenuous influence on people, was an antithesis of the Russian desire for spiritual calmness and freedom. Russians peasant traditionally mistrusted machines that "seemed [to them] as something anti-natural, a fruit of worthless thinking, which was invented in Europe and is entirely unsuitable for Russians."[140] It is important to note, however, that this technological skepticism was based on a cultural predisposition, rather than on the lack of technological imagination or skills. Due to their unmaterialistic mind-set, Russian people saw technology as something tedious and worthless; however, under the pressure of necessity (coming mainly from the authorities), Russian minds were able to produce technological wonders. As an old Russian proverb states: "[If you] beat a Russian—[he] would even make you a clock" (*pokoloti russkogo - on tebe i chasy sdelayet*).[141] This generally unfavorable perception of technology in general, and military technology in particular, has been a part of the Russian military reality since Peter the Great (the first tsar who actively tried to reform the Russian military through technological modernization). Technology has always been the Achilles heel of the Russian military, which was usually compensated for by a heavy reliance on the enormous human resources. And while it could be seen that the Russian military tradition is simply outnumbering the enemies, this not true, as "first and foremost it [Russian military] was concerned with overcoming the enemy qualitatively—morally."[142]

Notwithstanding, the major characteristics of the Russian mentality—collectivism, endurance, and patience—have all lain at the core of the perception of the moral superiority of the Russian forces over their counterparts throughout Russian history. While the notion of collectivism was translated into mutual help and assistance between soldiers, to a level that could barely be understood by their European counterparts, the endurance and patience were transformed into heroic readiness to withstand unbearable conditions and make enormous sacrifices.[143] Consequently, it is not surprising that Russian military culture considered new military technologies an important, but not obligatory, supplements to its military power, which was fundamentally based on spiritually superior "flesh," rather than sophisticated "war machines."[144] When an American, hitting his fingers, will seek a way to improve the hammer, a Russian will discuss intangible theories of a better way to strike the nail, but, in practice, neither being able to produce a better hammer, nor to change the way he holds the nail—he will endure the pain and continue striking.

In addition to the traditional technological drawbacks, the Soviet centralized economy and authoritarian conservative military tradition shielded the Soviet military from technological adventures and from Western tech-euphoria concepts. Building on the cultural fact that, faced with a problem, the Russian instinct does not immediately seek a technological solution, Soviet military decision-makers approved technology that was considered to be the minimum required, significantly compensating for technological inferiority through operational and conceptual creativity. Soviet technological achievements were mainly used for *pokazuha* (literally—show-off; a staged event for officials, international observers, and also for internal propaganda),[145] rather than for actual employment. Interestingly enough, this tendency has survived the Soviet Union's dissolution, being transferred to contemporary Russian forces. The 2008–2020 Plan for the Reorganization of the Armed Forces of Russia is intended, by its definition, to give "a new image" to the Armed Forces, rather than deal with the core problems and defining characteristics.[146]

Russian Military Thought and NLW

The Russian approach to waging wars has always been different from the Western one. Built on different cultural predispositions, Russian generals repeatedly produced solutions that were considered not only impossible, but even unthinkable: Suvorov's march across the Alps in 1799, or Kutuzov's decision to abandon Moscow in 1812. Although, directly after the Bolshevik Revolution, the Soviets struggled to separate between political ideology and military professionalism,[147] early Soviet military theoreticians commonly believed that scientifically developed military thought was a required foundation for crafting strategy and executing tactics.[148]

Even though it has been more than 20 years since the dissolution of the Soviet Union, it seems right to argue that contemporary Russian military thought is deeply rooted in the Soviet scientific approach, constructed after the Great Patriotic War, and it has preserved the ability to produce ideas and concepts "outside the box" that characterized the Soviet military scientific thinking. Moreover, while observing their Western counterparts and understanding their own technological inferiority, Russian military thinkers have always known how to produce innovative and creative strategies and tactics. The main secret of their intellectual success was that, due to purely Russian cultural predispositions, Russian military thinkers have never been deterred by "the gap between the theoretical and feasible."[149] In other words, Russian military thought has never been restricted by the framework of feasibility, allowing military thinkers and strategists to produce and debate strategic and tactical concepts, despite their potential impracticability.

As with many other technologies, the story of Russian military thinking about NLW began with the observation of emerging technologies and concepts in the West. The first Russian publications on military implementation of NLW appeared in mid-2000s, and always began with an analysis of the Western experience as the major point of reference for the subsequent discussion.[150] These publications cautiously advocated for the possible usefulness of NLW for Russian military as "unlike Russia, several NATO members already employ or very close to employ such high-technological weapons as NLW."[151]

An examination of Russian military publications on NLW shows that the *Conception to the Development and Employment of Weapons with Non-Lethal Action in the Armed Forces of the Russian Federation*, approved by the Russian Minister of Defense in 2009, served as an important milestone in the process of conceptualization of military employment of NLW. Since 2009, Russia's major military publication, *Voeniaia Misl'* (*Military Thought*), has been flooded with articles theorizing and conceptualizing the present and the future of NLW on the battlefield.[152] These articles, in addition to many others, published in *Strategicheskaya Stabil'nost'* (*Strategic Stability*),[153] *Zashcita i Bezopatnost'* (*Protection and Security*),[154] *Voprosi Oboronnoi Tekhniki* (*Questions of Defense Technology*)[155] and other military-professional journals, systematically discussed the potential strategic and tactic implementations of NLW on the battlefield. Similar to the American case, this theoretical discussion has been led by military, as well as academic, scholars and thinkers. On the academic forefront of the NLW discourse, a special place is reserved for Victor Selivanov, the head of the High-Precision Airborne Devices Department at the Bauman Moscow State Technical University (BMSTU) and member of the Russian Academy of Rocket and Artillery Sciences, and Denis Levin, an Associate Professor at BMSTU. On the military side of the debate, the main contributors are Colonel General Anatoliy Nagovitsyn, Major General Vasilii Burenok, and Colonels Vladimir Moiseev and Dmitri Zaitsev. All parts of the Russian military-science apparatus have been actively engaged in the analysis of NLW—from central military research institutes to military academies—sorting through NLW and producing a very comprehensive vision of their possible military employment. According to the affiliations of the authors, the centers of the military thinking and conceptualization about NLW have been: the 12th Central Scientific Research Institute of the Ministry of Defense of the Russian Federation (the 12th CSRI), the 46th Central Scientific Research Institute of the Ministry of Defense of the Russian Federation (the 46th CSRI), the Combined Arms Academy of the Armed Forces of the Russian Federation (CAAAF), and the Military Academy of the General Staff of Armed Forces of the Russian Federation (MAGS). Interestingly enough, as the first two are responsible for the research and development of new military technologies, and the latter two are the centers of military thinking, the combination of the research produced by them has covered the whole picture of NLW possibilities.

According to the group of Russian military officers led by Major General Vasilii Burenok, head of the 46th CSRI, NLW are characterized by:

> (1) high effectiveness of affecting factors on man-power, equipment and infrastructure; (2) ability to incapacitate targets for a period of time, required by special forces to accomplish their missions; (3) high level of information-

> psychological impact on the enemy; (4) selectiveness in affecting parts of the chosen targets; (5) ability to provide the required time before the realization of the chosen affecting factor and the regulation of its impact according to the operational and tactical situation; (6) compatibility and feasibility of integration with existent and future armaments.[156]

The systematic methodology of Russian military science helped to identify the role of NLW on the future battlefield, including mathematical simulations of the possible contribution of NLW to military success. Analyzing the possible implementations of NLW, Colonel Evgeni Efremov from the 12th CSRI, an institution that is responsible for the development of the requirements from military technologies, reported:

> The experience of local wars and armed conflicts shows that a mathematical modulation of combat activity is a required condition for the scientific forecast of the process and the outcome of military activity . . . [and] the development of relevant requirements from tactical-technical weapons' characteristics . . . including Weapons with Non-Lethal Action.[157]

Interestingly enough, one of the main conclusions of the proposed mathematical simulations was that employment of NLW, intended to minimize civilian casualties, will decrease the enemy's reinforcement from the civilian population, thus leading to a decrease in casualties of military personnel that employ these NLW.[158]

On the conceptual forefront of debate, the Russian military thinkers also delved deeper into possible concepts of operations and employment of NLW. As Colonels Vladimir Moiseev and Vladimir Orliyanskii reported in their imperative article, "Weapons with Non-Lethal Action and the Principles of Tactics":

> Rapid and massive employment of a variety of Weapons with Non-Lethal Action can allow the neutralization of entire enemy units, as they lack the information about the characteristics of these weapons, scales of their employment, damaging factors and ways of protection.[159]

While, similar to their American counterparts, Russian military thinkers predetermined the suitability of NLW more to law enforcement types of military missions (e.g., counter-terror or peacekeeping operations), they also delved into the tactical potential of NLW during these types of conflicts, suggesting that properly employed NLW can provide a force with significant tactical advantages. The major possible advantage of NLW on the battlefield, according to Russian publications, is a tactical surprise that can be achieved "not by the fact of the employment of the new weapons, but by skillful ways of their employment and avoiding repetitive patterns."[160]

In an attempt to organize the existing knowledge, emerging technologies, and the experience of their employment, different databases were created[161] and comprehensive plans of future development were discussed.[162] Economic perceptiveness,[163] legal aspects,[164] and even measures to protect friendly civilian populations against NLW employed by enemy forces[165]—all these aspects were extensively discussed in the Russian military literature.

The most interesting part of their research, however, is their discussions of the real events involving Russian Armed Forces in Chechnya and how the employment of NLW could have brought a better tactical solution, secure troops on the ground, and save enemy civilian lives.[166] There are two main conclusions these examinations brought Russian researchers. The first is that "arming the troops with different types of NLW significantly enhances their ability to impact the enemy in the place and time, where and when he expects it least."[167] And the second is that:

> Arming the troops with the most advanced types of NLW, and their decisive and extensive employment in combination with lethal weapons—will allow forcing the enemy to stop the armed struggle and will create conditions for the peaceful resolution of the existent conflict.[168]

Due to different cultural and historical predispositions, technological virtuosity has never been the center of gravity of the Russian military tradition. Some cultural aspects, however, have released Russian military thinkers and strategy crafters from the burden of being immediately relevant, allowing them to grasp distant, unfeasible, and unthinkable concepts.[169] Modelling themselves on the Soviet-built methodology of military science, contemporary Russian military researchers have produced analyses of NLW that are significantly more comprehensive than the product of their American counterparts. The extensive spectrum of the analysis, the different perspectives and assessments, offer a very detailed proposal for the military employment of NLW. Interestingly enough, according to the Russian point of view, NLW are intended to be employed by military units, not because of political concerns to avoid civilian casualties, but because, as several Russian military thinkers claimed, the massive employment of lethal weapons "will promote the growth of the population's aggressiveness, the destruction of economic and social infrastructures, and ultimately will lead to a further escalation of the conflict."[170] This understanding of the initial role of NLW corresponds to an even more significant breakthrough in Russian military conceptualizing: NLW have a potential to complement lethal means on conventional battlefields against conventional enemies' forces. According to Russian military thinkers, non-lethal capabilities have to be integrated within conventional lethal weapons, thus increasing their effectiveness:

> Variations of Weapons with Non-Lethal Action have to be compatible with the existing ordinary weaponry and complement its capabilities. It is important that the standard existing means of delivery (firearms, artillery, aircrafts, missiles, etc.) are used as means of the delivery [of non-lethal effects], and oversized systems are mounted on the standard already employed military transport platforms. The technical maintenance and resupply of the combat employment of different Weapons with Non-Lethal Action have to be a part of the existing system of the technical maintenance and resupply of the Armed Forces of the Russian Federation.[171]

Subsequently, according to Russian understanding, NLW can "minimize the enemy's capabilities to maneuver its reserves and second echelons" as well as "disorganize the coordination of command, control, intelligence and information."[172] This shows that, as per usual with the Russian military science, the fact that current NLW are far away from being able to fulfill these far-seeing visions does not stop Russian military thinkers from conceptualizing their potential influence on the future of warfare.

CONCLUSIONS: THE FAILURE OF THE RMA OF NLW IN RUSSIA

While the peak of Russian enthusiasm for NLW occurred almost two decades after the outset of the military revision of these weapons in the United States in the early 1990s, the current employment of NLW within the Russian military does not much differ from its American counterpart, presented in the previous chapter. It seems that the Russian military concepts approved in 2009 served as a trigger for the Russian military community to start engaging the topic of NLW, but without powerful leadership this eagerness rapidly attenuated, burying NLW within the Russian bureaucracy machine.

The history of Russian political and military reforms and modernizations offers a logical explanation for the sudden rise and fall of interest in NLW in the Russian political-military discourse. Historically, most of Russian political, economic and, especially, military reforms have been inspired by the West. Intuitively Russian strategy decision-makers closely observe Western military technological and conceptual developments in an attempt to "Russianize" them.[173] Consequently, it is not surprising that, after a decade since its development in the United States, the idea of NLW migrated to Russia, creating a sudden vast interest in these weapons. It is also not surprising that this interest was short-lived. Russian historical and cultural predispositions demand a strong personality to lead any reform or modernization: "there's a modernizer—there's modernization, there's no modernizer—do not expect even attempts to organize changes."[174] And, as discussed, Russian politicians have neither international nor domestic political pressures to minimize the lethality of their Armed Forces and demand military reform that will lead to the employment of NLW.

Since the mid-2000s, the Kremlin has implemented a very effective defensive foreign policy, built on Russia's growing economic status, the revitalization of the Russian role in international affairs, and the critical interpretation of Western actions. As long as Western militaries provide enough cases of brutal lethality during their military activity and do not employ NLW, Russian politicians see no reason to invest resources in the development of military operationally useful NLW. In other words, as long as the Kremlin will be able to justify enemy civilian casualties by referring to the level of lethality caused during U.S.-led operations, there will be no pressure to decrease the lethality of the Russian troops and consequently employ NLW. Moreover, the Russian cultural attitudes toward war and the Russian true belief that the main purpose of war is pure survival of one's own nation, rather than an attempt to defend values or high moral standards, prevent the formation of any domestic political pressure to minimize enemy civilian casualties.

Russian military thought has traditionally tended to concentrate on the big abstract ideas, often leading Russian military thinkers to develop state-of-the-art doctrines and concepts that were incompatible with the material capacity of the Russian military establishment—and NLW are not an exception. In 2012, in his famous article "Being Strong: National Security Guarantees for Russia," Vladimir Putin advocated the importance of the development of the so-called "Weapons Based on New Physical Principles" (*oruzhiya na novykh fizicheskikh printsipakh*).[175] While in the beginning, the meaning of the term was unclear, later, the Russian Ministry of Defense clarified that "Weapons Based on New Physical Principles" (also called "Non-Traditional Types of Weapons") are "new types of weapons with assaulting factors based on processes and phenomena that were not previously used for military purposes" and listed six types of these weapons: Directed Energy Weapons, Electromagnetic Weapons, Geophysical Weapons, Radiological Weapons, Genetic Weapons, as well as Non-Lethal Weapons.[176]

This very ambitious and abstract concept of a whole set of entirely new weaponry demonstrates best the paradoxical gap between the ability of Russian military thought to produce sophisticated concepts, and the feasibility of their implementation. On the one hand, Putin himself advocated these weapons, stating in 2012 that: "in the long term . . . the Weapons Based on New Physical Principles . . . alongside nuclear weapons, will allow us to obtain imperatively new instruments to achieve political and strategic goals,"[177] and Lieutenant General Stanislav Suvorov, the head of the Military-Scientific Committee of the Armed Forces, stated three years later, during the International Military-Technical Forum "Army 2015," that there were some new developments in this field as "a separate part of the exhibition was designated for the Non-Traditional Types of Weapons."[178] On the other hand, the definition of these weapons undermines the very feasibility of this new weaponry, clearly stating that "many types of these weapons fall under 1978 Environmental Modification Convention and other international treaties,"[179] and "given the unpredictability of the consequences of the use of certain types of these weapons . . . the international community is trying to prevent their testing or employment."[180]

While the inability of American military thought to produce a comprehensive operational concept for NLW limited their development and employment, the Russian military's traditional way developed an overwhelming and unfeasible conceptualization that is just impossible to materialize. Consequently, it is not surprising that the employment of NLW was limited to the Collective Rapid Reaction Force. The 2010 training operation "Vzaimodeystviye-2010" was the best example of Russian *pokazuha*—a good staged demonstration of technological achievement (in this case of NLW) to produce an image of fulfilment and please the high command. It seems that this experimentation with NLW was a reaction to the sudden attention from the strategy-makers and policy-designers. Once this pressure to present non-lethal capabilities was satisfied and consequently debilitated, the military, as well as industries, lost their interest. Notwithstanding the loss of official interest, Russian military thought has not ceased to exert itself, developing a comprehensive tactic and strategic understanding of the possible employment of NLW, as the lack of technological feasibility has never stopped Russian military thinkers from further conceptualizing and theorizing.

While analyzing the published products of their activities, it is important to highlight two main aspects. The first one is an interesting, but not surprising, fact—full compliance with the political course of complete and firm respect for the international law system. While some Western researchers gingerly suggest changes in the modern system of the Law of Armed Conflict (LoAC),[181] such actions have never been proposed in Russia, neither by military thinkers, nor by scholars. In the eyes of Russian specialists, the very conceptualization of future warfare (with or without NLW) has to be legally defensible within the existing system of international laws, as the international legal system is one of the major practicalities of the Russian foreign policy and Russia has no intention (or power) to change them.

The second aspect of the Russian NLW theoretical debate that should be highlighted is the conceptual idea of the employment of NLW on conventional battlefields. While the general discourse in Russian military literature is generally similar to the American appreciation of the advantageous characteristics of NLW during law enforcement operations, several Russian military theoreticians analyze the tactical usefulness of NLW against a conventional enemy.[182] While neither Russia, nor the United States, or any other country, have fully developed NLW that can suit this theory, it does not stop

Russian military thought from developing novel concepts for waging future wars.

Unfortunately, despite very progressive tactical and strategic concepts for the possible employment of NLW, and the fact that Russian military thinkers believe that "the creation of Weapons with Non-Lethal Action is an extremely important task for Russia,"[183] it seems that the Russian military is not going to lead in this field in the visible future. Strong political pressure is required to lift off an employment of a novel military technology within the context of the unmaterialistic Russian military culture, especially if it is not employed by the Western military first or counteracts an existing Western capability. As long as NLW do not pass this test of political necessity, they are doomed to remain a thought-provoking, yet idle tool in the Russian military arsenal.

NOTES

1. The Ministry of Interior of the USSR, *Ob utverzhdenii Nastavleniya po primeneniyu spetsial'nykh sredstv "Cheremukha" [Decree to Approve the Manual for the Use of Special Means "Cheremuha"]*, Moscow, September 24, 1970; The Ministry of Interior of the USSR, *Ob utverzhdenii Instruktsii po primeneniyu spetsial'nykh sredstv individual'noy zashchity Lichnogo sostava i drugikh mer bezopasnosti v organakh vnutrennikh del i Podrazdeleniya vnutrennikh voysk SSSR [Decree to Approve the Instructions for the Use of Special Means for the Individual Self-Defense of the Personnel and Other Security Measures by the Law Enforcement Agencies and Internal Forces of the Ministry of Interior of the USSR]*, Moscow, December 26, 1978; The Ministry of Interior of the USSR, *Ob utverzhdenii Instruktsii o poryadke primeneniya rabotnikami militsii khimicheskogo Sredstva spetsial'nogo naznacheniya "Cheremukha-10" [Decree to Approve the Instructions for the Use of Chemical Special Means "Cheremuha-10" by Police Officers]*, Moscow, November 19, 1981.

2. The Federal Law of the Russian Federation № 3-ФЗ, *O Politsii [About Police]*, Moscow, March 1, 2011, Articles 21–22.

3. The Ministry of Defense of the Russian Federation, *Kontseptsiya sozdaniya i primeneniya oruzhiya neletal'nogo deystviya v Vooruzhennykh Silakh Rossiyskoy Federatsii [The Conception to the Development and Employment of Weapons with Non-Lethal Action in the Armed Forces of the Russian Federation]*, Moscow, 2009.

4. V. Selivanov, "Oruzhiye neletal'nogo deystviya kak sredstvo bor'by s terrorizmom, obespecheniya mirotvorcheskikh operatsiy i operatsiy po pravoprinuzhdeniyu [Non-Lethal Weapons as Means for Struggle against Terrorism, Provision on Peacekeeping and Law Enforcement Operations]," *Vestnik Rossii'skoi' Akademii Estestvennih Nauk*, No. 4, 2005, 1–11; V. Selivanov, D. Levin, and S. Lyuwin, "Arkhitektura i realizatsiya bazy dannykh 'Oruzhiye Neletal'nogo Deystviya' [The Architecture and Realization of the Data-base 'Weapons with Non-Lethal Action']," *Voprosi Oboronoi Tehniki—Seriya 16*, no. 7–8 (2012): 63–76; V. Lyaschenko, *Torgovlya Oruzhiyem: oruzhiye neletal'nogo deystviya, sozdaniye, proizvodstvo, primeneniye, operatsii na mirovom rynke [Trade in Weapons: Weapons with Non-Lethal Action, Development, Production, Employment, Operations on the World Market]* (Moscow: Economika, 2011).

5. This document is classified, so it is difficult to say what its goals and definitions are; however the analysis of the situation with NLW prior this policy and during the five years after shows that there is not much difference. See also: V. Selivanov, D. Levin, and I. Ilyin, "Metodicheskiye osnovy programmno-tselevogo upravleniya razvitiyem oruzhiya neletal'nogo deystviya [Methodological Basics of Management by Objectives of Non-Lethal Weapons Development]," *Voprosy Oboronnoy Tekhniki, Seriya 16*, no. 3–4 (69–70) (2014): 36–50.

6. This influence can be easily traced in the Russian publications about military employment of NLW. For example, Selivanov, "Oruzhiye neletal'nogo deystviya [Non-Lethal Weapons]"; S. Savrasov, "Kontseptsiya primeneniya oruzhiya neletal'nogo deystviya v boyevykh operatsiyakh sukhoputnykh voysk VS SSHA [The NLW Deployment Policy of the U.S. Army's in Combat Operations]," *Zarubezhnoe voennoe obozrenie*, no. 10 (2009): 37–44; V. Antipov and S. Novichkov, "K voprosu o razrabotke i primenenii neletal'nykh sredstv porazheniya na khimicheskoy osnove [Regarding the Question of Development and Employment of Non-Lethal Means Based on Chemicals]," *Voenaya Misl'*, no. 9 (2009): 54–61; V. Vladimirov and G. Chernikh, "Sostoyaniye i osnovnyye napravleniya razvitiya oruzhiya neletal'nogo deystviya, sredstv i sposobov zashchity ot nego [The Current Situation and Main Directions for the Development of NLW, the Means and Ways of Protection against Them]," *Strategiya Grazhdanskoy Zashchity': Problemy i Issledovaniya* 2, no. 1(2) (2012): 13–22.

7. Both documents are still classified, however both of them are widely discussed in the open publications.

8. *The XXI Century Encyclopedia—Russian Arms and Technologies, Volume XII: Ordnance and Munition* (Moscow: Arms and Technologies, 2006), pp. 754–65, (Russian/English).

9. I. Petrov, "Oruzhiye nesmertel'nogo deystviya [Weapons with Non-Lethal Action]," *Vecherniaia Moskva*, no. 190 (2004, October 8); D. Zaitsev et al., "Perspektivy primeneniya kompleksov oruzhiya neletal'nogo deystviya na osnove KVCH-izlucheniya v spetsial'nykh operatsiyakh po presecheniyu masovykh besporyadkov [The Prospects for the Application of Millimeter-Wave NLW in Operations for the Suppression of Mass Riots]," *Strategicheskaya Stabil'nost'*, no. 2(55) (2011): 54–59.

10. E. Stupitskiy and L. Kuznetsova, "Issledovaniye vozmozhnosti sozdaniya sredstv elektroshokovogo neletal'nogo vozdeystviya na bioob'yekty na bol'shikh rasstoyaniyakh [The Research of the Possibility to Create Means for Non-Lethal Electroshock Impact on Biological Objects from Significant Distance]," *Voenaya Misl'*, no. 4 (2013): 38–46.

11. A. Nagovitsyn, A. Grudzinskii', and A. Sporykhin, "Oruzhiye neletal'nogo deystviya i perspektivy yego ispol'zovaniya v interesakh sil Organizatsii Dogovora o kollektivnoy bezopasnosti [Weapons with Non-Lethal Action and the Perspectives of Their Employment in the Interest of Collective Security Treaty Organization]," *Voenaya Misl'*, no. 3 (2011): 51–59.

12. Collective Security Treaty Organization (CSTO), *Soglasheniye o Kollektivnykh silakh operativnogo reagirovaniya Organizatsii Dogovora o kollektivnoy bezopasnosti [Agreement on the Collective Rapid Response Forces of the Collective Security Treaty]*, Moscow, June 14, 2009, Article 2.

13. Nagovitsyn, Grudzinskii', and Sporykhin, "Oruzhiye neletal'nogo deystviya [Weapons with Non-Lethal Action]"; A. Nagovitsyn, "Ispol'zovaniye oruzhiya neletal'nogo deystviya kollektivnymi silami ODKB [The Employment of Weapons with Non-Lethal Action by CSTO Forces]," *Zashchita i Bezopasnost'*, no. 3(58) (2011): 2–3.

14. Nagovitsyn, Grudzinskii', and Sporykhin, "Oruzhiye neletal'nogo deystviya [Weapons with Non-Lethal Action]"; Nagovitsyn, "Ispol'zovaniye oruzhiya neletal'nogo deystviya [The Employment of Weapons with Non-Lethal Action]."

15. Selivanov, Levin, and Ilyin, "Metodicheskiye osnovy [Methodological Basics]."

16. V. Korchak and S. Smirnov, "Regulirovaniye normativno-pravovykh otnosheniy pri sozdanii oruzhiya neletal'nogo deystviya [The Legislative Regulation Regarding the Development of Weapons with Non-Lethal Action]," *Kompetentnost'*, no. 4–5(45–46) (2007): 3–10.

17. A. Silnikov, "Vidy spetsial'nykh sredstv sostoyashchikh na vooruzhenii OVD, i ikh klassifikatsiya [Types and Classification of Special Equipment Adopted by the Russian Interior Ministry Units]," *Vestnik Sankt-Peterburgskogo universiteta MVD Rossii*, no. 3(51) (2011): 66–70.

18. *The XXI Century Encyclopaedia—Russian Arms and Technologies, Volume XV: Security and Law Enforcement Facilities* (Moscow: Arms and Technologies, 2007), 117–40, (Russian/English).

19. *The XXI Century Encyclopaedia—Russian Arms and Technologies, Volume XII*, 754–65.

20. "Voyennaya politsiya VVO poluchila partiyu travmaticheskikh pistoletov 'Osa' [Military Police Received Traumatic Pistols 'OSA']," *Ria Novosti*, February 13, 2014, http://ria.ru/defense_safety/20140213/994703801.html [accessed May 10, 2018].

21. A. Bogaturov, "Tri pokoleniya vneshnepoliticheskikh doktrin Rossii [The Three Generations of the Russian Foreign Policy Doctrines]," *Mezhdunarodnyye Protsessy* 5, no. 1(3) (2007): 54–69; N. Rabotyazhev and E. Solov'yev, "Ot 'Vozvrashcheniya v Mirovuyu Tsivilizatsiyu' k 'Suverennoy Demokratii': Evolyutsiya vneshnepoliticheskikh kontseptsiy Rossiyskikh 'Partiy Vlasti' [From the 'Comeback to the World Civilization' to the 'Sovereign Democracy': An Evolution of the Foreign Policy Conceptions of the Russian 'Parties in Power']," *Politiya*, no. 1(68) (2013): 36–54.

22. *Interview with K. Kosachev, the Chairman of the State Duma Committee on Foreign Affairs*, published in *Mezhdunarodnaya Zhizn,'* no. 4 (2010): 4–9, (Russian).

23. Bogaturov, "Tri pokoleniya [The Three Generations]."

24. Bogaturov, "Tri pokoleniya [The Three Generations]"; V. Zhurkin, M. Nosov, and D. Danilov, "Voprosy mezhdunarodnoy bezopasnosti [Questions of International Security]," in N. Shmelev (ed.), *Rossiya v mnogoobrazii tsivilizatsiy [Russia in the Diversity of Civilizations]*, (Moscow: Ves' Mir, 2011), 175.

25. Bogaturov, "Tri pokoleniya [The Three Generations]."

26. O. Shishkina, "Post-Soviet Foreign Policy Foundations: State Interests vs. Identities," *Moscow State Institute of International Relations (MGIMO) Blog - Govoryat Eksperty MGIMO*, June 9, 2014, https://mgimo.ru/about/news/experts/253273/ [accessed May 10, 2018].

27. Bogaturov, "Tri pokoleniya [The Three Generations]."

28. Vladimir Putin in Vladimir Solov'yev, *President* (Masterskaya Movie Company, 2015) (Russian).

29. T. Shakleina, *Rossiya i SSHA v mirovoy politike [Russia and the USA in the World's Politics]* (Moscow: Aspekt Press, 2012), 160–61.

30. Bogaturov, "Tri pokoleniya [The Three Generations]."

31. Åslund Anders, "An Assessment of Putin's Economic Policy," *CESifo Forum*, no. 9(2) (2008): 16–21.

32. K. Rogov, "Sverkhbol'shinstvo dlya sverkhprezidenstva [Supermajority for Super-presidency]," *Pro et Contra* 17, Issue 3–4 (2013): 107.

33. Vladimir Putin, *2006 Speech at Meeting with the Ambassadors and Permanent Representatives of the Russian Federation*, http://kremlin.ru/events/president/transcripts/23669, [accessed May 10, 2018], (Russian); Bogaturov, "Tri pokoleniya [The Three Generations]."

34. Vladimir Putin, *2007 Speech at the Munich Security Conference*, http://www.stringer-news.com/publication.mhtml?Part=50&PubID=7070 [accessed May 10, 2018], (Russian).

35. T. Shakleina, "Obshchnost' i razlichiya v strategiyakh Rossii i SSHA [Common Ground and Divergences in the Strategies of Russia and the U.S.A.]," *Mezhdunarodnyye Protsessy* 11, no. 2(33) (2013): 6–19.

36. "Interview with S. Ryabkov, the Deputy Foreign Minister of the Russian Federation," published in *Mezhdunarodnaya Zhizn'*, no. 1 (2014): 4–16, (Russian); Vladimir Putin, "2014 Speech at Meeting the Ambassadors and Permanent Representatives of the Russian Federation," published in *Mezhdunarodnaya Zhizn'*, no. 7 (2014): 3–12, (Russian); Putin, *2006 Speech at Meeting with the Ambassadors*; Sergey Lavrov, "Kak okonchatel'no zavershit' kholodnuyu voynu? [How to finally end the Cold War?]," *Mezhdunarodnaya Zhizn'*, no. 5 (2009): 3–12.

37. E. Primakov, "Vneshnyaya politika vo vse bol'shey stepeni napravlyayetsya na vosstanovleniye Rossii v kachestve derzhavy mirovogo klassa [Foreign Policy is More Directed to the Restoration of Russia as a World-Class Power]," *Mezhdunarodnaya Zhizn'*, no. 1–2 (2008): 48–51.

38. Dmitry Medvedev, *Russia Forward!*, 2009, http://www.kremlin.ru/transcripts/5413 [accessed August 8, 2014], (Russian).

39. A. Smirnov, "Kontseptsii vneshney politiki Rossiyskoy Federatsii: sravnitel'nyy analiz [The Foreign Policy Concept of the Russian Federation—Comparative Analysis]," in S. Smul'skii, O. Abramova, and V. Buyanov (ed.), *Vneshnyaya politika Rossii: teoriya i praktika [The Foreign Policy of Russia: Theory and Practice]* (Moscow: Kniga i Biznes, 2013), 58–83.

40. Ministry of Foreign Affairs of the Russian Federation, *The Foreign Policy Concept of the Russian Federation-2000* (Moscow, 2000) (Russian).

41. Ministry of Foreign Affairs of the Russian Federation, *The Foreign Policy Concept of the Russian Federation-2008* (Moscow, 2008) (Russian).

42. Ministry of Foreign Affairs of the Russian Federation, *The Foreign Policy Concept of the Russian Federation-2013* (Moscow, 2013) (Russian).

43. V. Averkov, "Prinyatiye vneshnepoliticheskikh resheniy v Rossii [Foreign Policy Decision-Making in Russia]," *Mezhdunarodnyye Protsessy* 10, no. 2(29) (2012): 59–70.

44. Sergey Lavrov, "Vneshnepoliticheskaya filosofiya Rossii [The Foreign Policy Philosophy of Russia]," *Mezhdunarodnaya Zhizn'*, no. 4 (2013): 2.

45. Zhurkin, Nosov, and Danilov, "Voprosy mezhdunarodnoy bezopasnosti," 175.

46. Vladimir Putin, *2012 Address to the Federal Assembly*, http://kremlin.ru/events/president/news/17118 [accessed May 10, 2018], (Russian).

47. "Interview with Sergey Lavrov, the Foreign Minister of the Russian Federation," published in *Mezhdunarodnaya Zhizn'*, no. 9 (2012): 20, (Russian).

48. Lavrov, "Vneshnepoliticheskaya filosofiya [The Foreign Policy Philosophy]," 2.

49. For example, in 2005, Vladimir Putin described Ivan Ilyin as a great Russian philosopher; see Vladimir Putin, *2005 Address to the Federal Assembly*, http://kremlin.ru/events/president/transcripts/22931 [accessed: May 10, 2018], (Russian); and Dmitry Medvedev wrote a foreword to Ivan Ilyin, *Puti Rossii [Russian Ways]* (Moscow: Vagrius, 2007).

50. Vladimir Putin, *2012 Speech at Meeting of Ambassadors and Permanent Representatives of the Russian Federation to the International Organizations*, published in *Mezhdunarodnaya Zhizn'*, no. 7 (2012): 4 (Russian).

51. Lavrov, "Vneshnepoliticheskaya filosofiya [The Foreign Policy Philosophy]," 2.

52. Ibid., 7. This narrative of "counterbalancing power" is based on Russian Imperial history (mainly the period between 1815 and 1914), rather than on Soviet legacy; see Ivan Ilyin, *Natsional'naya Rossiya, nashi zadachi, [National Russia, Our Goals]* (Moscow: Eskimo Algoritm, 2011), 84–95.

53. V. Baranovskii, "Osnovnyye parametry sovremennoy sistemy mezhdunarodnykh otnosheniy—chast' III [The Major Parameters of the Modern System of International Relations—Part III]," *Polis: Politicheskii Issledovaniya*, no. 5 (2012): 148–56; V. Buyanov, "Rossiya v globaliziruyushchemsya mire: politicheskiy aspekt [Russia in the Globalizing World: Political Aspect]," in Smul'skii, Abramova, and Buyanov, *Vneshnyaya politiku Rossii [The Foreign Policy of Russia]*, 106–29; "Interview with Sergey Lavrov."

54. Lavrov, "Kak okonchatel'no zavershi [How to finally end]," 6.

55. Shakleina, *Rossiya i SSHA [Russia and the USA]*, 174.

56. Vladimir Putin, *Speech of the President of the Russian Federation, 18 March 2014*, http://kremlin.ru/events/president/news/20603/videos [accessed May 10, 2018], (Russian).

57. E. Voronin, "Problema legitimnosti vooruzhennogo vmeshatel'stva. Liviyskiy Casus Belli [The Problem of Legitimacy of Armed Intervention. Libyan Casus Belli]," *Mezhdunarodnaya Zhizn'*, no. 8 (2012): 107–18; V. Kotlyar, "'Otvetstvennost' pri zashchite' i 'Arabskaya Vesna,' ['The Responsibility to Protect' and the 'Arab Spring']," *Mezhdunarodnaya Zhizn'*, no. 9 (2012): 99–114; S. Chernichenko, "Dvoynyye Standarty [Double Standartds]," *Mezhdunarodnaya Zhizn'*, no. 12 (2008): 63–71; "Interview with Sergey Lavrov"; Putin, *Speech of the President of the Russian Federation, 18 March 2014*.

58. "Interview with Sergey Lavrov," 4, 16.

59. "Interview with S. Ryabkov, The Deputy Minister of Foreign Affairs of the Russian Federation," published in *Mezhdunarodnaya Zhizn'*, no. 8 (2014): 2–12 (Russian).

60. "Press-conference with Sergey Lavrov, Russian Foreign Minister," published in *Mezhdunarodnaya Zhizn'*, no. 2 (2013): 11 (Russian).

61. Dmitry Medvedev, *Statement in Relation to the Recognition of the Independence of Abkhazia and South Ossetia*, Moscow, August 26, 2008, http://news.kremlin.ru/news/1223 [accessed May 10, 2018], (Russian).

62. Dmitry Medvedev, "Why I had to Recognise Georgia's Breakaway Regions," *Financial Times*, August 26 2008, https://www.ft.com/content/9c7ad792-7395-11dd-8a66-0000779fd18c [accessed May 10, 2018].

63. A. Moiseev, "O nekotorykh mezhdunarodno-pravovykh pozitsiyakh po ukrainskomu voprosu [About Some International Law Positions on the Ukrainian Issue]," *Mezhdunarodnaya Zhizn,'* no. 5 (2014): 53–65; Putin, *Speech of the President of the Russian Federation, 18 March 2014.*

64. Vladimir Putin in Andrey Kondrashev, *Krim. Put' na Rodinu [Crimea. The Way Home]*, (Vserossiyskaya Gosudarstvennaya Televizionnaya i Radioveshchatel'naya Kompaniya, 2015), (Russian).

65. Putin in Kondrashev, *Krim. Put' na Rodinu [Crimea . The Way Home]*.

66. Sergey Lavrov, "Speech at the First Forum of Young Diplomats," April 25, 2014, published in *Mezhdunarodnaya Zhizn',* no. 5 (2014): 5–20, (Russian); Putin, *Speech of the President of the Russian Federation, 18 March 2014.*

67. Operation to Force Georgia to Peace—is the Russian official name for the 2008 Russo-Georgian War that supports the Russian legal argument for Russian "Jus ad bellum." See the Independent International Fact-Finding Mission on the Conflict in Georgia, *Report*, Volume II, 2009, 188.

68. See the replies of the Russian side to the questions about the humanitarian law in the Independent International Fact-Finding Mission on the Conflict in Georgia, *Report*, 2009.

69. Dmitry Medvedev, *Press Conference after a Meeting with French President Nicolas Sarkozy*, September 8, 2008, http://kremlin.ru/events/president/transcripts/1330 [accessed May 10, 2018], (Russian).

70. Sergey Lavrov, quoted in A. Zemlyanko and V. Rozhkov, "Poteri v lokal'nykh boyevykh deystviyakh [Losses in Local Armed Operations]," *Zashchita i Bezopasnost',* no. 4(47) (2008): 18.

71. V. Ovchinnikov, "Lokal'nyye vooruzhennyye konflikty i bezopasnost' Rossii [Local Armed Conflicts and the Security of Russia]," *Zashchita i Bezopasnost',* no. 4(47) (2008): 9–11; *Interview with A. Starunskii, Former Deputy Commander of the Joint Group of Forces in the North Caucasus (2002–2003)*, published in *Zashchita i Bezopasnost',* no. 4(47) (2008): 6–8, (Russian); Rozhkov and Zemlyanko, "Poteri v lokal'nykh boyevykh deystviyakh."

72. Kondrashev, *Krim. Put' na Rodinu [Crimea . The Way Home].*

73. Georgii Florovskii, "Yevraziyskiy soblazn [Eurasian Temptation]," in L. Novikova and I. Sizemskaya (ed.), *Mir Rossii - Yevraziya: Antologiya [The World of Russia - Eurasia: An Anthology]* (Moscow: Vysshaya shkola, 1995).

74. Alexander Panarin, *Rossiya v tsivilizatsionnom protsesse: mezhdu atlantizmom i yevraziystvom [Russia in the Civilisation Process: Between Atlanticism and Eurasianism]*, (Moscow: Istitut filosofii RAN, 1995), 27–28.

75. Vladimir Pantin, "Istoricheskaya rol' rossii mezhdu zapadom i vostokom [The Historical Role of Russia between the West and the East]," *Istoriya i sovremennost',* no. 2 (2012): 93–112.

76. Ivan Ilyin, *Sushchnost' i svoyeobraziye Russkoy kul'tury [The Essence and Peculiarity of Russian Culture]* (Moscow: Russkaya Kniga-XXI Vek, 2007), 15.

77. For those who are interested in publications in English see for example, Oleg Kharkhordin, *The Collective and the Individual in Russia* (Berkeley: University of California Press, 1999); Yale Richmond, *From Nyet to Da: Understanding the Russians* (Yarmouth, ME: Intercultural Press, 1996); Isaiah Berlin, *The Soviet Mind: Russian Culture under Communism* (Washington, DC: Brookings Institution Press, 2004); Richard Harrison, *The Russian Way of War: Operational Art*, 1904–1940 (Lawrence, KS: University Press of Kansas, 2001); Jack Snyder, *The Soviet Strategic Culture: Implications for Limited Nuclear Operations* (Santa Monica, CA: The RAND Corporation, 1977); Dima Adamsky, *The Culture of Military Innovation: The Impact of Cultural Factors on the Revolution in Military Affairs in Russia, the US and Israel* (Stanford: Stanford University Press, 2010).

78. For example, Ilyin, *Sushchnost' i svoyeobraziye [The Essence and Peculiarity]*; Ilyin, *Natsional'naya Rossiya nashi zadachi [National Russia: Our Goals]*; Nikolay Lossky, *Kharakter Russkogo naroda—Kniga 1 [The Character of the Russian People—Book 1]* (Frankfurt Main: Possev-Velrlag, 1957); Ivan Solonevich, *Narodnaya Monarkhiya [People's Monarchy]* (Moscow: Algoritm, 2011); Nikolai Berdyaev, *Novoye Srednevekov'ye - Razmyshleniya o sud'be Rossii i Yevropy [New Middle Ages: Reflections on the Destiny of Russia and Europe]*,

(Moscow: Feniks-HDS Press, 1991); Nikolai Berdyaev, *Sud'ba Rossii [The Destiny of Russia]*, reprint edition from 1918 publication (Moscow: Filosofskoye Obshchestvo SSSR, 1990).

79. For example, in 2005, Vladimir Putin described Ivan Ilyin as great Russian philosopher (see Putin, *2005 Address to the Federal Assembly*), and in 2007 Dmitry Medvedev wrote a foreword to Ilyin, *Puti Rossii [Russian Ways]*.

80. The foreword of Dmitry Medvedev to Ilyin, *Puti Rossii [Russian Ways]*.

81. For example, Panarin, *Rossiya v tsivilizatsionnom protsesse [Russia in the Civilization Process]*; Pantin, "Istoricheskaya rol' rossii [The Historical Role of Russia]"; V. Lapkin and V. Pantin, "Dinamika obraza Rossii i tsikly reform-kontrreform [The Dynamics of Russia and the Reform/Counter-Reform Cycles]," *Rossiya i musul'manskiy mir*, no. 5 (2008): 15–22; V. Gubanov, *Russkiy natsional'nyy kharakter v kontekste politicheskoy zhizni Rossiii [Russian National Character in the Context of the Russian Political Life]* (St. Petersburg: Izdatel'skiy Tsentr SPbGMGU, 1999).

82. Ilyin, *Sushchnost' i svoyeobraziye [The Essence and Peculiarity]*, 112.

83. Ivan Sikorskii, *Cherty iz psikhologii slavyan - rech' na torzhestvennom zasedanii Slavyanskogo blagotvoritel'nogo obshchestva 14 maya 1895 goda [Lines from the Psychology of Slavs- A Speech at the Ceremonial Meeting of the Slavic Charitable Society on 14 May 1895]* (Kiev: Tipografiya imeni Chokolova, 1895); V. Gubanov, *Russkiy natsional'nyy kharakter [Russian National Character]*, 39–58; Ilyin, *Sushchnost' i svoyeobraziye [The Essence and Peculiarity]*, 17–22.

84. Gubanov, *Russkiy natsional'nyy kharakter [Russian National Character]*, 42–47.

85. Daniel Boorstin, *The Americans: The National Experience* (New York: Vintage Books, 1965), 115–23.

86. Veniamin Alekseyev, "Ekaterinburg," *Rodina*, no. 8 (2003): 100–145.

87. Peter Alabin, *Trekhvekovaya godovshchina goroda Samary [Three-Century-Old Anniversary of the City of Samara]* (Samara: Gubernskaya Zemskaya Tipografiya, 1887).

88. K. Umbrashko, "Istoricheskiy obzor: 'novonikolayevskaya guberniya—novosibirskaya oblast': lyudi, sobytiya, fakty,' [Historical Overview: 'Novonikolayevsk Province - Novosibirsk Region: People, Events, Facts']," published in the proceedings to the *2nd Russian Scientific-Practical Conference Dedicated to the 90th Anniversary of the Novonikolayevsk Province*, Part 1, Novosibirsk, October 25, 2011, 6–28.

89. In fact, this proverb originated from the short poem "Zabytaya derevnya [Forgotten Village]" by Nikolay Nekrasov.

90. This power of the rural commune was abolished only in 1906; see Ilyin, *Sushchnost' i svoyeobraziye [The Essence and Peculiarity]*, 47–48.

91. Berdyaev, *Sud'ba Rossii [The Destiny of Russia]*, 4–5.

92. Lossky, *Kharakter Russkogo naroda [The Character of the Russian People]*, 48.

93. Berdyaev, *Sud'ba Rossii [The Destiny of Russia]*, 14–15; Ilyin, *Sushchnost' i svoyeobraziye [The Essence and Peculiarity]*, 45–47; Lossky, *Kharakter Russkogo naroda [The Character of the Russian People]*; Gubanov, *Russkiy natsional'nyy kharakter [Russian National Character]*, 46–57.

94. Medvedev, *Russia Forward!*

95. Ilyin, *Sushchnost' i svoyeobraziye [The Essence and peculiarity]*; Lossky, *Kharakter Russkogo naroda [The Character of the Russian People]*; Solonevich, *Narodnaya Monarkhiya [People's Monarchy]*; Berdyaev, *Sud'ba Rossii [The Destiny of Russia]*.

96. Berdyaev, *Sud'ba Rossii [The Destiny of Russia]*, 5.

97. Fyodor Dostoyevsky, *Dnevnik pisatelya [*A *Writer's Diary]* (Kaliningrad: Yantarnii Rasskaz, 2005), 28.

98. Gubanov, *Russkiy natsional'nyy kharakter [Russian National Character]*, 44–45.

99. Ilyin, *Sushchnost' i svoyeobraziye [The Essence and peculiarity]*, 29.

100. I. Dzyaloshinskiy, "Kul'tura, zhurnalistika, tolerantnost' [Culture, Journalism, Tolerance]," in I. Dzyaloshinskiy and M. Dzyaloshinskaya (ed.), *Rossiyskiye SMI: kak sozdayetsya obraz vraga [Russian Mass Media: How the Image of Enemy Is Created]* (Moscow: Moskovskoe Byuro po Pravam Cheloveka "Academia," 2008), 24–25.

101. E. Shetopal, "Tsennostnyye kharakteristiki rossiyskogo politicheskogo protsessa i strategiya razvitiya strany [The Characteristics of Values of the Russian Political Process and the Country's Development Strategy]," *Polis: Politicheskii Issledovaniya*, no. 2 (2014): 61–71.

102. Gubanov, *Russkiy natsional'nyy kharakter [Russian National Character]*, 47–52.

103. *Brockhaus and Efron Encyclopedic Dictionary*, Petersburg, 1907, Vol. XVI, 836–39 (Russian).

104. Gubanov, *Russkiy natsional'nyy kharakter [Russian National Character]*, 172–73.

105. Ilyin, *Sushchnost' i svoyeobraziye [The Essence and Peculiarity]*, 115–16.

106. Ibid., 17–19.

107. Levon Petrosyan, "Konflikt mentalitetov kak osnovaniye protivostoyaniya politicheskikh kul'tur: Rossiya i severoatlanticheskiy soyuz v mire posle kholodnoy voyny [The Conflict of Mentalities as a Basis for the Political Cultures Stand-Off: Russia and the North Atlantic Treaty Organization in the Post-Cold War World]," PhD thesis, Moscow State Pedagogical University, 2003, 26–27.

108. Gubanov, *Russkiy natsional'nyy kharakter [Russian National Character]*, 146–47.

109. Ilyin, *Sushchnost' i svoyeobraziye [The Essence and Peculiarity]*, 108.

110. Georgii Fedotov, *Sud'ba i grekhi Rossii [The Fate and the Sins of Russia]*, Volume II, (St. Petersburg: Sofiya, 1991), 324–25.

111. Ibid., 252.

112. Ilyin, *Sushchnost' i svoyeobraziye [The Essence and Peculiarity]*, 110.

113. Gubanov, *Russkiy natsional'nyy kharakter [Russian National Character]*, 167.

114. Clifford Rogers, "'As if a New Sun Had Arisen': England's Fourteenth-Century RMA," in Williamson Murray and MacGregor Knox (ed.), *The Dynamics of Military Revolution 1300–2050* (Cambridge: Cambridge University Press, 2001), 24.

115. Despite the fact that in modern Russian language "*soldier*" (soldat) is more prevalent, it was introduced only in the seventeenth century during the military reforms of Peter the Great. Today, "*voin*" is still in use, although it has higher and poetic meaning.

116. Gubanov, *Russkiy natsional'nyy kharakter [Russian National Character]*, 167–70.

117. Ilyin, *Sushchnost' i svoyeobraziye [The Essence and Peculiarity]*, 109.

118. Ivan Solonevich, "Tak chto zhe yest' demokratiya? [So, What is Democracy?]," *Nasha Strana*, no. 35 (1-2), no. 36 (1-2), no. 38 (1-2), 1950; Ivan Solonevich, "Doklad o mezhdunarodnom polozhenii [Report about the International Situation]," *Nasha Strana*, no. 60 (1950): 1–2.

119. Gubanov, *Russkiy natsional'nyy kharakter [Russian National Character]*, 147.

120. Fyodor Dostoyevsky, "Vlas," *Polnoye sobraniye sochineniy v 30 tomakh. Tom 21-Dnevnik pisatelya 1873 god [The Omnibus in Thirty Volumes. Volume 21—Writer's Diary 1873]*" (Leningrad: Nauka, 1980), 36.

121. Vladimir Lenin, *Sotsializm i voyna [Socialism and War]* (Moscow: Partizdat TsK VKPb, 1937).

122. E. Rudnitskaya and N. Lisovoi, "Mirotvorcheskaya paradigma russkoy mysli [The Peacemaking Paradigm of the Russian Thought]," in E. Rudnitskaya (ed.), *Mirotvorchestvo v Rossii - Tserkov,' Politika, Mysliteli, ot rannego srednevekov'ya do rubezha XIX-XX stoletiy [Peacemaking in Russia: The Church, Politics, Thinkers: From the Early Middle Ages to the Boundary between the 19th and 20th Centuries]* (Moscow: Nauka, 2003), 4.

123. Lev Gudkov, "Pamyat' o voyne i massovaya identichnost' rossiyan [Memory of War and the Russian Identity]," *Neprikosnovennyy Zapas*, no. 2–3(40–41) (2005).

124. Aleksey Levinson, "Voyna, voyny, voyne . . . [War, Wars, to War . . .]," *Neprikosnovennyy Zapas*, no. 2–3(40–41) (2005); Lev Gudkov, "Tsinizm 'neperekhodnogo' obshchestva [Cynicism of 'Non-transitional' Society]," *Vestnik obshchestvennogo mneniya. Dannyye. Analiz. Diskussii* 2, no. 76 (2005): 55.

125. Levinson, "Voyna, voyny, voyne . . . [War, Wars, to War . . .]"; Gudkov, "Tsinizm 'neperekhodnogo' obshchestva [Cynicism of 'Non-transitional' Society]."

126. Dmitri Trenin, "Putin's Biggest Challenge Is Public Support," *Carnegie Moscow Center*, January 15, 2015, http://carnegie.ru/2015/01/15/putin-s-biggest-challenge-is-public-support/hzl5 [accessed May 10, 2018].

127. Anatol Lieven, *Chechnya: Tombstone of Russian Power* (New Haven, CT: Yale University Press, 1999), 169–71.
128. Dmitri Trenin and Aleksei Malashenko, *Russia's Restless Frontier: The Chechnya Factor in Post-Soviet Russia* (Washington, DC: Carnegie Endowment for International Peace, 2004), 50.
129. Lev Gudkov, "God chechenskoy voyny v obshchestvennom mnenii Rossii [A Year of Chechen Way in the Russian Public Opinion]," *Druzhba Narodov*, no. 2 (1996).
130. Vladimir Putin in Solov'yev, *President.*
131. Trenin and Malashenko, *Russia's Restless Frontier*, 50.
132. Trenin, "Putin's Biggest Challenge."
133. Alexander Vladimirov, *Osnovy Obshchey Teorii Voyny, [Fundamentals of General Theory of War]*, Volume 1 (Moscow: Moskovskiy Finansovo-Promyshlennyy Universitet "Sinergiya," 2013), 163.
134. V. Selivanov, V. Leonov, and N. Bagdasaryan, "Social Aspects of NLW Development," in *Proceedings of the 3rd European Symposium on Non-Lethal Weapons* (Ettlingen: Fraunhofer-Institut für Chemische Technologie ICT, 2005).
135. Oleg Khlobustov, "Chechnya, Terrorizm i Kontrterroristicheskiye Operatsii v Zerkale Obshchestvennogo Mneniya [Chechnya, Terrorism and Counter-Terrorism Operations in the Mirror of Public Opinion]," *Vlast'*, no. 5 (2003): 54–59.
136. Ibid.
137. For example, Christopher Donnelly, *Red Banner: The Soviet Military System in Peace and War* (London: Jane's Information Group, 1988); Adamsky, *The Culture of Military Innovation.*
138. Adamsky, *The Culture of Military Innovation*, 41, 47.
139. Donnelly, *Red Banner*, 61–65, 125–34.
140. Ilyin, *Sushchnost' i svoyeobraziye [The Essence and Peculiarity]*, 29.
141. Ibid., 29–30.
142. Adamsky, *The Culture of Military Innovation*, 43.
143. Gubanov, *Russkiy natsional'nyy kharakter [Russian National Character]*, 169.
144. Adamsky, *The Culture of Military Innovation*, 44–45.
145. Ibid., 45–48, 51.
146. A. Arbatov and V. Dvorkin, "Voyennaya reforma: proschety i puti ikh ispravleniya [Military Reform: Miscalculations and the Ways to Their Correction]," in M. Lipman (ed.), *Povestka dnya dlya novogo prezidentstva [Agenda for the New Presidency]*, (Moscow: Moscow Carnegie Centre, Rossiyskaya Politicheskaya Entsiklopediya, 2012), 41–52; P. Baev, "Trayektoriya voyennoy reformy v Rossii [The Trajectory of the Military Reform in Russia]," *Pro et Contra* 16, Issue 4–5 (2012): 171–84.
147. Andrei Kokoshin, *Soviet Strategic Thought, 1917–91* (Cambridge, MA: MIT Press, 1998), 11–39.
148. Adamsky, *The Culture of Military Innovation*, 47.
149. Ibid., 52–53.
150. For example, V. Burenok, V. Korchak, and S. Smirnov, "Oruzhiye neletal'nogo deystviya - sostavnaya chast' sistemy vooruzheniya budushchego [Weapons with Non-Lethal Action—The Constitutive Part of the Future Weaponry System]," *Vestnik Akademii Voyennykh Nauk*, no. 4 (2007): 117–27; Savrasov, "Kontseptsiya primeneniya oruzhiya [The NLW Deployment Policy]"; Antipov and Novichkov, "K voprosu o razrabotke [Regarding the Question of Development]"; Selivanov, "Oruzhiye neletal'nogo deystviya [Non-Lethal Weapons]."
151. Burenok, Korchak, and Smirnov, "Oruzhiye neletal'nogo deystviya [Weapons with Non-Lethal Action]," 119.
152. For example, V. Moiseev and V. Orlyanskii, "Oruzhiye neletal'nogo deystviya i printsipy taktiki [Weapons with Non-Lethal Action and the Principles of Tactics]," *Voennaia Misl'*, no. 6 (2011): 26–33; E. Efremov, "Matematicheskoye modelirovaniye primeneniya oruzhiya neletal'nogo deystviya v mirotvorcheskikh operatsiyakh [The Mathematical Simulation of the Employment of Weapons with Non-Lethal Action during Peacemaking Operations]," *Voennaia Misl'*, no. 4 (2012): 39–43; S. Panteleev, "Attraktory vrazhduyushchikh storon v lokal'nykh konfliktakh s primeneniyem sredstv neletal'nogo vozdeystviya [Attractors of the Fighting

Sides in Local Conflicts with the Employment of Weapons with Non-Lethal Action]," *Voennaia Misl',* no. 4 (2012): 11–16; A. Pronin, A. Leonov, and L. Kaplyarchuk, "Osnovnyye kriterii voyenno-ekonomicheskoy otsenki oruzhiya neletal'nogo deystviya [The Major Criteria of Military-Economic Evaluation of Weapons with Non-Lethal Action]," *Voennaia Misl',* no. 10 (2012): 43–50; D. Zaytsev, D. Soskov, and A. Foteev, "Oruzhiye neletal'nogo deystviya na osnove izlucheniy - fizicheskiye osobennosti i perspektivy primeneniya [Weapons with Non-Lethal Action Based on Radiation: Physical Characteristics and Employment Perspective]," *Voennaia Misl',* no. 4 (2013): 30–37; L. Ilyin and V. Rikhlin, "Inkapasitanty kak oruzhiye neletal'nogo deystviya [Incapacitants as Weapons with Non-Lethal Action]," *Voennaia Misl',* no. 9 (2014): 37–42; Kuznetsova Stupitskiy, "Issledovaniye vozmozhnosti sozdaniya [The Research of the Possibility to Create]"; Nagovitsyn, Grudzinskii', and Sporykhin, "Oruzhiye neletal'nogo deystviya [Weapons with Non-Lethal Action]."

153. For example, A. Burutin et al., "Sistematizatsiya tipovykh skhem primeneniya oruzhiya neletal'nogo deystviya [The Systematization of Typical Scenarios of the Implementation of Weapons with Non-Lethal Action]," *Strategicheskaya Stabil'nost',* no. 2(51) (2010): 18–26; D. Zaitsev et al., "Rol' i mesto oruzhiya neletal'nogo deystviya v konfliktakh nizkoy intensivnosti [The Role and the Place of Weapons with Non-Lethal Action in the Low-Intensity Conflicts]," *Strategicheskaya Stabil'nost',* no. 4(61) (2012): 27–35; D. Zaitsev et al., "Porazhayushchiye faktory oruzhiya neletal'nogo deystviya - fizicheskiye osobennosti, biologicheskiye effekty i taktika primeneniya [The Hitting Factors of Weapons with Non-Lethal Action: Physiological Features, Biological Effects and Employment Tactics]," *Strategicheskaya Stabil'nost',* no. 3(64 (2013): 2–11; A. Harvardt et al., "Perspektivy primeneniya oruzhiya neletal'nogo deystviya v urbanizirovannom rayone [The Perspectives of the Employment of Weapons with Non-Lethal Action in Urban Areas]," *Strategicheskaya Stabil'nost',* no. 1(61) (2014): 64–71.

154. Nagovitsyn, "Ispol'zovaniye oruzhiya neletal'nogo deystviya [The Employment of Weapons with Non-Lethal Action]."

155. Selivanov, Levin, and Lyuwin, "Arkhitektura i realizatsiya [The Architecture and Realization]"; Selivanov, Levin, and Ilyin, "Metodicheskiye osnovy [Methodological Basics]."

156. V. Burenok, A. Leonov, and A. Pronin, *Voyenno-Ekonomicheskiye i Innovatsionnyye Aspekty Integratsii Netraditsionnykh Vidov Oruzhiya v Sostav Sistemy Vooruzheniya [Military-Economic and Innovative Aspects of the Integration of Non-traditional Weapons]* (Moscow: Izdatel'skaya Gruppa Granitsa, 2014), 44–45.

157. Efremov, "Matematicheskoye modelirovaniye [The Mathematical Simulation]," 39.

158. Panteleev, "Attraktory vrazhduyushchikh storon [Attractors of the Fighting Sides]," 16.

159. Moiseev and Orlyanskii, "Oruzhiye neletal'nogo deystviya [Weapons with Non-Lethal Action]," 29–30.

160. Ibid.

161. Selivanov, Levin, and Lyuwin, "Arkhitektura i realizatsiya [The Architecture and Realization]."

162. Selivanov, Levin, and Ilyin, "Metodicheskiye osnovy [Methodological Basics]."

163. Pronin, Leonov, and Kaplyarchuk, "Osnovnyye kriterii [The Major Criteria]."

164. Smirnov Korchak, "Regulirovaniye normativno-pravovykh otnosheniy"; Artem Muranov, "Administrativno-pravovoy rezhim oruzhiya neletal'nogo deystviya: problemy formirovaniya i realizatsii [The Regulatory System of Weapons with Non-Lethal Action: Problems of Formulation and Implementation]," PhD thesis, The All-Russian Research Institute of the Ministry of Internal Affairs of the Russian Federation, 2008.

165. V. Vladimirov, "Sovremennaya voyna i grazhdanskaya oborona [Contemporary War and Civilian Defense]," *Strategiya Grazhdanskoy Zashchity: Problemy i Issledovaniya* 2, no. 1 (2012): 471–81; Vladimirov and Chernikh, "Sostoyaniye i osnovnyye napravleniya [The Current Situation and Main Directions]."

166. Burenok, Korchak, and Smirnov, "Oruzhiye neletal'nogo deystviya [Weapons with Non-Lethal Action]"; Zaitsev et al., "Rol' i mesto oruzhiya [The Role and the Place]"; Harvardt et al., "Perspektivy primeneniya oruzhiya [The Perspectives of the Employment of Weapons]"; Burutin et al., "Sistematizatsiya tipovykh skhem [The Systematization of Typical Scenarios]."

167. Moiseev and Orlyanskii, "Oruzhiye neletal'nogo deystviya [Weapons with Non-Lethal Action]," 30.

168. Zaitsev et al., "Rol' i mesto oruzhiya [The Role and the Place]," 35.
169. Adamsky, *The Culture of Military Innovation*, 50–52.
170. Zaitsev et al., "Rol' i mesto oruzhiya [The Role and the Place]," 31.
171. Pronin, Leonov, and Kaplyarchuk, "Osnovnyye kriterii [The Major Criteria]," 48–49.
172. Moiseev and Orlyanskii, "Oruzhiye neletal'nogo deystviya [Weapons with Non-Lethal Action]," 32.
173. Vladislav Inosemzev, *Poteryannoye desyatiletiye [The Lost Decade]* (Moscow: Moskovskaya shkola politicheskih issledovanii, 2013), 553–69; Lapkin and Pantin, "Dinamika obraza Rossii [The Dynamics of Russia]."
174. Inosemzev, *Poteryannoye desyatiletiye [The Lost Decade]*, 556.
175. Vladimir Putin, "Byt' sil'nymi: garantii natsional'noy bezopasnosti dlya Rossii [Being Strong: National Security Guarantees for Russia]," *Rossiyskaya gazeta*, No. 5708, February 17, 2012.
176. The Ministry of Defense of the Russian Federation, the online version of the *Encyclopedia of the Strategic Missile Troops* (*Entsiklopediya Raketnykh Voysk Strategicheskogo Naznacheniya*), http://encyclopedia.mil.ru/encyclopedia/dictionary/details_rvsn.htm?id=13770@morfDictionary; also the online version of the *Military Encyclopedic Dictionary* (*Voyennyy Entsiklopedicheskiy Slovar'*), http://encyclopedia.mil.ru/encyclopedia/dictionary/details.htm?id=7781@morfDictionary, [accessed May 10, 2018].
177. Putin, "Byt' sil'nymi [Being Strong]."
178. Quoted in "Minoborony: Na Forume 'Armiya-2015' Pokazhut Netraditsionnyye Vidy Oruzhiya [Ministry of Defense: Non-Traditional Types of Weapons Will be Demonstrated at 'Army-2020' Show']," *RiaNovosti*, April 21, 2015, http://ria.ru/defense_safety/20150421/1059915256.html, [accessed May 10, 2018].
179. The Ministry of Defense of the Russian Federation, the online version of the *Military Encyclopedic Dictionary*.
180. The Ministry of Defense of the Russian Federation, the online version of the *Encyclopedia of the Strategic Missile Troops*.
181. For example, Frederic Merget, "Non-Lethal Weapons and the Possibility of Radical New Horizons for the Laws of War: Why Kill, Wound and Hurt (Combatants) at All?," *Social Science Research Network*, July 1, 2008, http://papers.ssrn.com/sol3/papers.cfm?abstract_id=1295348 [accessed May 10, 2018]; David Fidler, "'Non-Lethal' Weapons and International Law: Three Perspectives on The Future," in Nick Lewer (ed.), *The Future of Non-Lethal Weapons: Technologies, Operations, Ethics and Law* (London: Frank Cass Publishers, 2002), 26–38.
182. For example, Moiseev and Orlyanskii, "Oruzhiye neletal'nogo deystviya [Weapons with Non-Lethal Action]."
183. Burenok, Korchak, and Smirnov, "Oruzhiye neletal'nogo deystviya [Weapons with Non-Lethal Action]," 124.

Chapter Four

"Never Again Shall Masada Fall!"

Israel and Enemy Civilian Casualties

Following the American discourse on NLW, the current academic and professional literature on NLW in Israel has been employing similar terminology, calling NLW *neshek ul-hereg* (non-killing weapons)[1] or *neshek pahot katlani* (less-lethal weapons).[2] Yet the official term used by the Israeli Defense Forces (IDF) for NLW is *emtzaim le-pizur hafganot* (crowd dispersal means).[3] One of the most prominent explanations for this term is the fact that NLW has been employed by the IDF as riot control weapons long before the contemporary understanding of NLW was formed.

The first employment of NLW by the Israeli military was during the 1970s, encountering disturbances and other forms of civilian violence, generally organized by youths and students in towns and cities in the Territories (Judea, Samaria, and Gaza). This employment, however, was limited in scale and restricted to tear gas and rubber bullets, usually employed by the Border Police under the command of IDF officers, rather than military units themselves.[4] The first significant peak in the IDF's interest in NLW, as a law enforcement tool to deal with riots and civilian disobedience, occurred during the First Intifada (December 1987–September 1993). The problem of the "gap between verbal warning and deadly force"[5] that the U.S. military encountered in Somalia in 1995, the IDF had to deal with as early as 1988, when the then Defense Minister Yitzhak Rabin had stated that "soldiers found themselves in a position of either fire or do nothing."[6] In an attempt to bridge this gap, the IDF had to employ a vast variety of NLW to fulfill its law enforcement missions in the Territories.[7]

The second major peak of interest in NLW occurred during the Second Intifada (Al-Aqsa Intifada, September 2000–February 2005). The IDF had found itself, yet again, encountering massive riots and the quest for new, improved NLW was given high priority.[8] Since then, the IDF's interest in NLW dwindled and intensified according to the political situation in the West Bank. Consequently, it seems right to argue that the IDF employs NLW on a much larger scale and has a much richer experience than its American or Russian counterparts due to its intensive involvement in law enforcement missions in the Territories.

Interestingly enough, despite this vast employment of NLW and the fact that the problem of civilian casualties during military operations has been at the core of the IDF's professional discourse for many years,[9] the IDF has always thought of NLW as a tool for riot and crowd control only, and the idea of their employment on the battlefield has never been considered as a viable option. While it is easy to criticize the IDF for its lack of vision,[10] it seems that the reasons for the very limited employment of NLW in the Israeli military, similar to the American and Russian cases, are rooted more in political and cultural contexts, rather than in the IDF's professional misconduct.

THE STATUS OF NON-LETHAL WEAPONS IN ISRAEL

With the end of the Six-Day War and the establishment of Military Government in the Territories, the IDF found itself dealing with local civilian disobedience in the form of stone-throwing, tire-burning, and road-blocking. As Major General Shlomo Gazit, the first IDF coordinator of government activities in the Territories, noted: "civil disobedience in the West Bank broke out at an extremely early stage, within weeks of the cessation of hostilities in the Six-Day War."[11] Unlike the Jordanians, who controlled Judea and Samaria before 1967, and who did not hesitate to use live ammunition against rioters, the IDF, dealing with the protesters, started to experiment with NLW. These sporadic outbreaks of violence were regarded as typical "current security" (*Bitachon Shotef*, or *Batash*) problems and were solved decisively and by relatively limited measures, as "in the aftermath of the 1967 war, the IDF enjoyed a significant deterrent profile in the Territories and the local population . . . was reluctant to take personal risks." When the employment of NLW was required, it was limited to rubber bullets and tear gas, and it was mainly executed by the Border Police (*Mishmar Hagvul*, or *Magav*) under the close supervision of IDF commanders.[12]

The outbreak of the First Intifada in December 1987 created an entirely different situation, when "Palestinian residents of the West Bank and the Gaza Strip attacked any Israeli who crossed their path . . . with rocks, knives, home-made firebombs and, eventually, gunfire."[13] The IDF found itself completely unprepared to face civilian violence on a large scale, accompanied by massive riots and demonstrations. Ehud Barak, the IDF Deputy Chief of Staff at that time, assessed that the new situation left the soldiers "with only two possibilities: to run away or to shoot. Both possibilities were bad . . . We were not technically prepared to deal with the violent popular riot on this scale."[14]

This technological gap was bridged by the IDF in two different ways. The first one was a vast employment of "off-the-shelf" law enforcement NLW—riot batons, rubber bullets, and tear gas. The second one was the rapid research and development process to improve the effectivity of the existing NLW and introduce new ones according to the new challenges posted by rioters. For example, when the existing rubber bullets proved ineffective at a distance of more than 20 meters, the IDF quickly introduced an improved version containing a quantity of lead.[15] When the effectiveness of this improvement also diminished, as the local population had learnt how to offset its effects (mainly by maintaining further distance from the forces), the IDF introduced plastic bullets. Yitzhak Rabin, then the Defense Minister, explained the purpose of this new NLW:

> It was demonstrated that the use of clubs, tear gas and rubber bullets was insufficient when confronting stone-throwers at ranges of 30–50 meters. . . . Therefore we added plastic bullets, which are intended to reach the stone throwers, organizers and agitators.[16]

While the newly introduced plastic bullets were intended to be used against point targets (stone throwers, organizers, etc.), the IDF was keen to introduce a weapon that would be more effective against a big mass of demonstrators. The first such attempt was made when the IDF introduced *Hatzatzit* (from *hatzatz*—gravel)—a system that produced a torrent of small size gravel and was carried by a small truck.[17] The second was when the Israeli Air Force, in a very short time, introduced several systems for massive employment of non-lethal means (tear gas and smoke grenades, lachrymatory and/or painting liquids, and even gravel) carried and dispersed by helicopters.[18]

With the end of the First Intifada, and especially after 1994, the same year the Oslo Agreements were signed, the IDF's interest in NLW dwindled. The plastic bullets were rejected, as they proved as significantly more lethal than expected—i.e., during the first 18 months of the Intifada the number of fatalities caused by plastic bullets stood at 82.[19] The *Hatzatzit* systems, which were regarded as odd, bizarre, and not "sophisticated" enough weapons, were completely abandoned. As Deputy Commissioner Alexander (Alik) Ron stated during the Or Commission (commission of inquiry into the clashes between law enforcement forces and Israeli citizens, in October 2000): "these systems have been rotting somewhere for many years, since the Intifada we have not used them."[20]

The IDF's interest in NLW was reignited in 2000. Despite the experience of the First Intifada, the Israeli military found itself unprepared and the quantitative and qualitative lack of NLW were first experienced during the withdrawal from Lebanon, when the IDF found itself encountering massive demonstrations led by Hezbollah fighters. This shortage became to be more significant with the start of the Second (Al-Aqsa) Intifada and following operation Defensive Shield that reinforced the need for more effective NLW.

The main reason why the Al-Aqsa Intifada required NLW that were different from the previously employed systems is the fact that, at that time, the Israeli activity in the Territories had a military, rather than law enforcement, nature. While Israeli infantry units, tanks, and armored personnel carriers (APCs) were deployed to fight terrorists, they frequently found themselves encountering stone-throwers and unarmed rioters. Without NLW, the Israeli troops faced the difficult choice between shooting the rioters (usually children and youths) and doing nothing—decisions that several times led to situations when military equipment (including machine-guns) was stolen from manned tanks and APCs.[21]

Desperate to bridge this gap, the IDF introduced non-lethal flash-bang shells for tanks[22]—the first and, so far, unique NLW intended to be fired from the tank's main cannon. Another, newly developed NLW was *Bo'esh* (Skunk)—a sticky liquid produced from an organic and non-toxic blend of baking powder, yeast, and other ingredients, which leaves a powerful odor similar to rot or sewage.[23] In 2002, during the Siege of the Church of the Nativity in Bethlehem (April 2–May 10) and the series of sieges of the Arafat compound in the Mukataa in Ramallah, *Bo'esh* was considered by the IDF, and rejected, as a possible non-lethal tool to resolve the siege situations.[24] Despite these very innovative technologies, with the end of the Al-Aqsa Intifada in 2005, similar to the end of the First Intifada, the IDF's interest in NLW dwindled again. The non-lethal tanks' shells were taken out of service,[25] and *Bo'esh* was in the development process for three more years and was operationally employed for the first time only in 2008, against demonstrators in Na'alin village in the West Bank.[26]

Since the end of the Second Intifada, the IDF's interest in the employment of NLW changed according to political developments in the region. On the one hand, due to the frequent violent demonstrations and riots organized by Palestinians and Israeli ultra-left movements against the Israeli West Bank Wall ("Separation Wall" or "Security Barrier"—depending on one's political views), most of the IDF units have been trained to fulfill crowd control missions and went through NLW basic training.[27] On the other, facing political-military turbulences, which required other than lethal resolution, the IDF found itself unprepared. For example, the lethal outcome of the military operation against the Gaza Freedom Flotilla, on May 31, 2010, resulted in vast criticism, reigniting the interest in NLW and exposing a detrimental gap in the IDF's employment of NLW.[28] Another example was in 2011, when anticipating the Palestinian unilateral declaration of independence in the UN, and a potential "Third Intifada" that would come afterwards, the IDF found itself desperately searching for new NLW and investing about NIS75 million to purchase mainly off-the-shelf weapons.[29]

Interestingly enough, despite its vast experience and technological novelty, the IDF has never formed a persisting professional institution responsible for the development and employment of NLW and their operational concepts. During the First Intifada, it was Deputy Chief of Staff Major General Ehud Barak who led the development of NLW and their fielding.[30] During the Al-Aqsa Intifada, the IDF established a short-lived Forum of Brigadier Generals to define the operational gaps versus the threats. This forum had already been disbanded by mid-2001, and the responsibility for NLW was relocated to the head of the IDF Operations Directorate.[31] In 2006, after the end of the Second Intifada, the operational and conceptual development of NLW was in the hands of a specially established committee, The NLW Forum at the General Officer Commanding Army Headquarters, which was responsible for the evaluation of NLW and writing operational concepts for their employment. While this committee was disbanded in 2009, just three years after its establishment, one of its main achievements was the establishment of the Department of Riot Control at the main training base of the IDF Central Command. This accomplishment, however, was limited, as the responsibilities of this department were mainly restricted to the supervision of riot control training of the IDF units before their deployment in the Territories, and did not include the development of operational concepts for NLW or their technological requirements.[32]

To summarize the current NLW situation in the IDF, it is important to emphasize two main factors. The first one is the technological level of the employed NLW and the scale of their employment. On the one hand, the range of the routinely employed NLW within the IDF does not differ much from the NLW employed by its American and Russian counterparts, and mainly consists of different means to employ tear gas and kinetic rubber bullets.[33] On the other, during the peaks of civil unrest in the Territories, the IDF introduced several out-of-the-box non-lethal technologies, such as *Hatzatzit*, *Bo'esh*, and non-lethal shells for tanks. Moreover, due to the constant deployment of Israeli military units in the Territories as law enforcement forces, they have extensive experience in riot control missions, and, thus, the employment of NLW.

The second is the IDF's conceptual perspective on NLW. Despite the vast employment of NLW, and despite significant NLW technological innovations, the IDF has never considered their possible employment on the battlefield as a tool to minimize civilian casualties during military operations.[34] NLW have been continuously considered by the IDF as less prestigious weapon systems intended for riot control during *Batash* missions ("current security"—i.e., law enforcement deployment in the Territories and along Israel's borders), rather than military operations. Moreover, NLW were frequently conceptualized as systems that "might harm the [IDF's] ability to prepare for a [future] battle, and even sabotage the ability to achieve victory in it."[35] While it seems easy to criticize the IDF for its lack of vision, will, or ability to re-evaluate its assessment of NLW,[36] it seems that the reasons for the IDF's denial to employ NLW during military operations, similar to the American and Russian cases, are rooted more in the political and cultural contexts, rather than in the IDF's professional misconduct.

ISRAELI FOREIGN POLICY AND INTERNATIONAL CRITICISM

When the Israeli government appointed in 1968 the freshly retired Lieutenant General Yitzhak Rabin as Israel's ambassador to the United States, Rabin asked the foreign ministry: "What does the government expect Israel's Ambassador to the United States to achieve?" He was then told: "Diplomatic Objectives? We have no idea."[37] This anecdote provides a hint of the Israeli approach to its foreign policy, suggesting that either Israel does not care about its foreign affairs or the Israeli foreign policy aims are so solid that they are obvious. Indeed, to outsiders, the Israeli foreign policy decision-making may seem reactionary and quite irrational. The best example of this was the visit of U.S. Vice President Joe Biden to Jerusalem in March 2010, who in an attempt to re-launch peace negotiations requested Israel to freeze construction in the West Bank. When he landed in Tel Aviv, he was taken aback by Israel's announcement of 1,600 new apartments to be built in east Jerusalem—a move that led to what was reported as one of the most serious rows between the two allies in recent decades.[38]

Israelis, however, have a clear view of their foreign policy behavior, which is firmly based on a set of Israeli traditional cultural codes—"existential anxiety; upright defiance; a *dugri* (frank) speech culture and a fear of seeming to be a *fraier* (sucker)."[39] While some of these characteristics will be discussed in the next part of this chapter, the main focus here is on foreign policy mechanisms that assist Israel's political leadership to withstand international criticism; in other words, how this cultural code has been translated into foreign policy narratives intended to fend off international criticism regarding Israel's military activities and their outcomes (especially in terms of civilian casualties).

There is a general agreement between scholars that Israel's foreign policy is deeply rooted in numerous Jewish traditions, and its analysis has to start, at the very least, from the beginning of modern Zionism and the work of Benjamin Ze'ev Herzl.[40] This comprehensive examination, however, is beyond this chapter, and it seems reasonable to begin the following examination from 1948, the year of the establishment of the state of Israel, and go back to the pre-state roots of the Israeli foreign policy narratives when it is required.

The discussion of the American and Russian foreign policies, presented in the previous chapters, primarily focused on the mechanisms intended to fend off international criticism following the military activity of these states, or, in other words, how these states have maintained the legitimacy of their military operations. In the case of Israel, however, this question of legitimacy has always been a much more complicated issue. Unlike the United States and Russia, which have had to defend the legitimacy of their military deployments that were criticized for causing civilian casualties and collateral damage, Israel, from its very beginning, has had to defend, not only the legitimacy of its military operations, but also its own legitimacy as a Jewish state.[41] Consequently, in Israeli foreign policy, these two different types of legitimacy seem to be highly interconnected, as according to many Israeli scholars and politicians, "after more than 60 years of the independence of the State of Israel, there are still many question marks regarding its sovereignty and its right to be an equal member of the international community."[42]

As the establishment of the State of Israel in 1948 was an outcome of a UN resolution, the initial attitude of Israeli leadership toward the international community was very positive and was characterized by a "mutual trust built on mutual expectations and their realization."[43] Based on the traditional political Zionism expressed by Chaim Weizmann in 1907 that "[we have] to make the Jewish question an international one. It means going to the nations and saying to them: We need your help to achieve our aim,"[44] the initial foreign policy of Israel was based on a "knock on every door" policy.[45] For David Ben-Gurion, the first Israeli Prime Minister, the crucial basis for Israeli foreign policy was his conclusion that Jews "are the weakest nation on earth . . . we lack any declared friendship anywhere in the world . . . [and] we must take all possible steps to find understanding, if not friendship, anywhere in the world."[46] Many scholars define this period as the "Golden Age" of Israel, when it enjoyed the broadest international approval in its history. In addition to the openness of Israel's leaders towards the world, its military victory in the War of Independence was seen by the world as a stand of "the few against the many," as a "David facing Goliath," and, therefore, Israel was endorsed by the majority of countries. Moreover, the successful immigration absorption, the establishment of kibbutz communities, the development of the Negev desert, and the creation of an independent military industrial complex—all these created vast empathy and support Israel in the eyes of the international community.[47]

This period, however, was very short-lived. In an effort not to be seen as taking sides in the emerging Cold War confrontation between the two giants, Israeli leaders begun to realize that that despite Israel's non-committed foreign policy, "both the United States and the Soviet Union tended to judge Israel by its pro-Communist or pro-Western proclivity and pressured Israel to define its position much more clearly."[48] Ben-Gurion, who realized that Israel could become a hostage of the Soviet-American struggle, had, already in 1952, started to emphasize the independence of the Israeli political decision-making process:

> There is no external force—even as most powerful, aggressive and rich as it could be imagined—that will define the needs and the values of Israel. The foreign policy of Israel is organized according to the values and the basic needs of the Jewish nation, and not by any other decision-maker.[49]

During the 1950s, two different developments in the global arena forced Israel to harden this narrative. The first one was the choice of the USSR to support Arab countries against Israel. After 1954, when the Kremlin for the first time vetoed the UN Security Council resolution in favor of an Arab country (Syria) over Israel, the Soviet bloc countries had ceased their support of Israel. The second development was the decolonization process that added a significant number of new sovereign Arab and Muslim states to the international community, and, therefore, to the UN.[50] Following these changes on the international map, and subsequent rising criticism of Israel, Israeli leadership adopted a narrative that could be described as: "we're on our own because the world is against us"[51]—a statement that expressed a fundamental distrust of the goodwill of the international community. This was the time when the "mutual trust" between the UN and Israel was transformed in one of the most common narratives of the Israeli foreign policy—*Um Shmum* (the UN is nothing)—a term coined by David Ben-Gurion in 1955.[52]

The Israeli military victories in the Six-Day War, in 1967, and the Yom Kippur War, in 1973, completely changed the way Israel was perceived by the international community. From the small "David," which barely won its independence in 1948, Israel twice thrashed significantly superior Arab militaries, emerging as a regional "Goliath" itself. Utilizing the bipolar tension of the Cold War, the Arab countries succeeded in recruiting significant support in the U.N. against Israel, which was expressed in a chain of anti-Israeli resolutions. The peak of the Arab states' efforts to delegitimize Israel was in 1975 when the UN General Assembly passed a resolution that determined that "Zionism is a form of racism and racial discrimination."[53] The response of Israeli Prime Minister Yitzhak Rabin was simple: "The UN lost its ethical and political validity and became a battlefield for wrangling and harassment that have no connection, whatsoever, with the principles and ideals for which it was established."[54]

Since then, the general approach of Israeli politicians toward the UN was best described by the prominent Israeli diplomat Abba Eban: "Any UN decision against Israel automatically enjoys major support, and if Israel would propose that the Earth is round, the majority in the UN will vote against."[55] In other words, facing criticism from the international community, the Israeli political reaction was based on its disdain for what it saw as other powers' biased political action, disconnected from the political or military situation on the ground.

An additional facet of this *Um Shmum* narrative was the increasing unilateralism in the Israeli foreign policy. The disdain for and mistrust of international organizations led Israeli politicians to emphasize the independence of their decisions; as Ben-Gurion famously put it, "It's not important what the gentiles would say; it is important what the Jews would do."[56] Interestingly enough, this political unilateralism has allowed Israel to persist with its actions regardless of the vast criticism coming, not only from Israel's natural enemies (i.e., the Arab and Muslim world), but also from its allies. For example, despite the fact that the construction of the "Separation Wall" was ruled illegal by the International Court of Justice in Hague and was highly criticized by the international community (and more cautiously by the United States), Israel erected the "Security Barrier" as planned.[57] As Raanan Gissin, the then Israeli prime minister's spokesperson, put it: "I believe that after all the rancor dies, this resolution will find its place in the garbage can of history. The court has made an unjust ruling denying Israel its right of self-defense."[58]

While U.S. foreign policy creates an image of *a good protector against a bad evil* to legitimize American actions, and Russians argue for *double standards* to fend off international criticism, Israelis simply state: "we care nothing for what UN members think or vote" and "when it comes to Israel's survival, we must always remain the masters of our own fate."[59] Nevertheless, it would be wrong to argue that this Israeli traditional *chutzpah* (insolence or audacity), in its attitude toward the international community, is the only factor that enables them to withstand criticism of its political-military decisions. After all, Israel's economic, political and military power and independence cannot be compared to those of Washington or Moscow. This additional factor, which has been allowing Israel to keep afloat in a rising tide of international criticism, was best expressed by Prime Minister Benjamin Netanyahu in his speech to the U.S. Congress, on July 10, 1996:

> You, the people of America, offered the fledgling Jewish state succour and support. You stood by us time and time again. . . . The United States has given Israel—how can I tell it to this body? The United States has given Israel, apart from political and military support, munificent and magnificent assistance in the economic sphere. With America's help, Israel has grown to be a powerful, modern state.[60]

One of the main foundations of the Zionist movement was a quest for the support of the Great Powers. With the end of World War II and the surprising support of both of the newly emerging superpowers—the United States and the USSR—Israel preferred to adopt a political line that was defined at that time as "non-identification." This Israeli attempt to play on both sides of the field in the context of the rising tensions of the Cold War proved to be unfeasible for two main reasons. The first was the fact that after a short love affair with the Soviets, during the first few years of independence, Israel began to understand that Moscow had too little to offer to the newly established state which was facing a massive Arab political-military-economic boycott. The second was the rising American political, economic, military, and strategic weight in the world, as well as Washington's increasing interest in Middle Eastern affairs and its approach toward Israel—all of which were considered by Israel as far more beneficial than whatever the Kremlin could offer. Nevertheless, in the months leading up to the Sinai Campaign (October 1956), it was clear to all that Israel's choice, whether due to its own initiative or circumstantial, was made in favor of the West.[61]

During the first years after 1948, however, Washington was hesitant in its official support of Israel, and the initial American involvement in the 1960s and 1970s in the Middle East, in general, and in Israel's affairs, in particular, was mainly due to its own economic (i.e., Arab oil) and strategic (i.e., the Cold War) interests.[62] During the following decades, and especially after the Yom Kippur War, American patronage has become one of the most dominant narratives of Israel's foreign policy—"despite the fact that no formal alliance exists between the two nations, the United States has evolved into the ultimate guarantor of Israel's security, its principal diplomatically and its foremost economic supporter."[63] This special relationship between two countries was best emphasized in the *Economist* as "the mighty edifice of support" that Washington provides Israel each time it faces a severe criticism from all corners of the globe.[64]

On the one hand, many scholars and politicians regard this unconditional American support for Israel as a result of the powerful role of the Jewish/Israeli lobby in American domestic politics—as Ehud Olmert, former Israeli prime minister, stated in 2006: "Thank God we have AIPAC [the American Israel Public Affairs Committee], the greatest supporter and friend we have in the whole world."[65] On the other hand, there are scholars who reject this "theory of everything" (i.e., the Jewish/Israeli control of U.S. foreign policy).[66] While some of them claim that the United States supports Israel due to its own strategic interests and positions in the Middle East and "Israel is America's most reliable ally in the region,"[67] others simply blame American public opinion for overwhelmingly taking Israel's side.[68] Nevertheless, regardless of the roots of American diplomatic protection, Israeli political leaders take it for granted, even if "sometimes they [themselves, i.e., Israelis] do not act like a [most] loyal ally."[69]

In analyzing Israel's foreign policy, it is important to highlight its almost instinctive reliance on U.S. support. After all, between 1972 and 2015, Washington vetoed 43 UN Security Council resolutions that were critical of Israel (more than half of all American vetoes during these years).[70] Consequently, this unprecedented protection by the United States, "the most powerful state in the world,"[71] combined with the Israeli traditional narrative of *Um Shmum* (i.e., disdain for international criticism), allows Israeli politicians to minimize the influence of international pressure on their decision-making process regarding Israel's military activities. In other words, facing international criticism, Israel's leaders not only explicitly reject it as biased against Israel, but also count on American protection against truly meaningful UN Security Council decisions.

While in the previous chapters, the analysis showed how American and Russian foreign policies fend off general international criticism regarding their military actions, and then the narratives discussed were projected on the criticism of enemy civilian casualties, the case of Israel is different. On the one hand, Israeli politicians have been continuously trying to link the legitimacy of Israel as a Jewish state and the legitimacy of its military actions, even at the price of misusing anti-Semitism terminology and abusing history.[72] One of the best examples of such abuse occurred in 2015, when the Israeli Prime Minister Benjamin Netanyahu argued that Adolf Hitler had only wanted to expel Jews from Europe, and it was the Grand Mufti of Jerusalem, Haj Amin al-Husseini, who told him to "Burn them."[73]

On the other hand, international criticism of Israel during the past decade has continuously focused on the collateral damage and civilian casualties of Israel's military activity, rather than on the legitimacy of the State of Israel. The best examples of this are the international investigations of Israel's repetitive military operations in Gaza. Concentrating on these military operations, the following analysis offers a better understanding of how the Israeli foreign policy narratives described above were used to fend off international criticism, allowing Israeli political decision-makers to be relatively relaxed about enemy civilian costs during these operations.

Since the Israeli unilateral disengagement from Gaza in 2005 and Hamas's subsequent rise to power in 2007, Israel launched three major military operations against the terrorist organizations there—Operation Cast Lead (December 2008–January 2009), Operation Pillar of Defense (November 2012), and Operation Protective Edge (July–August 2014). While it is difficult to establish an objective number of Palestinian casualties in each of these operations, as this number is highly politicized,[74] it seems right to argue that Gaza, and its inhabitants, suffered much greater damage than Israel did. Consequently, it is not surprising that these operations had exposed Israel to severe international criticism that was followed by international investigations of IDF's conduct. The two best known of these investigations were the United Nations Fact Finding Mission on the Gaza Conflict (the Goldstone Report) and the United Nations Fact Finding Mission on the 2014 Israel–Gaza conflict.[75]

Analyzing the reactions of Israeli scholars and politicians to the findings of these reports, it is possible to point to four main areas that have been perceived by Israelis as international criticism of their actions: (1) Israel's actions were disproportional; (2) Israel did not try hard enough to avert enemy civilian casualties; (3) Israel's use of force was not merely a last-resort option; and (4) The Israeli occupation makes all military actions immoral/illegal.[76] The main argument, however, that is emphasised by these four claims is that "Israel uses excessive force and endangers (or kills) too many civilians in its response to Palestinian aggression."[77] To fend off this criticism, Israel's foreign policy has been concentrating on five major narratives.

The first narrative of Israel's foreign policy for dealing with international criticism is the Israeli traditional *Um Shmum*, i.e., the international community is biased against Israel and, therefore, Israel can disdain its criticism. For example, Shimon Peres, the President of Israel at the time of Operation Cast Lead, stated that the Goldstone Report "makes a mockery of history"[78]; and Avigdor Lieberman, the Minister of Foreign Affairs at the time, claimed that "the Goldstone Report wishes to take the UN back to the dark ages, where it was also determined, through the leadership of utilitarian countries, that Zionism is racism."[79] Continuing the narrative that *the world is always against Israel*, Israel rejected the Goldstone Report by claiming that "both the mandate of the Mission and the resolution establishing it prejudged the outcome of any investigation and gave legitimacy to the Hamas terrorist organization."[80] Israel's Prime Minister Benjamin Netanyahu put it more directly: "this report is tendentious and biased, that its end result was determined before it started. This is a kangaroo court that decided to convict Israel in any case."[81]

The second narrative, as discussed above, is Israel's reliance on the American "generous and unconstrained support,"[82] which is, usually, not long in coming. Less than three months after the initial release of the Goldstone Report, the U.S. Congress passed, by 344 votes to 36, a condemning resolution stating that the report is "irredeemably biased and unworthy of further consideration or legitimacy."[83] Moreover, in 2015, the United States was the only country that voted in the Human Rights Council against the report of the Fact Finding Mission on the 2014 Israel–Gaza Conflict.[84] When Netanyahu's statement in the U.S. Congress that "Israel is not what is wrong about the Middle East; Israel is what's right about the Middle East"[85] was cheerfully applauded, it seems only right to argue that Israel can rely on its

American ally to stand by it and protect Israel from what the U.S. Congress defines as "anti-Israel bias at the United Nations."[86]

The third narrative that has developed quite recently within the specific context of the conflict in Gaza is a purely legal one. To justify the IDF's activity as a part of the emerging *lawfare* (a method of political struggle where law is used as a means of influence on military objectives)[87] around the hostilities in Gaza, Israel has been very constantly arguing that all its military activities were carried out according to international law. Rejecting the accusations emphasized by international investigations, Israel has continuously argued that it was entitled to attack Gaza, based on the right of self-defense according to Article 51 of the UN Charter, and that the IDF took all precautions required by law to minimize civilian casualties.[88] Specifically referring to these requirements, the Israeli Ministry of Foreign Affairs stated that:

> The IDF not only implemented a range of precautions related to targeting and munitions, but also used an extensive system of graduated warnings to civilians, including both general advance warnings through media broadcasts and widespread leafleting, regional warnings to alert civilians to leave specific areas before IDF operations commenced, and specific warnings to civilians in or near military targets, through telephone calls and warning shots with light weapons.[89]

This legal argument ultimately leads to the fourth narrative that helps Israel to withstand international criticism—a *holier-than-thou* approach. Not only that, according to Israeli leaders, all military activities have been conducted within the framework of international law, but also that "the IDF is the most moral army in the world."[90] To fend off the accusations of disproportional harm to civilians, Israel argues that "the Israel Defense Forces demonstrate higher military and moral standards"[91] than other military forces, pointing to the fact that the IDF harms fewer civilians than other armies do when acting in similar conditions.[92]

On the one hand, these foreign policy narratives undoubtedly help Israeli leadership fend off international criticism and feel relatively free to deploy military force as and how it deems necessary. On the other, it will be unfair to claim that Israel enjoys the same freedom of political maneuvering on the international arena as the United States or Russia do. Following the reliance of Israeli foreign policy on American protection, in the Israeli political discourse, international pressure is commonly translated as American pressure to cease hostilities. In other words, Israel's disdainful approach to international political institutions and its dependence on U.S. protection turn Washington into "the only foreign actor whose policies constitute a critical input into Israeli decision-making."[93] Israeli leaders, however, understand that given the broader American agenda, the opinion of the international community cannot be entirely downplayed (even if they want to) and a broader consensus of support among other countries would reduce pressure on Washington, thus providing Israel with a greater freedom.

In an attempt to win the support (of, or, at least, to minimize criticism by) other countries, Israel's foreign policy utilizes its fifth narrative. As Israel inherently does not trust international organizations, its diplomatic activity to achieve broader international support focuses on a direct approach based on a separate bilateral relationship with each of its counterpart states, the support of which Israel is looking for. While in doing so, Israeli leaders many times "craft diplomatic scenes that forsake common diplomatic language, manner and ceremony"[94] by bypassing international organizations, it seems that this repudiation has been serving Israel well. One of the most fruitful examples of this narrative was Israel's success in obtaining the support of the European Union during Operation Cast Lead via the Czech Republic, relations with which are often described as "the strongest between Israel and any [other] European country."[95]

At the time of the beginning of operation Cast Lead, the European presidency was held by France, and following a meeting of the foreign ministers held on December 30, 2008, the European Union condemned both Israel and Hamas, calling for an "immediate and permanent ceasefire" and "an unconditional halt to rocket attacks by Hamas on Israel and an end to Israeli military actions."[96] On January 3, 2009, three days after the Czech Republic took over the role of the presidency of the European Union, Jiri Putuznik, spokesman for the new Czech presidency, stated that "at the present time and at the light of the events of these last days, we estimate that this measure [Israeli military operation] constitutes a defensive action, not offensive."[97] While this statement attracted vast criticism from many European countries, forcing the Czechs to apologize for the "misunderstanding," arguing that "it was a personal error,"[98] this example demonstrates perfectly the role of bilateral relations within Israeli foreign policy.

Another good example of the value of Israel's direct approach in foreign policy, at the expense of multilateral international organizations, was the reaction of Israeli leaders to the Goldstone Report. Mistrusting their ability to influence the decision-making process at the UN Human Rights Council, Israel immediately turned to a direct approach, separately asking different countries "to express disagreement with the report and to oppose any use of it as the basis for anti-Israeli resolutions at other international institutions."[99] This approach was best emphasized in a *Ha'aretz* article published on the day the Goldstone Report was released:

> Following the report's release, Netanyahu, Lieberman, President Shimon Peres and Defence Minister Ehud Barak will telephone many of their counterparts around the world. . . . Israel's diplomatic efforts will focus on the five perma-

> nent members of the UN Security Council—the United States, Russia, China, Britain and France—but will also give priority to members of the European Union, because of their influence in the UN Human Rights Council in Geneva.[100]

Summarizing Israel's foreign policy narratives that have been structured by the Israeli leaders in order to fend off international criticism of its military actions, it is important to separate between the political rhetoric and the actual deeds. While from a first impression one might assume that Israeli leadership simply ignores international organizations, accusing them of being anti-Israeli and biased while trusting the United States to protect Israel against them, the real situation is quite different. In fact, it seems that Israeli leadership truly understands its weak political position on the international arena, on the one hand, and its economic and military dependence on its international counterparts, on the other (mainly but not only the United States, as the European Union is the Israeli biggest trade partner).[101] One of the most prominent Israeli strategists, Yehoshafat Harkabi, stated in 1979:

> The biggest change that has happened in the world is the creation of the international system and the international public opinion that influence the behavior of the states and the decisions of their leaders. The international system . . . can tip the scales and determine success or failure. . . . The international system is most important for the small states, as their dependence on the big states has been growing and the support of the big states, especially in terms of weapons, becomes vital.[102]

Israel is a small state and, therefore, according to many Israeli scholars it "has [always] had to take international factors into account in all of its military engagements with its neighbors"[103] much more seriously than anyone of the previously examined countries. The political rhetoric of *Um Shmum* or the *holier-than-thou* attitude have been very useful in fending off criticism, but definitely not enough to get a free rein for military operations. To maneuver through the deep waters of international criticism, Israel has adopted two supplementary ways to minimize international criticism and provide itself with a limited window of political opportunity for its military actions.

The first of Israel's ways to create a more supportive atmosphere in the international arena is the policy of restraint (i.e., toleration of offensive activities done by Israel's enemies). According to Israel's viewpoint, this restraint shows that Israeli governments are not eager for battle and do everything possible to avoid confrontation and, therefore, eases the political efforts to explain Israel's military actions as self-defense.[104] The second way is an attempt to choose the most convenient timing to achieve as large a margin for maneuvers as possible. The best example of this was Operation Cast Lead, which: (1) started at the end of the calendar year, a period marked by a near-freeze of activity in international diplomacy due to the Christmas season; (2) in the interregnum between the Bush and Obama administrations, which gave Israel a great deal of freedom[105]; and (3) the incoming Czech presidency in the European Union, which, as noted above, promised to ease pressure from Europe. It is important to note that these two ways are highly interconnected, as restraining from action and waiting for the right timing in international affairs, Israeli leaders are at risk of losing domestic support, especially in such situations where Israeli civilians are threatened by continuous enemy bombardments. This dilemma, however, seemed to have been resolved after the successful introduction of the Iron Dome Anti-Missile Shield as "the public's sense of protection by Iron Dome gives time and space for the government to make [more] calculated decisions on how to proceed in response to the rocket fire,"[106] taking into consideration international, domestic, and operational factors.

Israeli politicians and diplomats, undoubtedly, toil much harder than their Russian or American counterparts to defend their state's military engagements in the international arena, and they are much more sensitive to international pressure. When it comes to military actions, "Israel faces criticism regarding both the decision to react forcibly and the scope of force employed"[107] and it understands that it acts on borrowed time. Interestingly enough, unlike in the cases of Russia or the United States, most of the criticism of Israel's military operations specifically points to enemy civilian casualties, and Israeli political-military decision-makers do indeed pressure the IDF to minimize the collateral damage of its actions. After all, as the former commander of British forces in Afghanistan Colonel Richard Kemp put it: "No army in the world acts with as much discretion and great care as the IDF in order to minimize damage. The US and the UK are careful, but not as much as Israel."[108] The main question, however, is why this pressure has not pushed for the military employment of NLW. To answer this question, a better understanding of Israeli culture and its attitude toward the enemy is required.

ISRAELI CULTURE AND ENEMY CIVILIAN CASUALTIES

In 1971, Sir Ernest John Barnes, the British ambassador in Israel, provided a significant insight into the Israeli character. In his report to the British Foreign Office, he wrote:

> Her [Israel's] attitudes spring from deeper historical roots too. A people, which has been pushed around the world for two thousand years, does not intend to be pushed any more. Nor is it only ancient history, which produces the Masada complex of total resistance, death rather dishonor. The memories

> of persecution, and above all of Hitler's holocaust, are much nearer the surface. . . . Be that as it may, many Israelis feel oppression in their bones . . . [109]

Almost 40 years later, when another Briton, an English student who studied in Israel, was asked as a part of sociological survey to describe Israeli society, he said:

> No Englishman in London will ever say the sentence that you [Israelis] always say—"Who knows, may be in one generation, we will not be here at all." The possibility that London and England will not stand for next generations or for the next hundred years does not appear in our minds. But for you, it comes in every second sentence. You live in a constant anxiety. Before I came here, I did not understand the depth of your trauma. [110]

In their observations of the Israeli culture, or *Israeliut* (Israeliness) as it is often called by scholars,[111] these two outsiders not only accurately pointed at several characteristics ("constant anxiety," "Masada complex," etc.), but also suggested that any attempt to understand Israel's culture has to begin with the long Jewish history. In other words, "if you want to understand Israelis, you have to start with their past, very distant past."[112]

The significant influence of the distant Jewish historical experience on the Israeli mind-set, and especially such constitutive events as the Siege of Masada, the Bar-Kokhba Revolt, the Holocaust, etc., has been greatly discussed by many scholars and researchers in the fields of sociology, history, politics, and strategy.[113] Professor Yair Gad, one of the most prominent contemporary Israeli sociologists, in his book *The Code of Israeliness*, identified the ten most pivotal narratives of the Israeli character: (1) existential anxiety; (2) proud standing and deliberate audacity (*chutzpah*); (3) the feeling of entitlement to and the ownership of the Land of Israel (*Eretz Yisrael*); (4) a *give-and-take* style contract with the state; (5) a fear of being a *fraier* (somebody who has been taken advantage of); (6) "*I speak means I exist*"; (7) creative improvisation (*iltur*) and lack of seriousness; (8) anti-hierarchy and equality; (9) collectivism and meddlesomeness; and (10) intimacy. In the conclusion to his book, he writes:

> In each and every Jew living in Israel, deep in their minds and stomach, there is a shared culture—culture that is based on a shared history and a shared destiny. . . . The Israeli culture embeds in us [Israelis] a deep historical consciousness, as well as horrific future scenarios. . . . In our every movement, in our every expression and reaction, it is possible to find the roots of this epic story.[114]

While it seems that the contemporary Israeli mind-set is deeply rooted in the long Jewish history, the following analysis does not intend to reproduce a complete picture of the Israeli culture. It seeks, instead, to understand a much narrower phenomenon—how the Jewish, and the more recent Israeli, experiences have shaped what Israelis think about war and enemy civilian casualties. Those who are interested in a broader picture are advised to focus more on the works of the leading Israeli scholars that rest at the core of this analysis.

Israeli Culture and the Value of Human Life

Much has been thought and written about Israeli culture. Different scholars have discussed it from different perspectives, analyzing its different aspects, yet all agree on the fact that "Israel is what is known in psychological literature as a post-disaster community."[115] While the long Jewish history provides many traumatic experiences: the enslavement in Egypt, and the Babylonian captivity; the exile after the destruction of the Second Temple by the Romans, and the Alhambra Decree ordering the expulsion of Jews from Spain; the Holocaust is, undoubtedly, the experience that shaped most of the contemporary Israeli character.[116] This rich apocalyptic history of attempted annihilation, best represented by the Jewish/Israeli calendar that is strewn with related holidays and memorials, lies at the core of one of the most prominent characteristics of Israeli national mind-set—the existential anxiety.[117]

On the one hand, the initial ideological and political goal of the *Yishuv* (pre-state Jewish settlement in Israel) leadership was, in accordance with the Zionist ideology, to establish a "normal" state. In other words, the leading vision was to establish a state that would become a "normal" member of the family of nations and "that traumatic memories of the helpless state of the Jews during World War II would give way to rational considerations."[118] On the other, the destruction and persecution did not stop with the establishment of the State of Israel, and, in fact, quite the opposite was true. The War of Independence, which immediately followed the declaration of independence and the subsequent repetitive conflicts with the neighboring Arab countries, as well as the Palestinians, have recreated and enforced in the Israeli mind a strong sense of continuously being under existential threat. As Yair Gad put it:

> In the Israeli head, there is a well-established doctrine of near-annihilation. In the Six-Day War we thought that we would have to take a long swim to Cyprus; in the Yom Kippur War we already imagined concentration camps; in the First Gulf War, in 1990, we saw in our mind a destroyed Tel Aviv; in preparation for the rerun in 2003, we saw ourselves burnt by mustard gas; and since the Second Lebanon War and operations in Gaza we know that rockets are an existential threat in all parts of Israel.[119]

The interpretation of any event and political development through the lens of the long Jewish history of persecution, the Holocaust, and the more contemporary experience of Israeli conflicts, ultimately has led the Israelis to think that "all the world is against us."[120] And, while one might assume that such a long history of oppression would generate more universal sympathy for "all oppressed," it seems that this is not the case.

The trauma of the Holocaust has been perceived by the Israeli mind as a singular and very unique experience. Though the majority of the Jewish population in Israel is not direct heirs of the survivors of the Holocaust,[121] "the self-understanding of individuals in Israel [is] as survivors, heirs, and kin to the victims."[122] Consequently, this Israeli collective memory of the Holocaust, preoccupied by an all-consuming fear of a repetition of the trauma, shapes the Israeli attitude toward others, creating a very distinctive divide between enemies and friends. When it comes to friends, Israelis easily open their hearts and come to their aid; however, when it comes to enemies, the Israeli mind is insensible and offensive. For example, Israel would easily send a generous rescue team to an Asian country struck by an earthquake; yet, Israelis' hearts are hardened to the suffering of their neighbors in Gaza, who have been locked for years in the Gaza Ghetto—"We [Israelis] even will not allow them to call it a 'Ghetto,' because 'Ghetto' is ours and ours only."[123]

In analyzing this Israeli narrative of existential anxiety within the context of valuing human life, it is important to highlight the relationship between the collective and the individual in Israeli culture. While, as discussed in the previous chapters, the American character is predisposed toward individualism, and the Russian one toward collectivism, the Israeli mind-set, based on Jewish traditions, paradoxically comprehends both. On the one hand, due to a long history of life in exile, usually in hostile environments, Jewish communities developed a very strong sense of collectivism. Community was essential not only for physical survival, but also for the survival of Jewish traditions—"for the individual, the collectivity was a community of fate—one's destiny was inseparable from that of the group."[124] Very similar to the Russian concept of *krugovaya poruka* (collective responsibility and punishment), the Jewish communities in the diaspora shaped the Jewish mind around an idea that the community is responsible for one's actions and the only prospect for help lay within community. However, while in the Russian case *krugovaya poruka* was imposed by law,[125] in the Jewish communities this collectivism was an inherent part of Talmudic teaching, as Rabbi Hillel said: "Do not separate yourself from the community."[126] In other words, the Israeli mind based on this Jewish mentality is less oriented to oneself and "tends to be focused more on the pursuit of collective goals and well-being of the society."[127]

On the other hand, simultaneous with the strong sense of collectivism, the Jewish tradition also cultivated a very deep respect for the uniqueness of the individual, as Rabbi Hillel also said: "If I am not for myself, who will be for me?"[128] Within Jewish communities individualism was encouraged from an early age and its values were well protected by the Talmudic tradition, just as one of the most prominent Jewish maxims states: "To save a single soul out of Israel is to save an entire world." While this combination of appreciation of the community and the importance of individuals within it helped Jewish people to survive two thousand years of diaspora, it also ultimately left a significant footmark on the Israeli mind-set, shaping the relations of Israelis between themselves, and between Israelis and others.

The Israeli character, bred on the traditional view of the Jewish community as one collective with one shared destiny, tends to understand others in a very straightforward way, automatically and very easily labelling others either "friend" or "foe." Unable to comprehend the complexity of a "gray" political pattern, the Israeli mind-set almost intuitively rushes to utilize the collective memory of the Jewish history of persecution, pointing out its uniqueness: "We are the Chosen People, Holy People. There is no nation that is more ethical than us, and everybody who claims differently will be forced to realize his mistake."[129] Interestingly enough, in their reaction to any attempt to undermine this claim, Israelis "shoot down" allegations targeting even their friends. Professor Dan Kaspi summarized best this predisposition of the Israeli character by describing the traditional Israeli response to a speech of President Barack Obama that was perceived as not sympathetic enough:

> Hussein Obama is a belittler [*katan aleinu*]! We are Chosen and Enlightened People! Without any prejudices towards the skin color of any one of the United States presidents, he just does not know history. Because one who survived Pharaoh and Nazis, damn them, will also succeed against a greenhorn president with brownish skin.[130]

While this predisposition of the Israeli character to react to external threats is based on its collectivistic nature, as an old Talmudic maxim says "all of Israel stand surety for one another," for inspiration for their internal affairs Israelis traditionally tend to draw from their inherent individualism. The Jewish religious tradition, unlike many other religions, is far from dogmatic; innovative explanations, variant opinions, and general independence of thought have traditionally been highly prized: "no institution or dynasty has a monopoly on authority, let alone truth."[131] This inborn self-confidence lies at the core of the Israelis' predisposition to argue and sincerely believe that they have the rightest answer to any question: "every Israeli has an opinion, an assessment, an evaluation and a decisive conclusion on any given subject: governmental policy, military operation, scientific discovery, criminal news item or a singer's haircut."[132]

The paradoxical relations between the collective and the individual in the Israeli mind-set also explain best the complex relations between the State of Israel and its citizens. On the one hand, Israelis see the state in the traditional terms of one big Jewish community, and therefore instinctively rush to contribute to the common good by serving in the IDF, by volunteering, donating, etc. On the other, they expect and demand from the state an unprecedented individual approach to each one of its citizens, especially in times of distress. In the mind of an Israeli, there is an unwritten contract with the state, as Yair Gad put it:

> When I [an Israeli] am in distress, it [Israel] will come to rescue me. To dress me up. To give me emergency money. To fly me back home, and fast. And why all these? Because I contributed. Because I deserve it. Because I am an Israeli. Because it is my country, and it is an alternative for God.[133]

This expectation from the state, combined with the traditional manner of arguing (or demanding), had created another paradox in the Israeli character. On the one hand, externally, Israelis are very sensitive to protecting their state, expressing high levels of patriotism and concern for its image. On the other, internally, they fully exercise the right to criticize it, the right that is preserved for them and them alone, because they are Israelis and they contributed to the state. In other words, Israeli society represents a very close-knit community, when it comes to outsiders, but within itself it is entirely opposite. As Ben Gurion observed:

> Many of our inhabitants, including Israeli youth, have not learned how to respect their fellow-citizens and treat them with politeness, tolerance and sympathy. Elementary decency is lacking among us.[134]

Interestingly enough, while Ben Gurion made his observation on an individual level, many scholars agree with the fact that the Israeli collective, as a whole, behaves in a similar manner.[135] This strong sense of community based on the long history of persecution and consequential existential anxiety shaped in the Israeli mind another pivotal cultural narrative: never be a passive victim again. Benjamin Netanyahu expressed it best:

> Nevertheless, today we must ask the question: have the lessons of the Holocaust been learned? I believe that there are three lessons: fortify your strength, teach good deeds and fight evil.

And if somebody was unsure what "good deeds" mean, he clarified:

> Good deeds have a complementary side, and that is the third lesson of the Holocaust: fight evil. It is not enough to simply do good and be tolerant. A free society must ask itself what it will do when faced with the destructive forces of evil that seek to destroy and trample man and his rights.[136]

While for thousands of years, stateless Jews had to accept the doctrine of submission to the powerful, a submission that more than several times almost ended with total annihilation, since the establishment of their own state and their own military forces the Israeli mind is not ready to accept any type of patronizing. Some might even argue that this behavior is a psychological compensation for past humiliations, or a post-traumatic syndrome.[137] Either way, building and exercising military strength seems to be the prime lesson learned by Israelis from their history.

As discussed above, the Israeli mind-set tends to divide the world categorically into "friends" and "foes." Subsequently, it is not surprising that the appreciation of human life in the Israeli mind also differentiates between "our lives," "the lives of our friends," and "the lives of our enemies." One of the most important influences of the traumatic Jewish history on the Israeli culture is the commitment "to ensure the future of the Jewish people."[138] Consequently, Israelis appreciate not only their own lives, but also the lives of all Jewish people around the world. Moreover, when it comes to a universal value of life, the Israeli memory of the Holocaust forces the Israeli mind to demonstrate the best side of its character. Moral outrage toward bystander nations for their lack of help during the Holocaust created in the Israeli mind "a need to demonstrate that Israelis have higher humanitarian values than other countries,"[139] and generous Israeli rescue teams sent all around the world are proof of that.

A closer examination, however, shows that this humanitarian spirit is very limited, and the universal appreciation of human life is not as all-inclusive as it seems. When it comes to something or somebody that is perceived as a threat, the Israeli mind-set demonstrates much less sensitivity (if any). One of the best examples of this is the way in which Israel deals with African refugees in Israel. Unwilling to grant them refugee status, the Israeli authorities simply offered them a choice between $3,500 in cash and a one-way ticket home, or imprisonment in a specially built detention facility.[140] Another obvious example is the way in which the Israeli mind-set appreciates (or not) the lives of Palestinians. As Palestinians, unlike refugees from Sudan, Eritrea, and other African countries, are perceived by the Israeli mind-set as enemies, at least as far as the conflict is unsolved, they are treated according to the Israeli perception as an enemy in wartime.

War and the Israeli Character

While Americans perceive war as a temporary interruption, and Russians understand it as an integral part of the human condition, it seems right to argue that the Israeli perception of military confrontation presents a type of hybrid of these two. In an attempt to understand this Israeli characteristic, it is important to differentiate between a "state of war" and a "state of conflict" in the Israeli historical-cultural context. Baruch Kimmerling, one of the most prominent Israeli sociologists, who tended to question the official narrative of Israel's creation, defines the "state of war" as a period of active combat between Israel and its one, or more, enemies. All other periods, during which active combat is dormant, can be defined as a "state of conflict," since Israel's enemies carry on warfare by other means (economic, political, terror, etc.).[141]

On the one hand, the Israeli mentality, similar to the Russian character, perceives conflict as a permanent condition or destiny of Israeli society, as Moshe Dayan put it:

> We are a generation of settlers; without a helmet or a gun we will be unable to plant a tree or build a house. . . This is the destiny of our generation. The only choice we have is to be armed, strong and resolute, else the sword will fall from our hands and the thread of our lives will be severed.[142]

On the other, war, similar to the American view, is perceived by Israelis as a temporary interruption. In contrast to the Americans, however, the Israeli mind-set, steeped in the traditional existential anxiety, intuitively assumes that the only possible predominant goal of war is "to ensure the very existence of society," and it might be achieved "only through optimal mobilization of virtually all available human and material resources of the system."[143] Consequently, as waging wars demands such an enormous mobilization of society, wars have to be waged as decisively as possible and be concluded as fast as possible.

Due to the Israeli history of confrontations, the "state of conflict" has become a routine, something that military specialists call "current security" (*bitachon shotef*, or *batash*), punctuated by "states of war," i.e., Israeli-Arab wars (Six-Day Way, Yom Kippur War, etc.) or large military operations in Gaza and the Western Bank (Operation Defensive Shield, Operation Cast Lead, Operation Protective Edge, etc.). Since the focus of this research is on the latter, it is important to understand three main pillars of the Israeli cultural approach toward wars: *harta'a* (deterrence), *hatra'a* (early warning) and *hachra'a* (humiliation on the battlefield).[144] While the first pillar is intended to deter the enemy from attacking by maintaining considerable military power, and the purpose of the second one is to alert Israeli leadership when the first fails, the third, "humiliation on the battlefield," is the most crucial, as it intends not only to bring an ultimate victory, but also to restore deterrence. In the Israeli mind, the main objective of war is not a terminal knockdown of the enemy, as it seems impossible in light of the imbalance of resources vis-a-vis the Arab states and international political pressures. For Israelis, the major strategic goal of war is deterring the enemy from attacking Israel again for as long a period as possible.[145]

According to the Israeli understanding, "all of Israel's wars and military campaigns . . . were fought because the deterrence Israel had hoped to establish by prior threats or actions, broke down,"[146] and therefore, once deterrence failed and war began, the enemy has to be punished so severely that it will be deterred from doing it again. This idea of deterrence by punishment was formulated best by Moshe Dayan in 1959:

> We cannot guard every water pipeline from explosion and every tree from uprooting. We cannot prevent every murder of a worker in an orchard [or] a family in their beds. But it is in our power to set a high price on our blood, a price that will be too high to pay for the Arab community, the Arab army, or the Arab government.[147]

While one might have thought that this sort of "massive retaliation" during military operations belongs to the past, the opposite is true. One of the best examples was during the Second Lebanon War, when Prime Minister Ehud Olmert requested to determine "such a price tag, that nobody will wish to trouble us again,"[148] Defense Minister Amir Peretz claimed that "the central goal is to cause Hezbollah to feel bitten [and] persecuted . . ."[149] and Chief of Staff Lieutenant General Dan Halutz stated the need to "turn back the clock in Lebanon by 20 years."[150]

Consequently, it is not surprising that many scholars describe the Israeli way of war as a matter of emotional revenge.[151] As Niccolo Petrelli concluded:

> Mainstream Yishuv thinking stressed the importance and effectiveness of offensive retaliatory actions for the purposes of taking revenge, exacting punishment and, most of all, deter Arab violence.[152]

The perception of military operations as an act of revenge is deeply rooted in the Israeli culture as retaliatory raids against Arab villages and the destruction of their fields or herds were a part of Israeli official policy during the 1950s.[153] Moreover, in the Israeli mind, short periods of war that punctuate long periods of "state of conflict" are intended not only to restore the failed deterrence, but also are seen as an opportunity to *lishbor kelim* (to break the rules) of *havlaga* (self-restraint) that usually precedes wars. The way in which Israelis measure the success of war demonstrates best this traditional desire for revenge. During the early days of the Second Lebanon War, support among the Jewish public reached 82 percent; however, by the end of the war, polls reflected a dramatic turnaround, as polls taken on August 10, four days before the end, found that only 20 percent of the overall responders felt that "Israel had won the war." According to the analysts, one of the main explanations for such a dramatic change was the Israeli public refusal to accept the fact that "after thirty-three days of air and ground warfare, the IDF was unable to make even a dent in Hezbollah's capacity to attack Israel."[154] Despite the fact that as an outcome of this war, the Israeli northern border has been enjoying the longest period of peace in Israel's history, this war has been perceived as a failure, mainly because "the best military in the Middle East" was humiliated by a "para-military organization with several thousand fighters," and the IDF was not able to humiliate it in turn.[155] In other words, as an Israeli proverb states: "Arabs understand only force" (*Aravim mevinim rak koah*), and the Israeli mind-set feels obligated to forcefully retaliate back each time it gets the opportunity.

While Israel's retaliation and revenge is aimed directly against military enemies, such as Hamas or Hezbollah, indirectly, it obviously harms enemy populations. Moreover, there are three main arguments that suggest that the Israeli public has very little objection to it. The first argument is the long and constitutive history of retaliatory raids and "collective punishment" from Israel's past. Interestingly enough, while in the mid-1950s the Israeli prime ministers developed doubts about the utility of retaliation as a method for deterring Arab infiltrators, or compelling Arab governments to prevent infiltration, they still tended to authorize reprisals to calm the Israeli public outrage that demanded violent reactions.[156]

The second argument is the Jewish traditional collectivism that perceives community as responsible for one's actions. For example, in the context of the Palestinian conflict, Israeli collectivistic culture projects its narratives onto Palestinian society, considering all Palestinians as responsible for the actions of Hamas. In 1994, despite the vast support for the peace negotiations, the majority of the Jewish population in Israel (between 54 and 71 percent) believed that if the Palestinians had a chance, they would annihilate Israel. In 2000, with the Al-Aqsa Intifada as the background, 68 percent of the Jewish population in Israel perceived all Palestinians as violent. A year into the intifada, 23 percent believed that all Palestinians support terror, 32.7 percent believed that the majority of Palestinians support terror, and 17.3 percent believed that half of the Palestinians supported terror (i.e., only 27 percent did not blame Palestinian people for the terrorist actions done by terror organizations).[157]

This leads to the third argument—an assumption that the enemy population is responsible for the action of its military, and, once it suffers, it will force its military factions to halt activity against Israel. According to the Israeli military culture of retaliation, the purpose of "extensive damage on the civilian infrastructure of Hezbollah in Beirut and southern Lebanon"[158] was to retaliate against Hezbollah and deter it from further actions, and the IDF violent response in Gaza "would lead to destruction and increased domestic pressure on the organization's leadership."[159] The Israeli public understands that such a retaliation will come with a significant toll of enemy civilian casualties; however, Israelis have little problem with that as (1) Hezbollah's and Hamas's rockets cover all parts of Israel, endangering all of Israel's population; and (2) in the Israeli mind, to repeat, the enemy population is responsible in any case.

It is impossible to discuss the cultural narratives of war of any nation without mentioning its founding myths. While Russians have their Patriotic wars and Americans have their War of Independence, the Civil War, and World War II, the Israeli cultural attitude toward war is undoubtedly brewed on the mythical narratives of the Siege of Masada (73–74 A.D.)[160] and the Bar Kokhba Revolt (132–135 A.D.).[161] Interestingly enough, these two not-so-successful (if not to say the least successful) events in Jewish history, which had very little appeal to Jewish communities in exile, were reconstructed by the Zionism movement, making Masada as "a symbol of military valor and national commitment" and turning Bar Kokhba into "a national hero."[162]

The Siege of Masada was the final defeat of a much bigger Jewish uprising (the Great Jewish Revolt against Rome, between the years 66–73 A.D.) that ended in a catastrophic disaster and the violent oppression of the Jews. According to the Israeli myth, which does not necessarily represent the true course of events, at the end of the Great Revolt Jewish rebels found themselves trapped on the top of Masada fortress, and as they realized that there was no more hope of holding against the Roman army, they chose to kill themselves rather than surrender and become slaves.[163] As the only available written source on the battle of Masada is Josephus's *The Jewish War*, for many centuries it was unknown outside the Christian Church, in general, and by the Jewish community, in particular. In the late nineteenth century, however, when Zionism's founders were re-examining Jewish antiquity in an attempt to establish a basis for the emerging national Hebrew culture, the Masada story "came to represent a resolute commitment to national freedom."[164] Since then, the Siege of Masada has represented, within Israeli culture, a narrative of determination to survive or die, a symbolic equivalent to the American Alamo, but with more multiplied influences on the national character. Like the besieged and desperate defenders of Masada, contemporary Israelis see themselves surrounded by hostile and numerically superior forces.[165] Concurringly, a verse from a popular poem, by Hebrew poet Yitzhak Lamdan—"Never again shall Masada fall!"—has become a national slogan and a patriotic vow that shaped generations of soldiers in the IDF.[166] Moreover, the preference to die rather than be captured, as it is represented in the myth of Masada, explains best the IDF's combat procedures in the case of kidnapping (*Noal' Khanibal*) that orders soldiers to prevent kidnapping by any cost, including killing the hostage.[167]

The Bar Kokhba Revolt, the last Jewish uprising, was also mythicized by Zionism's leaders at the same time as was the Masada myth and for the same reasons. While previously Jewish memory tended to consider this revolt as one of the most traumatic disasters in Jewish history, and Bar Kokhba himself was considered as a bold, but arrogant and short-tempered leader who led his people to doom and exile,[168] the emerging Hebrew nationalism saw it otherwise. The glorification of Bar Kokhba by Zionism stemmed from the conviction that he represents a spirit of heroism that was suppressed during centuries of Jewish exile, and the rebellion became a model for "breaking the yoke of foreign rule."[169] Moreover, the main narrative of the Bar Kokhba Rebellion in Zionism's interpretation, which has become embedded in Israeli culture, is that it represents "the Jewish people's capacity for survival, in spite of all hardship it has endured," and that Jewish existence cannot be taken for granted, but has to be taken by force.[170]

In the conclusion of this discussion about the Israeli cultural character, it is difficult to disagree with those who title Israel as "a nation in arms" or a "nation in uniform" that lives in a situation that can been labelled a "dormant war."[171] The Israeli mind—shaped by an existential anxiety, steeped in the myths of a prerequisite "fight for survival" (i.e., the Bar Kokhba Rebellion), and a readiness to sacrifice in a cul-de-sac situation (i.e., the Siege of Masada)—is predisposed to see enemy society and its military as one whole and has very little sympathy for the enemy population. Despite the vast theoretical discourse led by Jewish and Israeli, secular and religious, philosophers on the ethical and moral problems of enemy civilian causalities, this debate always ended with the question of legality.[172] Probably because, as J. David Bleich, a Jewish orthodox scholar who is considered an authority on Jewish law and ethics, controversially concluded:

> There exists no discussion in classical rabbinic sources that takes cognizance of the likelihood of causing civilian casualties in the course of hostilities legitimately undertaken as posing a halakhic or moral problem.[173]

Consequently, when civilians are killed during legitimately undertaken hostilities, their deaths are not just, but they are justifiable in the circumstances of war, i.e., "when the damage isn't disproportionate to the probable benefits of attacking the target" and "when this damage is the unintended side-effect of a legitimate military operation."[174] And it seems that in the Israeli culture, the IDF's military operations are always legitimate, because, after all, *Never again shall Masada fall!*

ISRAELI MILITARY CULTURE AND NLW

One of the most popular Israeli slogans regarding the relationship between the Israelis and their military is *am bone tzava bone am* (the nation builds the military builds the nation). While the degree to which this slogan reflects the real nature of socio-military relations in Israel is debatable,[175] it definitely points to complex and intense relations between Israel's culture and its military. Already in 1974, British Brigadier Nigel Bagnall (later Field Marshal and the Chief of the General Staff Sir Nigel Thomas Bagnall) in his analysis of the IDF's performance during the Six-Day War pointed to a strong connection between Jewish history, the emerging Israeli culture, Israel's geopolitical situation, and the IDF's way of war.[176] The same arguments have been supported by many Israeli and foreign scholars and military thinkers.[177]

The main interest of this research is the failure of the IDF to employ NLW during its military operations, and, therefore, a broad discussion of Israeli military culture is beyond its scope. As NLW is a military technology, the following analysis will narrow its focus on the role of technology in Israeli military culture in general, and what Israeli military thought has to say about NLW, in particular.

Israeli Military Culture and Technology

Since the very beginning of Israel, the idea of "qualitative edge" (*itaron echuti*) has been one of the foundation stones of Israeli military culture. Realizing that the IDF would never be able to counterbalance the numerical superiority of its adversaries, creating and maintaining this edge became a lodestar of Israeli military thinking and build-up in terms of material and human factors.[178] As Major General Israel Tal, one of the prominent Israeli military strategists, put it:

> Israel has to turn to all of its national resources in wartime, and to rely on the quality of its society in all areas. . . The qualitative difference between Israel and Arabs must be one not of degree but of kind.[179]

The role of technology in the Israeli military culture expanded gradually. In the early years after achieving independence, Israel experienced political difficulties with procuring advanced foreign weaponry, and the IDF had to rely mostly on pre-war and World War II technologies. Consequently, analyzing this early period of the IDF's relationship with technology, researchers argue that "the IDF had little understanding of technology"[180] and "in Israeli opinion, many complex items of Western inventiveness offer little advantage over the simpler equipments they are designed to replace."[181] However, driven by a fundamental sense of insecurity and anxiety, Israel strove to improve not only its general technological edge, but to attain autonomy in weapons production.[182] Therefore, about the time of the Six-Day War, and especially in 1970s, when the various political obstacles that prevented Israel from purchasing foreign weaponry were significantly relieved, Israel already had developed its own defense industry. By 1982 and the First Lebanon War, the IDF internally developed and fielded technology with "ultra-sophisticated capabilities."[183] While during the 1950s and 1960s "the IDF sought to achieve technological superiority where possible," and during the 1970s and 1980s "technology had come to occupy an ever more important role in the IDF's self-understanding," by the late 1990s, the IDF had not only "excellent equipment, much of it of indigenous manufacture," but also succeeded "to field military capabilities in advance even of the United States."[184]

Since the 1970s, the IDF has been considering the advanced technology as a lodestar for military successes, as a panacea not only for minimizing costs and duration of wars, but also casualties among the Israeli civilian population and military personnel.[185] While "offensive romanticism," which is deeply integrated in the Israeli military culture and operational mind-set,[186] led, for example, to the development of the sophisticated *Merkava* tank or attacking Unmanned Aerial Vehicles (UAVs), defense technologies have "become a crucial element of Israel's approach in defending the country,"[187] with the Arrow Anti-Missile System and the Iron Dome Anti-Missile Shield as good examples. Moreover, since the number of casualties among IDF's soldiers determines the level of the IDF's success in the eyes of the Israeli public,[188] technologies for military personnel's protection have also been enjoying a high priority. The number of casualties in the IDF Armor Corps during the Second Lebanon War created vast public criticism of the penetrability of the *Merkava* tanks.[189] While this criticism was not supported by facts as, in comparison to the Yom Kippur War, the percentage of penetrated tanks (by anti-tank missiles) was significantly lower (86 percent in 1973 vs. 47 percent in 2006) and the average number of killed soldiers in each penetrated tank was halved (from two to one),[190] the IDF rushed to introduce the *Meil' Ruach* (windbreaker) Active Protection System for Vehicles.[191] Interestingly enough, while the main criticism was against the poor training of the crews before the war and the ill employment of the tanks during the war,[192] the IDF preferred to invest in a technological solution, rather than improving the quality of training and operational concepts.

Analyzing the contemporary IDF approach to technology, Avi Kober pointed to the rising "cult of technology" stating that "in the recent years, technology started overshadowing the non-material aspects of Israeli strategy and tactics, becoming the main factor in military build-up and operations."[193] He stresses the fact that the IDF has become heavily technologically oriented and identify military quality with high-technological capabilities. Interestingly enough, Kober explains this shift in the Israeli military culture as overexposure to American military thinking. While his argument that the American fascination with technology "has been received in the IDF enthusiastically with little skepticism, and has affected IDF commanders' thinking and modus operandi"[194] points at a change in the Israeli military way of thinking, it seems that very little has changed in the IDF's culture of development of the technology itself.

Describing the way in which the IDF develops its operational concepts, Eliot Cohen, Michael Eisentstadt, and Andrew Bachevich stated that: "Israeli operational concepts resemble keys carefully crafted to fit particular locks, rather than a general approach to the problem of opening doors."[195] This description, however, also describes best the IDF's attitude toward the development of new weaponry. Despite the warning of the former director of the Research and Development (R&D) Directorate at the Ministry of Defense, Major General Yitzhak Ben-Israel, that the IDF's approach to R&D processes has to be based on "relative edge" (*itaron yahasi*), rather than "bridging the gaps" (*sgirat pa'arim*), this is not the case in the IDF, as he concluded himself that the IDF's disproportional investment in the Air Force was of the latter, rather than the former.[196] Having no resources for great technological leaps in all of the military realm, the IDF has always preferred to invest in specific technologies to bridge certain gaps, at the huge expense of others. In 2014, during Operation Protective Edge, the oversophisticated *Merkava IV* tanks with the state-of-the-art *Meil' Ruach* Active Protection Systems were followed by the M113 armored personnel carriers built in the 1970s,[197] and Puma armored engineering vehicles that are based on the hulls of the British Centurion tanks built in the late 1950s.[198] This technological mishmash matches well the Israeli approach to military strategy, which is often described as "a conceptual salad, marked by a lack of vision."[199]

Despite IDF commanders' fascination with American techno-mania, the IDF's budget is a fraction of its American counterpart and, therefore, unlike the U.S. military, the IDF is "wringing the last iota of usefulness" out of any technology.[200] The *Merkava* project, the IDF Air Force, and the anti-missile defense systems are the outcomes of specified prioritizing and heavy investment.[201] In all other cases, technological advances rest on the Israeli cultural belief that "small technological advantages—marginal edges in range, accuracy, or maneuverability—can yield large differentials of combat power."[202] In other words, the IDF's fiscal constrains combined with the IDF's popular slogan *ze ma yesh, ve'im ze nenatzeah* (this is what we have, and with this we will win) has led to the situation in which the advancement of technologies, which are not the main focus, is based on small and steady adaptations and upgrades that answer specific deficiencies discovered during cumulative combat experience. When hitting his fingers, an American will seek a way to improve the hammer and a Russian will endure the pain; an Israeli, having neither American economic privileges, nor Russian readiness to endure, will invent a small and cheap gadget that will hold the nail while striking.

Israeli Military Thought and NLW

Much has been written on the lack of intellectual tradition in the Israeli military culture, the IDF's "anti-intellectualism," or even "false intellectualism."[203] The existing research proposes three main cultural explanations for this phenomenon. The first one is the incompetence of Jewish thought on war, driven from the central historical fact of the Jewish stateless condition, as "Jews had no 'high' politics, no politics of war and peace, from the time of Bar-Kokhba (135 A.D.) to the time of Ben-Gurion (1948)."[204] The second explanation is the fact that the Israeli military culture is rooted in the partisan-like culture of the *Palmach* (the striking companies of the Jewish military force during the British mandate) that favored heroic practitioners over theoreticians, leading the IDF to promote "heroic and charismatic generals [who] exhibited little intellectual curiosity in studying military science."[205] The third one is the previously discussed "cult of technology" that has led the Israeli generals "to think in terms of technological force multipliers, such as smart weapons, than to create force multipliers based on smart doctrines."[206]

Traditionally, the IDF's culture cultivated "doers," rather than "talkers" or "thinkers." The lack of intellectual vigor, however, has always been compensated by Israeli cultural emphasis on *iltur* (improvisation), when officers' abilities "to orient, to think, and to bounce ideas quickly," finding a better, "not-by-the-book" solution, are considered a hallmark of military performance.[207] On the one hand, this prominence of *iltur* in the IDF has often been connected to the lack of military professionalism.[208] One of the most recent criticisms was raised by the Winograd Commission (the Commission of Inquiry into the Events of Military Engagement in Lebanon, 2006) that highlighted "the connection between cultural organisation of improvisation and the lack of professionalism at the level of soldiers, as well as commanders."[209] On the other, creative improvisation has frequently compensated for anti-intellectualism and a deficit of professionalism, as "improvisers" are also "problem solvers" and, as the military is constantly busy with fighting, there is never time to "sit and study."[210]

The IDF, lacking the Russian cultural predisposition toward abstract thinking or the American extensive military education system that forces them to think (to a certain success), has never become a "bookish military." Consequently, while the IDF has always been good in fielding innovative technologies, "Israeli military thought lacked the ability to analyze, envision, and assess future [technological] developments and to generate knowledge in that concept."[211]

It is not surprising then that the theoretical discourse about possible usefulness of NLW in Israel is very limited (if, indeed, less than a dozen articles can be considered a discourse).[212] While the theoretical and conceptual milieu that surrounds NLW in the IDF generally suits the intellectual desert described above, a title of "oasis" might be bestowed on the work of Lieutenant Colonel Guy-Zahar Shtultz, who was responsible for the field of NLW and riot control at the IDF Infantry and Paratroopers Forces Headquarters. After a very comprehensive theoretical analysis of the political-military situation in which the IDF has been acting in Gaza and the West Bank, Shtultz produced a very concrete and practical list of recommendations, concluding that:

> A deployment of effective NLW by the [IDF's] fighting forces is a requirement of reality. A proficient employment of NLW will increase the legitimacy of our forces' activity in civilian environments, in general, and in Gaza and the West Bank, in particular, thus, providing the IDF with an advantage, which will frequently be of a strategic value. An effective employment of NLW will also provide our forces with a clear tactical advantage, as it will expand the range of the available means and solutions during operations in civilian environments, where the employment of usual lethal weapons is not optimal.[213]

Shtultz's wise analysis, however, fell on deaf ears, as the lack of enthusiasm for NLW in the IDF has been based on two paramount factors. The first one is the IDF's principal differentiation between *Bitachon Shotef* (*batash*, "current security," "state of conflict") and *Bitachon Yisodi* ("state of war," large military operations). On the one hand, Israeli forces have to prepare for war, or, at least, for a large military operation that is waged as a war. On the other, they constantly perform law enforcement missions in the West Bank, or routine patrolling of Israel's not-so-peaceful borders. As "an appointment of continuous *batash* missions on military units that were built to fight war disturbs their operational logic and harms their readiness for war,"[214] the IDF generals have cultivated an understanding that the accumulated experience in *batash* is not relevant for the IDF's preparations for war.[215] After the Second Lebanon War, the sense of *what is good for batash is not necessary good for war* was sharpened even more, due to the criticism that the IDF's poor performance was an outcome of the IDF's inability to switch from the mode of *batash* to the mode of war.[216] As discussed at the beginning of this chapter, NLW are primary considered by the Israeli military as weapons for *batash* and, therefore, their employment during a war is intuitively regarded as something that is unthinkable. When, as Brigadier General Oren Abman stated, the transition from *batash* to war "must be expressed in terms of aggressiveness and offensiveness, meaning the employment of all firepower that is available to the unit"[217] there is, understandably, no place for something that is not intended to kill (i.e., NLW).

The second reason for the lack of the IDF's conceptual interest in the employment of NLW on the battlefield is the fact that the reduction of enemy civilian population casualties has been treated by Israeli military culture as a purely legal issue. It will be unfair to claim that the IDF does not try to reduce casualties among the enemy civilian population, as even foreign military specialists assess the IDF's attempts to avert unnecessary harm to the local populace as one of (if not the) best in the Western world.[218] It seems, however, that the main leading factor behind these attempts is the IDF's obsessiveness (some would even argue over-obsessiveness)[219] with legitimacy, as Israeli generals understand that "every time, when the IDF finishes an operation, the UN Human Rights Council establishes an inquiry into the war crimes supposedly committed by Israel."[220] While this aspect of "lawfare" is relatively new to Israeli military culture, there is a traceable rising influence of the IDF International Law Department (known by its Hebrew acronym DABLA) on the strategic, operational, and tactical decision-making process.[221] In other words, instead of developing concepts of how to employ weapons that are intended to minimize civilian casualties (i.e., NLW), the Israeli military struggles to improve the compliance of the IDF with international humanitarian law, and, most importantly, the collection process of evidence for this compliance.[222]

One of the outcomes of this approach to reducing enemy civilian casualties was the introduction of Population-during-Warfare Officers (*ktzin uchlisiya be lehima*). As one of the lessons of Operation Cast Lead (and probably the results of the subsequent Goldstone Report), these officers have been attached to each deployed battalion and their duty is "to promote the balance between military necessity and the humanitarian need of the local populace," as well as to record the enemy's abuses of international law.[223] Interestingly enough, Lieutenant Colonel Guy-Zahar Shtultz, one who so arguably promoted the possible usefulness of NLW to reduce enemy civilian casualties, was promoted to be the Head of the Department of Civilian Affairs during Fighting (*rosh anaf ha'markiv ha'ezrachi be lehima*)—the department responsible for the development of training and operational procedures for Civilian-Population-during-Warfare Officers.[224] This promotion reflects best that the IDF's thinking about the problem of enemy civilian casualties is more inclined toward the legal aspects of this issue, rather than towards a search for a technological solution (i.e., NLW).

CONCLUSIONS: THE FAILURE OF THE RMA OF NLW IN ISRAEL

In summarizing the history of NLW in Israel, it is important to highlight the fact that compared with the American and Russian militaries (and probably any other in the world), the IDF seems to be the most experienced. The long history of deployment as law enforcement forces during routine *batash* operations forced the IDF to develop and employ NLW at an unprecedented scale. Moreover, the combination of the Israeli *iltur* culture combined with traditional out-of-the-box thinking, and a tendency to look for simple and cheap technological solutions, allowed the IDF to develop and successfully employ such extraordinary NLW as *Hatzatzit*, *Bo'esh*, or non-lethal shells for tanks. The IDF, however, is "a dynamic and agile military," and its culture tends to believe that "the wars of the past are irrelevant (or almost irrelevant) for the wars of the future."[225] It is not surprising then that once the immediate operational need for NLW dwindled, *Hatzatzit* was forgotten,[226] non-lethal shells for tanks were abandoned,[227] and, in fact, even the Civilian-Population-during-Warfare Officer was a forgotten lesson of the Al-Aqsa Intifada that was relearned after Operation Cast Lead.[228]

On the one hand, as it was discussed previously, the international criticism of Israeli military operations specifically focuses on enemy civilian casualties, burdening the Israeli political-military decision-making process. Despite its traditional foreign policy narratives of *Um Shmum*, *the world is always against us*; the *chutzpah* attitude to the international community and the traditional reliance on American protection, Israel is, undoubtedly, most vulnerable to external criticism, especially compared with the United States or Russia. Therefore, it could be assumed that the Israeli political-military decision-makers will do more than their counterparts in Washington or Moscow to pressure their military to look for less lethal technologies in an attempt to reduce this international pressure.

On the other hand, the cultural existential anxiety embedded in Israeli society, combined with the readiness to make sacrifices (the Masada myth) and a belief that Israel's existence cannot be taken for granted, but has to be taken by force regardless the price (the Bar Kokhba Syndrome), contradicts the very idea of reducing the punishment of a society that dares to aim a blow at Israel. While the traditional Jewish individualism increases the value of Jewish life, and the Holocaust experience explains Israeli concern about the suffering of others, the cultural collectivism shaped by two thousand years of persecution pays little respect to the lives of Israel's enemies and their societies. In 1983, in his criticism of Israeli culture, Yehoshafat Harkabi said that "to admire the Bar-Kokhba Rebellion is to admire rebelliousness and heroism detached of responsibility for their consequences."[229] It seems that 30 years later, very little has changed, as in 2015, surveys showed that a majority (53 percent) of the Jewish population in Israel agreed with the statement that "any Palestinian who has perpetrated a terror attack against Jews should be killed on the spot, even if he has been apprehended and no longer poses a threat."[230]

Israeli inherent predisposition towards violent solutions, regardless of the consequences, and their tendency to see a whole society as responsible for the actions of its individuals, leaves very little space (if any) for the appreciation of enemy civilian populations and, therefore, in the Israeli mind-set anything less than the employment of lethal force is just unthinkable. After all, in their evaluation of the IDF's performance during Operation Protective Edge "only 6 percent of the entire Jewish public saw the IDF as having used too much firepower."[231]

It seems right to argue that, maneuvering between these two almost opposite political vectors, the Israeli leadership pressured it military to deal with both of the problems by overemphasizing the legal issues surrounding its military operations. In other words, the IDF has been forced not only to pretend to be *holier-than-thou*, but to actually try and be that. To improve the IDF's compliance with international humanitarian law, the IDF introduced DABLA, which provides "operational legal advice" to the scale that "a military lawyer had to be present for every approval of targeting killing."[232] Moreover, the purpose of the above mentioned Population-during-Warfare Officers is to improve this compliance on the ground. According to many Israeli generals and military scholars, the IDF has become obsessed with the legitimacy and legality of its actions. Each time there is a suspicion of misconduct or overuse of firepower, the IDF does not hesitate to open its own investigations.[233] In other words, averting enemy civilian casualties, in the IDF's eyes, is a problem that has to be solved by an increased compliance with the law, rather than more suitable weaponry that will minimize civilian casualties and collateral damage (i.e., NLW).

The conceptual perception that NLW belong to *batash* and, therefore, they do not have a place on the battlefield reduces even more the chances for the RMA of NLW in Israel. Once a non-lethal effect is required, the IDF prefers to use lethal weapons in a non-lethal way, firing in the air or close to the target. The Israeli Air Force procedure *hakesh-be-gag* (knock-on-the-roof), is the best example of this; a first missile is shot to disperse civilians from the roof of a target, and then the second one (usually deadlier than the first) is shot to destroy the target.[234] As long as this perception stays prevalent, and as far as the political leadership does not pressure the IDF to look for technological solutions, rather than legal, to reduce enemy civilian casualties during Israeli military operations, NLW are doomed to play an important, yet very limited role during routine *batash* deployments.

NOTES

1. For example, the Hebrew version of Roi Ben-Horin, "Non-Lethal Weapons: Theory, Practice, and What Lies Between," *Strategic Assessment* 2, no. 4 (2001, January): 21–27 was titled as "Neshek Al-Hereg: Ha'alaha, Hama'ase ve-mashebenehem" and published in *Idkun Astrategi* 4, no.1 (2001): 18–24.

2. For example, Guy-Zahar Shtultz, "Haf'alat koah medureget be'imutim besviva ezrahit [The Employment of Graduate Force in Conflicts within Civilian Environment]," *Ma'arachot* 438 (2011): 8–17.

3. The State Comptroller of Israel, *Annual Report 54A for 2003 Year*, Jerusalem, 2003, 89–90 (Hebrew).

4. Aryeh Shalev, *The Intifada: Causes and Effects* (Jerusalem: Jerusalem Post, 1991), 56–62.

5. F. M. Lorenz, "Non-Lethal Force: The Slippery Slope to War?," *Parameters* XXVI, no. 3 (1996): 52.

6. "Interview with Defense Minister Yitzhak Rabin," published in *Jerusalem Post*, February 16, 1989.

7. Shalev, *The Intifada*, 106–11.

8. The State Comptroller of Israel, *Annual Report 54A*, 87–88.

9. For example, Gidon Avidor (ed.), *Hamashmauyot haben'leumiyot, hamishpatiyot vehamusriyot belehima bekerev uh'lusiya ezrahit [The International, Legal and Moral Consequences of Warfare in Civil Environment]* (Latrun: Zvi Meitar Institute for Land Warfare Studies, 2010).

10. Shtultz, 'Haf'alat koah medureget.'

11. Shlomo Gazit, *Hamakel vehagezer: hamimshal haisraeli be'yehuda veshomron [The Stick and the Carrot: The Israeli Government in Judea and Samaria]* (Tel Aviv: Zmora-Bitan, 1985), 269.

12. Shalev, *The Intifada*, 56–62.

13. Chaim Herzog, *The Arab-Israeli Wars: War and Peace in the Middle East*, 3rd edition, revised and updated by Shlomo Gazit (Barnsley: Frontline Books, 2010), 397–98.

14. Ehud Barak quoted in Ahron Bregman and Jihan El-Tahri, *The Fifty Years War: Israel and Arabs* (London: Penguin Books & BBC Worldwide, 1998), 189.

15. Shalev, *The Intifada*, 106–11.

16. Yitzhak Rabin quoted in *Ha'araretz*, October 5, 1988.

17. Efraim Lapid, "Haintifada harishona, ha'avanim she'banu et heskem Oslo [The First Intifada, the Stones that Built the Oslo Agreements]," *IsraelDefense*, December 8, 2013.

18. Dror Merom, "Hamasokim neged hafganot hamoniyot [Helicopters against Massive Demonstrations]," *Bit'aon Hel' Avir*, no. 67 (168) (1989): 14–15.

19. The State Comptroller of Israel, *Annual Report 54A*, 101; Shalev, *The Intifada*, 109.

20. Alexander (Alik) Ron in "Eduto shel' nitzav Alik Ron bephnei va'adat Or, 3 september 2001, helek shlishi [The Testimony of Major General Alik Ron to the Or Commission, 3 September 2001, Part 6]," *Ha'aretz Online*, 4 September 2001, http://www.haaretz.co.il/misc/1.1555776 [accessed May 10, 2018].

21. "Yaladim palestinim ganvu tziyud tzvai mitank [Palestinian Children Stole Military Equipment from a Tank]," *Yedioth Ahronoth*, January 2, 2002, http://www.ynet.co.il/articles/0,7340,L-1489428,00.html [accessed May 10, 2018].

22. Amos Ha'rel, "Tzal inase lefazer hafganot hamoniyot bepagazei tankim im k'liey plastik sh'osim rak ra'ash [The IDF Will Try to Disperse Massive Riots with Tanks' Shells with Plastic Warheads that Make Noise Only]," *Ha'aretz Online*, February 28, 2005, http://www.haaretz.co.il/misc/1.1505848 [accessed May 10, 2018].

23. Kobi Ben-Simhon, "Hatargil hamasriah; hamafginim mi'Na'alin tzrihim liknot deodorant [A Stinky Hoax; The Rioters from Na'alin Have to Buy Deodorant]," *Ha'aretz Online*, September 4, 2008 http://www.haaretz.co.il/misc/1.1347475 [accessed May 10, 2018].

24. Amir Oren, "Nishkal shigur 'Bo'esh' leknesiyat hamolad [It was Considered to Deploy 'Skunk' to the Church of the Nativity]," *Ha'aretz*, April 12, 2002, http://news.walla.co.il/item/208321 [accessed May 10, 2018].

25. Amir Rapaport, "Metach gavoha [High Tension]," *IsraelDefense*, April 26, 2011, http://www.israeldefense.com/en/node/9371 [accessed May 10, 2018].

26. Efrat Vais, "Lerishona: Magav Pizer Mafginei Smol' betotah sirahon [For the First Time: The Border Police Dispersed Leftist Demonstrators with a Stinky Cannon]," *Yedioth Ahronoth*, August 10, 2008, http://www.ynet.co.il/articles/0,7340,L-3580288,00.html [accessed May 10, 2018].

27. See Idan Seger, "Ei'n makom letauiyot: kah tzal' metkonen lemehumot [There Is No Place for Mistakes: This Is the Way in Which the IDF Prepares Itself for the Riots]," *Mako*, September 20, 2011, http://www.mako.co.il/pzm-magazine/Article-c7ddcfef4e18231006.htm [accessed May 10, 2018]; Yuval Azulai, "Technologiya hal-hereg: haim hayah nitan limnoa hasa'ara be 'marmara'? [Non-Lethal Technology: Whether Was It Possible to Prevent the 'Marmara' Storm?]," *Globes*, September 20, 2011, http://www.globes.co.il/news/article.aspx?did=1000683960 [accessed May 10, 2018].

28. Shai Levy, "Neshek (Hal) Hereg [(Non) Lethal Weapons]," *Mako*, http://www.mako.co.il/pzm-magazine/Article-c4b5b1a6475a531006.htm [accessed May 10, 2018]; Azulai, "Technologiya hal-hereg [Non-Lethal Technology]."

29. Ofer Shelah, "Mekavim letov, mitkonenim le r'a' [Hoping for Good, Preparing for Bad]," *NRG*, September 17, 2011, http://www.nrg.co.il/online/1/ART2/286/345.html [accessed 10 May 2018]; Moriya Ben-Yosef, "Muetzet Hait'tztaidut Beneshek Pahot Katlani [The Employment of Non-Lethal Weapons is Speeding Up]," *IsraelDefense*, June 5, 2011; Rapaport, "Metach gavoha [High Tension]"; Azulai, "Technologiya hal-hereg [Non-Lethal Technology]."

30. Lapid, "Haintifada harishona [The First Intifada]"; Merom, "Hamasokim neged hafganot [Helicopters against Massive Demonstrations]."

31. The State Comptroller of Israel, *Annual Report 54A*, 88–89.

32. Shtultz, "Haf'alat koah medureget [The Employment of Graduate Force]."

33. Sarit Michaeli, *Crowd Control: Israel's Use of Crowd Control Weapons in the West Bank* (Jerusalem: B'tselem, 2013).

34. Shtultz, "Haf'alat koah medureget [The Employment of Graduate Force]."

35. Ben-Shalom, Uzi, "'Faser lematzav ham'em': mah'shevot al neshek pahot katlani be'sde hakrav ['Faser switched to Shock': Thoughts on Less Lethal Weapons on the Battlefield]," *Be'yabasha*, no. 15 (2010, September).

36. Shtultz, "Haf'alat koah medureget [The Employment of Graduate Force]"; Ben-Shalom, "'Faser lematzav ham'em' ['Faser switched to Shock']."

37. Sasson Sofer, "Towards Distant Frontiers: The Course of Israeli Diplomacy," *Israel Affairs* 10, no. 1–2 (2004): 1–9.

38. "Ties between Israel and US 'worst in 35 years,'" *BBCNews*, March 15, 2010, http://news.bbc.co.uk/1/hi/8567706.stm [accessed May 10, 2018].

39. Gad Yair and Sharona Odom-Weiss, "Israeli Diplomacy: The Effects of Cultural Trauma," *The Hague Journal of Diplomacy*, no. 9 (2014): 1–23.

40. For example, Dov Ben-Meir, *Mediniyut Hutz [Foreign Policy]* (Tel Aviv: Yedioth Ahronot and Chemed Books, 2011); Aaron Klieman, *Israel & The World after 40 Years* (McLean: Pergamon-Brassey's International Defense Publishers, 1990); Sason Sofer, *Lidata shel ha'mah'shava hamedinit be'israel [The birth of the Political Thought in Israel]* (Jerusalem: Shoken, 2001); Uri Bialer, "Top Hat, Tuxedo and Cannons: Israeli Foreign Policy from 1948 to 1956 as a Field of Study," *Israel Studies* 7, no. 1 (2002): 1–80.

41. Klieman, *Israel & The World*, 7–17.

42. Ben-Meir, *Mediniyut Hutz [Foreign Policy]*, 25.

43. Arie Gronik, "Yahasei Israel ve'um—shalosh tkufot ve'shalosh she'elot [The Israel-U.N. Relations—Three Periods and Three Questions]," in Benjamin Noiberger and Arie Gronik (ed.), *Mediniut Hutz bein imut le'esderim—Israel 1948-2008 [The Foreign Policy between Conflict and Agreement—Israel 1948–2008]*, Vol. B (Raanana: The Open University of Israel, 2008), 1080.

44. Dan Leon and Yehuda Adlin (ed.), *Chaim Weizmann: Statesman of the Jewish Renaissance* (Jerusalem: Alpha Press, 1974), 139.

45. Uri Bialer, *"Our Place in the World"—Mapai and Israel's Foreign Policy Orientation 1947-1952* (Jerusalem: Magnes Press, Hebrew University, 1981), 10.

46. David Ben-Gurion quoted in Bialer, *"Our Place in the World,"* 11.

47. Jonatan Manor, "Kishalon ha'asbara or kishalon ha'mediniyut? [A Failure of the Hasbara or a Failure of the Policy?]," in Gronik Noiberger, *Mediniut Hutz bein imut le'esderim*; Sharon Pedro, *Utzma raka ke'nehes leumi [Soft Power as a National Asset]* (Jerusalem: The Jewish People Policy Planning Institute, 2004).

48. Bialer, "Top Hat, Tuxedo and Cannons," 28.

49. David Ben-Gurion, "Israel ben ha'umot [Israel between the Nations]," *Israel Government Year Book*, 1952, 23.

50. Gronik, "Yahasei Israel ve'um [The Israel-U.N. Relations]," 1086–87.

51. Uri Bialer, "Ben-Gurion ve'she'elat ha'orientatziya ha'beinleumit shel' Israel 1948–1956 [Ben-Gurion and the Question of the International Orientation of Israel 1948–1956]," *Cathedra*, no. 43 (1987): 148.

52. Yair and Odom-Weiss, "Israeli Diplomacy," 7.

53. UN General Assembly, *Resolution 3379: Elimination of All Forms of Racial Discrimination*, November 10, 1975, New York; Efraim Inbar, *Israel eina mevudedet [Israel Is Not Isolated]* (Tel Aviv: Begin-Sadat Center for Strategic Studies, 2013); Manor, "Kishalon ha'asbara [A Failure of the Hasbara]," 65–68.

54. Yitzhak Rabin, "Speech in the Knesset Regarding the U.N. General Assembly Resolution from 10 November 1975 about Zionism," November 11, 1975 (Hebrew), published in Inbal Telem, Shmuel Tzvaog, and Benjamin Noiberger (ed.), *Mediniyut ha'hutz shel Israel—kovetz mismachim [The Foreign Policy of Israel—Documents Collection]*, Vol. A., (Raanana: The Open University of Israel, 2004), 352.

55. Yohanan Bein, "Israel, ha'um ve'ha'ma'araha le'shalom [Israel, the U.N. and the Struggle for Peace]," in Moshe Igar, Yosef Guvrin, and Rei Oded (ed.), *Misrad ha'hutz—50 hashanim ha'rishonot [The Foreign Affairs Office—The 50 First Years]* (Jerusalem, Keter, 2002), 883.

56. Yair and Odom-Weiss, "Israeli Diplomacy," 7.

57. Jeremy Pressman, "Israeli Unilateralism and Israeli-Palestinian Relations 2001–2006," *International Studies Perspectives*, no. 7 (2006): 360–76.

58. "World Court Tells Israel to Tear Down Illegal Wall," *Guardian*, July 10, 2004, http://www.theguardian.com/world/2004/jul/10/israel3 [accessed May 10, 2018].

59. Yair and Odom-Weiss, "Israeli Diplomacy," 8, 12.

60. Benjamin Netanyahu, *Speech to US Congress,* July 10, 1996, http://mfa.gov.il/MFA/MFA-Archive/1996/Pages/PM%20Netanyahu-%20Speech%20to%20US%20Congress-%20July%2010-%201996.aspx [accessed May 10, 2018].

61. Bialer, "Top Hat, Tuxedo and Cannons," 23–33.

62. John Campbell, "The United States and the Arab-Israeli Conflict," in Robert Freedman (ed.), *World Politics and the Arab-Israeli Conflict* (New York: Pergamon Press, 1979).

63. Sofer, "Towards Distant Frontiers," 6.

64. "To Israel with Love," *Economist*, August 3, 2006, http://www.economist.com/node/7255198 [accessed May 10, 2018].

65. Ehud Olmert quoted in "To Israel with Love." See also John Mearsheimer and Stephen Walt, *The Israel Lobby and US Foreign Policy* (New York: Penguin Books, 2007).

66. Yossi Shain and Neil Rogachevsky, "Between JDate and J Street: US Foreign Policy and the Liberal Jewish Dilemma in America," *Israel Journal of Foreign Affairs* 3 (2011): 37–48.

67. Sofer, "Towards Distant Frontiers," 6.

68. An argument reflected in "To Israel with Love."

69. Mearsheimer and Walt, *The Israel Lobby*, 75.

70. UN Security Council, *Veto List*, http://research.un.org/en/docs/sc/quick [accessed: 27 October 2015].

71. Inbar, *Israel eina mevudedet [Israel Is Not Isolated]*, 14.

72. Norman Finkelstein, *Beyond Chutzpah: On the Misuse of Anti-Semitism and the Abuse of History*, second edition (London: Verso, 2008).

73. "Netanyahu Holocaust Remarks: Israeli PM Criticised," *BBCNews*, October 21, 2015, http://www.bbc.co.uk/news/world-middle-east-34594563 [accessed May 10, 2018].

74. The Intelligence and Terrorism Information Center (ITIC), *Estimate of the Number of Hamas Operatives Killed in Operation Protective Edge (Part Eleven)*, Tel Aviv, July 20, 2015.

75. UN Human Rights Council, *Human Rights in Palestine and Other Occupied Arab Territories: Report of the United Nations Fact-Finding Mission on the Gaza Conflict*, September 25,

2009; UN Human Rights Council, *Report of the Independent Commission of Inquiry Established Pursuant to Human Rights Council Resolution*, June 24, 2015.

76. Shmuel Rosner and Michael Herzog, *Jewish Values and Israel's Use of Force in Armed Conflict: Perspectives from World Jewry* (Jerusalem: The Jewish People Policy Institute, 2015), 21–25.

77. Ibid., 24.

78. Roni Sofer, "Peres: Goldstone Report Makes Mockery of History," *Yedioth Ahronoth*, September 16, 2009, http://www.ynetnews.com/articles/0,7340,L-3777814,00.html [accessed May 10, 2018].

79. Roni Sofer, "Lieberman: Goldstone Report Meant to Destroy Israel's Image," *Yedioth Ahronoth*, September 17, 2009, http://www.ynetnews.com/articles/0,7340,L-3778143,00.html [accessed May 10, 2018].

80. Israel Ministry of Foreign Affairs, *Israel's Analysis and Comments on the Gaza Fact-Finding Mission Report* (communicated by the Foreign Ministry Spokesperson), September 15, 2009, http://mfa.gov.il/MFA/PressRoom/2009/Pages/Israel_initial_reaction_Goldstone_Mission_15-Sep-2009.aspx [accessed May 10, 2018].

81. Israel Ministry of Foreign Affairs, *Goldstone Report: Interviews with PM Netanyahu on Israeli Television* (communicated by the Prime Minister's Media Adviser), September 17, 2009, http://mfa.gov.il/MFA/PressRoom/2009/Pages/Goldstone_Report_Interviews_PM_Netanyahu_17-Sep-2009.aspx [accessed May 10, 2018].

82. Mearsheimer and Walt, *The Israel Lobby*, 48.

83. "The Goldstone Report," *Journal of Palestine Studies* 39, no. 2 (Winter 2010): 109.

84. Elad Simchayoff and Udi Segal, "Doh ha'um al mivtza 'tzuk eitan' ushar, rak aratzot habrit hitnagda [The UN Report about Operation Protective Edge Was Approved, only the U.S. Was Against]," *Mako*, http://www.mako.co.il/news-military/israel-q3_2015/Article-268b8dec0445e41004.htm?sCh=31750a2610f26110&pId=786102762 [accessed May 10, 2018].

85. Benjamin Netanyahu, *Speech to US Congress,* July 10, 1996, http://www.theglobeandmail.com/news/world/americas/transcript-of-prime-minister-netanyahus-address-to-us-congress/article2032842/ [accessed May 10, 2018].

86. "The Goldstone Report," *Journal of Palestine Studies*, 109.

87. The term *lawfare* was developed by USAF Colonel Charles J. Dunlap, Jr. in "Law and Military Interventions: Preserving Humanitarian Values in 21st Conflicts," *Proceedings to the Humanitarian Challenges in Military Intervention Conference*, Carr Center for Human Rights Policy, Kennedy School of Government, Harvard University, Washington, 29 November 2001. For Israel's interpretations of *lawfare* see: Eyal Weizman, "Lawfare in Gaza: Legislative Attack," *Open Democracy*, https://www.opendemocracy.net/article/legislative-attack [accessed May 10, 2018]; Noam Noiman, "Lohamat mishpat: iumim ve'hizdamnuyiot [Lawfare: Threats and Opportunities]," *Ma'arachot*, no. 449 (2013): 16–23; Gerald Steinberg, "Lawfare," *Hamas, the Gaza War and Accountability under International Law, Updated Proceedings to an International Conference, 18 June 2009*, Jerusalem Center for Public Affairs and the Konrad Adenauer Stiftung, Jerusalem, 2013.

88. Israel Ministry of Foreign Affairs, *The 2014 Gaza Conflict, 7 July–26 August 2014: Factual and Legal Aspects*, Jerusalem, May 2015; Israel Ministry of Foreign Affairs, *FAQ: The Operation in Gaza-Factual and Legal Aspects*, August 16, 2009, http://mfa.gov.il/MFA/ForeignPolicy/FAQ/Pages/FAQ-Operation_in_Gaza-Legal_Aspects.aspx#13 [accessed May 10, 2018].

89. Israel Ministry of Foreign Affairs, *FAQ: The Operation in Gaza*.

90. Sofer, "Lieberman: Goldstone Report."

91. Israel Ministry of Foreign Affairs, *Israel Rejects General Assembly Resolution*, November 5, 2009, http://mfa.gov.il/MFA/PressRoom/2009/Pages/Israel_rejects_General_Assembly_resolution_5-Nov-2009.aspx [accessed May 10, 2018].

92. William Saletan, "Israel's Unprecedented Steps to Avoid Civilian Casualties," *The National Post*, July 16, 2014.

93. Mark Heller, "The International Dimension: Why So Few Constrains on Israel?," in Shlomo Brom and Meir Elran (ed.), *The Second Lebanon War: Strategic Perspectives* (Tel Aviv: Institute for National Security Studies, 2007), 209.

94. Yair and Odom-Weiss, "Israeli Diplomacy," 2.

95. Tamas Berzi, "European Reactions to Israel's Gaza Operation," Jerusalem Center for Public Affairs, *Articles on Europe & Israel* 8, no. 20 (2009).

96. Council of European Union, *EU Statement Following Israel's Operation Cast Lead in Gaza Calling for an Immediate and Permanent Ceasefire*, December 30, 2008.

97. "Gaza EU Statement a Misunderstanding—Czech Minister," *Reuters*, January 4, 2009, https://www.reuters.com/article/idUSL4734544 [accessed May 10, 2018].

98. Ibid.

99. Jack Khoury et al., "Israel Girds for Diplomatic War Over 'Biased' UN Gaza Report," *Ha'aretz*, September 15, 2009, http://www.haaretz.com/israel-girds-for-diplomatic-war-over-biased-un-gaza-report-1.7831 [accessed May 10, 2018].

100. Ibid.

101. Israel State Bureau of Statistics, *Israel's Foreign Trade in Goods, by Country—September 2015*, Jerusalem, October 25, 2015, http://www.cbs.gov.il/reader/newhodaot/hodaa_template_eng.html?hodaa=201516283 [accessed May 10, 2018].

102. Yehoshafat Harkabi, "Hir'urim al torat ha'bitachon ha'leumi [Thoughts about the National Security Doctrine]," *Ma'arachot*, no. 270–271 (1979): 44.

103. Oded Eran, "International Aspects of Operation Pillar of Defence," *INSS Insight*, no. 386 (2012); See also Yuval Bazak, "Responding to the Need for International Legitimacy: Strengthening the IDF Strike Force," *Military and Strategic Affairs* 3, no. 2 (2011): 3–17.

104. Zaki Shalom, "Is the IDF Prepared to Face a Regular War against the Arab States?," *Military and Strategic Affairs* 1, no. 1 (2009, April): 17–23.

105. Ibid.

106. Emily Landau and Azriel Bermant, "Iron Dome Protection: Missile Defense in Israel's Security Concept," in Anat Kurz and Shlomo Brom (ed.), *The Lessons of Operation Protective Edge* (Tel Aviv: Institute for National Security Studies, 2014); Yiftah Shapir, "Iron Dome: The Queen of Battle," in Shlomo Brom (ed.), *In the Aftermath of Operation Pillar of Defense: The Gaza Strip, November 2012* (Tel Aviv: Institute for National Security Studies, 2012).

107. Pnina Baruch, "Operation Protective Edge: Legality and Legitimacy," *INSS Insight*, no. 577 (2014).

108. Lahav Harkov, "Former British Commander in Afghanistan: No Army Acts with as Much Discretion as IDF does," *Jerusalem Post*, April 9, 2014.

109. John Barnes, "The Intransigent Israeli?," British Embassy, Tel Aviv, November 18, 1971, republished in Shmuel Bar, "The Intransigent Israeli?—A Document from the Archives of the British Foreign Office (FCO/17 1554)," *Israel Journal of Foreign Affairs* 4, no. 1 (2010): 103–13.

110. Yair Gad, *Tzofen ha'isareliyut: aseret ha'dibrot shel' shnot he'al'paiim [The Code of Israeliness: The Ten Commandments for the 21st century]* (Jerusalem: Keter Books, 2011), 15.

111. For example, Baruch Kimmerling, *The Invention and Decline of Israeliness: State, Society, and the Military* (Berkley: University of California Press, 2001); Motti Regev, "To Have a Culture of Our Own: on Israeliness and Its Variants," *Ethnic and Racial Studies* 23, no. 2 (2000): 223–47; Gad, *Tzofen ha'isareliyut [The Code of Israeliness]*.

112. Gad, *Tzofen ha'isareliyut [The Code of Israeliness]*, 17.

113. See Yehoshafat Harkabi, *The Bar Kokhba Syndrome: Risk and Realism in International Politics* (Chappaqua: Rossel Books, 1983); Moshe Lissak (ed.), *Israeli Society and Its Defense Establishment: The Social Political Impact of a Protracted Violent Conflict* (London: Frank Cass, 1984); Robert Wistrich and David Ohana (ed.), *The Shaping of Israeli Identity: Myth, Memory and Trauma* (London: Fank Cass, 1995); Edna Lomsky-Feder and Eyal Ben-Ari (ed.), *The Military and Militarism in Israeli Society* (New York: State University of New York Press, 1999).

114. Gad, *Tzofen ha'isareliyut [The Code of Israeliness]*, 187–88.

115. Raymond Cohen, *Culture and Conflict in Egyptian-Israeli Relations: A Dialogue of the Deaf* (Bloomington: Indiana University Press, 1990), 38; Daniel Bar-Tal, *Lechiot im*

ha'sichsuch: nituah psichologi-havrati shel' he'hevra ha'yehudit be'israel [Living with the Conflict: Socio-Psychological Analysis of the Jewish Society in Israel] (Jerusalem: Karmel, 2007); Dalia Ofer, "We Israelis Remember, but How?: The Memory of the Holocaust and the Israeli Experience," *Israel Studies* 18, no. 2 (2013): 70–85; Yair Gad, "Israeli Existential Anxiety: Cultural Trauma and the Constitution of National Character," *Social Identities: Journal of the Study of Race, Nation and Culture* 20, no. 4–5 (2014): 346–62; Gad, *Tzofen ha'isareliyut [The Code of Israeliness]*.

116. Gad, *Tzofen ha'isareliyut [The Code of Israeliness]*, 15–31.

117. Gad, "Israeli Existential Anxiety."

118. Ofer, "We Israelis Remember, But How?," 72; Ofira Seliktar, *New Zionism and the Foreign Policy System of Israel* (Carbondale: Southern Illnois University Press, 1986).

119. Gad, *Tzofen ha'isareliyut [The Code of Israeliness]*, 15.

120. Gad, "Israeli Existential Anxiety," 355.

121. Yechiel Klar, Noa Schori-Eyal, and Yonat Klar, "The 'Never Again' State of Israel: The Emergence of the Holocaust as a Core Feature of Israeli Identity and Its Four Incongruent Voices," *Journal of Social Issues* 69, no. 1 (2013): 125–43.

122. Ofer, "We Israelis Remember, But How?," 72.

123. Gad, *Tzofen ha'isareliyut [The Code of Israeliness]*, 30.

124. Cohen, *Culture and Conflict in Egyptian-Israeli Relations*, 31–32.

125. *Brockhaus and Efron Encyclopedic Dictionary*, Petersburg, 1907, Vol. XVI, 836–39, (Russian).

126. Quoted in Cohen, *Culture and Conflict in Egyptian-Israeli Relations*, 32.

127. Dima Adamsky, *The Culture of Military Innovation: The Impact of Cultural Factors on the Revolution in Military Affairs in Russia, the US and Israel* (Stanford: Stanford University Press, 2010), 110.

128. Quoted in Cohen, *Culture and Conflict in Egyptian-Israeli Relations*, 32.

129. Gad, *Tzofen ha'isareliyut [The Code of Israeliness]*, 45.

130. Dan Kaspi, "Shuv yesh lanu oiev meshutaf" ["We Have a Mutual Enemy Again"], *Ynet*, July 31, 2009, http://www.ynet.co.il/articles/0,7340,L-3753712,00.html [accessed May 10, 2018].

131. Cohen, *Culture and Conflict in Egyptian-Israeli Relations*, 33.

132. Gad, *Tzofen ha'isareliyut [The Code of Israeliness]*, 101.

133. Ibid., 74–75.

134. Quoted in Shlomo Avnery, *The Making of Modern Zionism* (New York: Basic Books, 1981), 277.

135. For example, Bialer, "Top Hat, Tuxedo and Cannons"; Yair and Odom-Weiss, "Israeli Diplomacy."

136. Benjamin Netanyahu, *Speech at the Holocaust Martyrs' and Heroes' Remembrance Day Ceremony*, April 11, 2010, http://www.pmo.gov.il/English/MediaCenter/Speeches/Pages/speechshoa1104010.aspx [accessed May 10, 2018].

137. For example, Cohen, *Culture and Conflict in Egyptian-Israeli Relations*, 36; Gad, *Tzofen ha'isareliyut [The Code of Israeliness]*.

138. Ofer, "We Israelis Remember, but How?," 71.

139. Klar, Schori-Eyal, and Klar, "The 'Never Again' State of Israel," 136.

140. William Booth, "Israeli Government to Refugees: Go Back to Africa or Go to Prison," *The Washington Post*, May 14, 2015, https://www.washingtonpost.com/world/middle_east/toughening-its-stance-toward-migrants-israel-pushes-africans-to-leave/2015/05/14/e1637bce-f350-11e4-bca5-21b51bbdf93e_story.html, [accessed May 10, 2018].

141. Baruch Kimmerling, "Making Conflict a Routine: Cumulative Effects of the Arab-Jewish Conflict Upon Israeli Society," in Lissak, *Israeli Society*.

142. Moshe Dayan quoted in Shabtai Teveth, *Moshe Dayan: The Soldier, the Myth, the Legend* (Steimatzky's Agency with Weidenfeld & Nicolson, 1972), 240.

143. Kimmerling, "Making Conflict a Routine," 14.

144. Although Adamsky translates *hachra'a* as "battlefield decisions," "humiliation on the battlefield" seems to be more appropriate, as he writes himself that "although not intended as a

terminal knockout, it was to bring the ultimate victory." Adamsky, *The Culture of Military Innovation*, 112.

145. Gabriel Siboni, "War and Victory," *Military and Strategic Affairs* 1, no. 3 (2009): 39–49.

146. Mark Heller, "Israeli Deterrence in the Aftermath of Protective Edge," in Brom Kurz, *The Lessons of Operation Protective Edge*.

147. Moshe Dayan, "Peulot Tzvayiot be'yemei shalom [Military Operations in Peacetime]," *Ma'arachot* 119 (1959): 56.

148. Ehud Olmert quoted in State of Israel, *The Partial Report of the Commission of Inquiry into the Events of Military Engagement in Lebanon 2006*, April 2007, (Hebrew), 81.

149. Amir Peretz quoted in State of Israel, *The Partial Report of the Commission of Inquiry*, 71.

150. Dan Halutz quoted in William Arkin, *Divining Victory: Airpower in the 2006 Israel-Hezbollah War* (Maxwell: Air University Press, 2007), 152.

151. See Oded Lowenheim and Gadi Heimann, "Revenge in International Politics," *Security Studies* 17, no. 4: 685–724; Gil Muciano, "A Matter of Honour: A Review of Israeli Decision Making during the Second Lebanon War," *The Atkin Paper Series*, The International Centre for the Study of Radicalization and Political Violence, March 2011.

152. Niccolo Petrelli, "Israel's Struggle Against Hamas," PhD Dissertation, Roma Tre University, 2014, 63.

153. Raanan Kuperman, "The Impact of Internal Politics on Israel's Reprisal Policy During the 1950s," *Journal of Strategic Studies* 24, no.1: 1–28; Petrelli, "Israel's Struggle," 24–89.

154. Yehuda Ben-Meir, "Israeli Public Opinion and the Second Lebanon War," in Brom and Elran, *The Second Lebanon War: Strategic Perspectives*.

155. State of Israel, *The Partial Report of the Commission of Inquiry*, 11.

156. Kuperman, "The Impact of Internal Politics."

157. Bar-Tal, *Lechiot im ha'sichsuch [Living with the Conflict]*, 237–38.

158. Yair Evron, "Deterrence and its Limitations," in Brom and Elran, *The Second Lebanon War: Strategic Perspectives*.

159. Avner Golov, "Rethinking the Deterrence of Hamas," in Kurz and Brom, *The Lessons of Operation Protective Edge*.

160. Barry Schwartz et al., "The Recovery of Masada: A Study in Collective Memory," *The Sociological Quarterly* 27, no. 2 (1986): 147–64; Nachman Ben-Yehuda, "The Masada Mythical Narrative and the Israeli Army," in Lomsky-Feder and Ben-Ari, *The Military and Militarism in Israeli Society*; Yael Zerubavel, "The Multivocality of a National Myth: Memory and Counter-Memories of Masada," in Wistrich and Ohana, *The Shaping of Israeli Identity*.

161. See Yael Zerubavel, "Bar Kokhba's Image in Modern Israeli Culture," in Peter Schafer (ed.), *The Bar Kokhba War Reconsidered: New Perspectives on the Second Jewish Revolt against Rome* (Tubingen: Mohr Siebeck, 2003); Harkabi, *The Bar Kokhba Syndrome*.

162. Schwartz et al., "The Recovery of Masada," 151; Zerubavel, "Bar Kokhba's Image," 279.

163. See Schwartz et al., "The Recovery of Masada"; Ben-Yehuda, "The Masada Mythical Narrative"; Zerubavel, "The Multivocality of a National Myth."

164. Zerubavel, "The Multivocality of a National Myth," 111.

165. Schwartz et al., "The Recovery of Masada."

166. Zerubavel, "Bar Kokhba's Image."

167. Michal Danieli, "'Noal' Khanibal: adif hayal harug ni'asher hayal hatuf? ['Khanibal' Procedure: Is a Dead Soldier Preferable to a Kidnapped Soldier?]," *Mako*, October 17, 2011, http://mobile.mako.co.il/pzm-magazine/Article-d4a692186511331006.htm&sCh=3d385dd2dd5d4110&pId=1093150966 [accessed May 10, 2018].

168. Zerubavel, "Bar Kokhba's Image."

169. Harkabi, *The Bar Kokhba Syndrome*.

170. Ibid., 107–10.

171. Avner Yaniv, "A Question of Survival: The Military and Politics under Siege" in Avner Yaniv (ed.), *National Security and Democracy in Israel* (Boulder: Lynne Rienner, 1993).

172. For example, Ehud Luz, "'Jewish Ethics' as an Argument in the Public Debate Over the Israeli Reaction to Palestinian Terror," *Israeli Studies* 7, no. 3 (2002): 134–56; Amichai Cohen and Stuart Cohen, "Israel and International Humanitarian Law: Between the Neo-Realism of State Security and the 'Soft Power' of Legal Acceptability," *Israel Studies* 16, no. 2 (2011): 1–23; Ziv Boher and Mark Osiel, "Proportionality in Military Force at War's Multiple Levels: Averting Civilian Casualties vs. Safeguarding Soldiers," *Vanderbilt Journal of Transnational Law* 46, no. 747 (2013): 747–822; Michael Walzer, "The Ethics of Warfare in the Jewish Tradition," *Philosophia* 40 (2012): 633–41.

173. J. David Bleich quoted in Michael Walzer, "The Ethics of Warfare," 637.

174. Walzer, "The Ethics of Warfare," 637.

175. Tal Peled, "Am bone tzava bone am—ha'umnam? [Nation Builds Military That Builds Nation—Is That So?]," *Ma'arachot* 442 (2012): 68–72.

176. Nigel Bagnall, *The Israeli Experience: A Study of Quality* (Surrey: Tactical Doctrine Retrieval Cell, Staff College, 1974).

177. For example, Michael Handel, *Israel's Political Military Doctrine* (Cambridge, MA: Harvard University Press, 1973); Israel Tal, *Bitachon leumi: meatim mul rabim [National Security: Few against Many]* (Tel Aviv: Dvir, 1996); Stuart Cohen (ed.), *Democratic Societies and Their Armed Forces: Israel in Comparative Context* (London: Frank Cass, 2000), part IV; Uri Bar-Joseph (ed.), *Israel National Security Towards the 21st Century* (London: Frank Cass, 2001).

178. Adamsky, *The Culture of Military Innovation*, 113.

179. Tal, *Bitachon leumi*, 70.

180. Adamsky, *The Culture of Military Innovation*, 113.

181. Bagnall, *The Israeli Experience*, 205.

182. Sharon Sade, "Israel's Defence Industry in the 21st Century: Challenges and Opportunities," *Strategic Assessment* 7, no. 3 (2004): 31–40; Alex Mintz, "The Military-Industrial Complex: The Israeli Case," in Lissak, *Israeli Society*.

183. Adamsky, *The Culture of Military Innovation*, 114.

184. Eliot Cohen, Michael Eisenstadt, and Andrew Bacevich, *Knives, Tanks & Missiles: Israel's Security Revolution* (Washington, DC: The Washington Institute for Near East Policy, 1998), 59, 47, 139.

185. See Moshe Sharvit, "Mekoma shel ha'technologia be'doktrina shel tzal [The Place of Technology in the IDF's Doctrine]," *Ma'arachot* 393 (2004): 81–83; Zvi Ofer and Avi Kober (ed.), *Ichut ve'kamut: dilemot be'bniyat ha'koah ha'tzva [Quality and Quantity: Dilemmas of Building a Military Force]* (Tel Aviv: Ma'arachot, 1985).

186. Adamsky, *The Culture of Military Innovation*, 126; Cohen, Eisenstadt and Bacevich, *Knives, Tanks & Missiles*, 44.

187. Landau and Bermant, "Iron Dome Protection."

188. Yehuda Ben-Meir, "Operation Cast Lead: Political Dimensions and Public Opinion," *Strategic Assessment* 11, no. 4 (2009): 29–34.

189. Hanan Grinberg, "Ha'til she'kofef et egrof ha'plada [The Missile that Bended the Iron Fist]," *Ynet*, http://www.ynet.co.il/articles/0,7340,L-3297389,00.html [accessed May 10, 2018].

190. Yitzhak Ben-Israel, *Milchemet tilim Harishona: Israel—Hizballa (kaitz 2006) [The First Missile War: Israel—Hezbollah (2006)]* (Tel Aviv: Tel Aviv University Press, 2007), 37.

191. Amos Dotan, "Technologiyat meil' ruach: yoter mi'mahapecha technologit be'lechimat shir'yon [The Technology of Meil' Ruach : More than a Revolution in Armour Warfare]," *Ma'arachot* 448 (2013): 68–74.

192. Moshe Kaplinsky, "The IDF in the Years before the Second Lebanon War," *Military and Strategic Affairs* 1, no. 2 (2009): 25–37; Grinberg, "Ha'til she'kofef [The Missile that Bended]"; Ben-Israel, *Milchemet tilim Harishona [The First Missile War]*.

193. Avi Kober, "What Happened to Israeli Military Thought?," *Journal of Strategic Studies* 34, no. 5 (2011): 724.

194. Avi Kober, "The IDF in the Second Lebanon War: Why the Poor Performance?," *Journal of Strategic Studies* 31, no. 1: 3–40.

195. Cohen, Eisenstadt, and Bacevich, *Knives, Tanks & Missiles*, 125.

196. Yitzhak Ben-Israel, "Torat ha'yahasut shel bniyat ha'koah [The Theory of Relativity of the Military Build-up]," *Ma'arachot* 352–53 (1997): 33–42.

197. Shai Levy, "Ason ha'nagmashim: lama tzal meshtamesh be'kli lo memugan mi-1971? [The Disaster of APCs: Why does the IDF uses vehicles with no armour from 1971?]," *Mako*, July 22, 2014, http://mobile.mako.co.il/pzm-israel-wars/operation-protective-edge/Article-2fab4a8f83e5741006.htm [accessed May 10, 2018].

198. The IDF General Officer Commanding Army Headquarters, *Puma*, http://mazi.idf.il/4986-5648-HE/IGF.aspx [accessed May 10, 2018], (Hebrew).

199. Adamsky, *The Culture of Military Innovation*, 111.

200. Cohen, Eisenstadt, and Bacevich, *Knives, Tanks & Missiles*, 139.

201. Lazar Berman, "Capturing Contemporary Innovation: Studying IDF Innovation against Hama and Hezbollah," *Journal of Strategic Studies* 35, no.1 (2012): 121–47.

202. Cohen, Eisenstadt, and Bacevich, *Knives, Tanks & Missiles*, 76.

203. See Adamsky, *The Culture of Military Innovation*; Kober, "What Happened to Israeli Military Thought?"; Avi Kober, "The Rise and Fall of Israeli Operational Art, 1948–2008," in John Olsen and Martin Creveld Van (ed.), *The Evolution of Operational Art: From Napoleon to the Present* (Oxford, Oxford University Press, 2010).

204. Walzer, "The Ethics of Warfare," 633.

205. Adamsky, *The Culture of Military Innovation*, 121.

206. Avi Kober, "The Intellectual and Modern Focus in Israeli Military Thinking as Reflected in *Ma'arachot* Articles, 1948–2000," *Armed Forces & Society* 30, no. 2 (2003): 145.

207. Adamsky, *The Culture of Military Innovation*, 117–19.

208. Kober, "The IDF in the Second Lebanon War."

209. State of Israel, *The Full Report of the Commission of Inquiry into the Events of Military Engagement in Lebanon 2006*, January 2008 (Hebrew), 425.

210. Amihud Shachar, "Ha'baya eina be'tzal [The Problem is Not in the IDF]," *Ma'arachot* 380–381 (2001): 88–89.

211. Adamsky, *The Culture of Military Innovation*, 122.

212. As for 2015, the list of articles that directly discuss NLW includes: Yitzhak Ben-Israel, "Neshel al-heger [Non Lethal Weapon]," *Ma'arachot* 363 (1999): 20–24; Ben-Horin, "Non-Lethal Weapons"; Shtultz, "Haf'alat koah medureget [The Employment of Graduate Force]"; Ben-Shalom, "'Faser lematzav ham'em' ['Faser Switched to Shock']." Also, there are several articles that loosely discuss NLW in the context of riot control: Yonatan Sher, "Hafsad lo ba' be'hashbon [A Riot Is Not an Option]," *Be'yabasha*, no. 24 (2010); Yehuda Albek, "Pizur hafganot seder: ha'etgar ve'hama'ane [Riot Control: The Challenge and the Solution]," *Ma'arachot* 449 (2013): 24–29.

213. Shtultz, "Haf'alat koah medureget [The Employment of Graduate Force]," 17.

214. Tamir Mantzur-Karmel, "Likrat hel hagana mirchavit [Towards a Territorial Defence Force]," *Ma'arachot* 459 (2015): 20.

215. Lieutenant Colonel Y. and Colonel Y., "Mischakei milchama memuchshavim be tzal [Computerised War-games in the IDF]," *Ma'arachot* 343 (1995): 29.

216. For example, State of Israel, *The Full Report of the Commission of Inquiry*, 550–54; Kober, "The IDF in the Second Lebanon War."

217. Oren Abman, "Ha'ma'avar le'lehima m'shigra ve'mi'batash [The Transition to Fighting from Routine and from Batash]," *Ma'arachot* 441 (2012): 21.

218. Harkov, "Former British Commander."

219. Rafael Rodnik, "Haderch le'hasagat legitimatziya [The Way to Achieve a Legitimacy]," *Ma'arachot* 449 (2013): 9–15.

220. Raz Sagi, "Mi'tzava ha'hagana le'tzava ha'hagana ha'mishpatit [From the Defence Force to the Legal Defence Force]," *Ma'arachot* 462, (2015): 58.

221. Cohen and Cohen, "Israel and International Humanitarian Law"; Noiman, "Lohamat mishpat."

222. Aviram Zino, "Ha'mishor ha'mishpati: keitzad ne'erechet Israel le'yom she'le'achar ha'mivtza be'haza [The Legal Dimension: How Israel Prepares Itself for the Day after the Operation in Gaza]," *MaarivOnline*, August 9, 2014, http://www.maariv.co.il/news/new.aspx?pn6Vq=E&0r9VQ=FMMFL [accessed May 10, 2018].

223. Noam Vitman, "4 shanim le'oferet yatzuka: ktzinei ha'uchlusiya hafchu le'helek integrali mi'hakoach ha'lochem [4 Years after the Cast Lead: Population-during-Warfare Officers Have Become an Integral Part of the Fighting Forces]," *IDF Website*, http://www.idf.il/1133-17987-he/Dover.aspx [access: 10 May 2018].

224. Ibid.

225. Gabi Grinfeld, "Lean halchu ha'lekhachim kulam? [Where Did All Lessons Go To?]," *Ma'aruchot* 380–381 (2001): 83; Shachar, "Ha'baya eina be'tzal," 89; Yaakov Tzur, "Haim hiskil tzal lehafik et lekhei milhamotav ve'le'yasmam? [Whether Did the IDF Succeed to Generate the Lessons from Its Wars and Learn from Them?]," *Ma'arachot* 378–379: 12–19.

226. Eduto shel' nitzav Alik Ron [The Testimony of Major General Alik Ron].

227. Rapaport, "Metach gavoha [High Tension]."

228. Vitman, "4 shanim le'oferet yatzuka [4 Years after the Cast Lead]."

229. Harkabi, *The Bar Kokhba Syndrome*, 105.

230. The Israel Democracy Institute, *The Peace Index: October 2015*, October 5, 2015, http://www.peaceindex.org/indexMonthEng.aspx?num=298 [accessed May 10, 2018].

231. The Israel Democracy Institute, *The Peace Index: August 2014*, August 12, 2014, http://www.peaceindex.org/indexMonthEng.aspx?num=283&monthname=August#.VkIiwdLhDcs [accessed May 10, 2018].

232. Cohen and Cohen, "Israel and International Humanitarian Law," 9, 15.

233. Sagi, "Mi'tzava ha'hagana le'tzava ha'hagana ha'mishpatit [From the Defence Force to the Legal Defence Force]"; Rodnik, "Haderch le'hasagat legitimatziya [The Way to Achieve a Legitimacy]."

234. Ben-Shalom, "'Faser lematzav ham'em' ['Faser switched to Shock']."

Conclusion

Enemy Civilian Casualties: Do We Really Care?

This book addressed the puzzling story of military employment of NLW in the United States, Russia, and Israel in an attempt to understand why the RMA of NLW did not occur. While American, Russian, and Israeli military organizations have approached NLW in different ways, it seems that minimizing enemy civilian casualties has never been defined by their political leaders as a significant challenge demanding a fundamental change in the conduct of war. Consequently, this conclusion will be divided into two main parts. First, to establish why, contrary to expectations, there was no RMA of NLW, it is important to resolve three main questions regarding the American, Russian, and Israeli cases:

a. to what extent does international political pressure to minimize enemy civilian casualties influence the political-military decision-making processes?
b. Why is there no domestic political pressure to minimize enemy civilian casualties?
c. If the challenge of minimizing civilian casualties has never been handed to military organizations, then why is there still an interest in NLW?

Secondly, in an attempt to take these conclusions beyond the American-Russian-Israeli case studies and produce a broader argument, it seems important to resolve an additional question: What does the absence of the RMA of NLW teach about the decline in violence in the twenty-first century?

INTERNATIONAL CRITICISM AND ENEMY CIVILIAN CASUALTIES

Causing civilian casualties during military operations has become a much-politicized topic in international relations since World War II. While countries, as well as non-state actors, use this argument to criticize their political adversaries in the international arena, the American, Russian, and Israeli governments are well prepared and politically equipped to fend off this criticism. Interestingly enough, each one of them has created its own foreign policy mechanisms to withstand international pressure, so that they feel relatively free to employ their military forces regardless of the extensive collateral damage and enemy civilian casualties. Moreover, it seems that their success in countering international criticism is rooted much more in their general political, economic, and military power than in their attempts to minimize the undesired outcome of their military activities. Consequently, it is not surprising that American politicians are much more successful in minimizing the outcome of international criticism than their Russian counterparts, and that the Israeli foreign policy is the most vulnerable of the three.

While Russians and Israelis are struggling to fend off the international criticism of their military actions, Washington adopts an entirely different strategy based on the prevention of the criticism, rather than on the reaction to it. This strategy of prevention is based on two main narratives of America's foreign policy in terms of its military operations. The first one is based on the Cold War heritage that bestows on the United States the position of *the great protector of the free world* against *a great evil.* As long as Washington enjoys this status (and American leadership puts a great effort in its preservation), U.S. military activity would be relatively free from international criticism, regardless of the undesired collateral damage and civilian casualties. When "small massacres" (or "big mistakes") committed by the U.S. military create an international outrage, such as in the case of the Kunduz hospital airstrike on October 3, 2015,[1] America "expresses its condolences" stating that it "goes to great lengths to limit the loss of life."[2] Moreover, as the history of the past few decades illustrates, the international outrage fades away very quickly and these events have very little (if any) effect on the Washington political-military decision-making process; after all, the U.S. military *protects the free world* against *the great evil.* The second mechanism of U.S. foreign policy to prevent international criticism is the American tradition of creating the "coalitions of willing." Based on its political and economic power, Washington builds on the self-interests of other states, "recruiting" them to participate in U.S.-led military activities. On the one hand, this practice helps Washington to preserve its freedom of decision-making, as in most of the cases the actual military contribution of the coalition's members is very limited. On the other, the fact that many states raise their flags in support of U.S. military operations (i.e., provide political support) prevents them from openly criticizing American actions, simultaneously making it more difficult for third states to criticize the whole coalition.

Unlike Washington, Moscow bases its political freedom for military actions on a much more defensive strategy. While the United States prevents international criticism by maintaining the "good" intentions of its military activity, Russia fends the criticism off by claiming that this criticism represents "double standards." During the last decade and a half the Kremlin has invested much effort rehabilitating not only its economic and military power, which were lost after the collapse of the USSR, but also its role in the international arena. Building on its status as the biggest nuclear power, a permanent member of the UN Security Council, its rehabilitated economy, and unprecedented domestic support, the Kremlin feels free to disdain international criticism as a political manipulation that is intended, according to President Putin, "to accuse Russia of growing ambitions . . . as though those who say that, have no ambitions [of their own]."[3] When Moscow decides to employ its military, it feels few constraints concerning possible undesired collateral damage and civilian casualties, at least as far as the levels of damage and lethality caused by Western operations would not be minimized significantly. Interestingly enough, the American techno-utopia in military affairs led Washington to base its criticism of the Russian intervention in the Syrian conflict in 2015 on Russian technology, rather than on the outcome of Russian activity. Thus, Americans claim that, while Western attacks "are the most precise in the history of warfare" and so "the amount of care that we [the U.S.] have taken to preserve civilian life, to preserve civilian infrastructure, is unprecedented"; Russian's munitions "lag so far behind those of Western militaries" that they "inadvertently kill innocent men, women and children."[4] It did not take long for the Kremlin to answer this narrowly technical criticism. Just five days later, it launched 26 guided missiles from a distance of 1500 km,[5] signalling to the U.S. not only the technological sophistication of Russian military, but also the fact that Moscow uses the same technology as the U.S. and, therefore, 'takes care' to avert civilian casualties at least on the same level as the U.S. does. In other words, Russian political leadership is powerful and confident enough to shrug off international criticism of its military activities and their undesired outcomes.

In comparison to the powerful positions of America and Russia in the international arena, which allow them to pay very little regard to the criticism of their military actions, Israel seems to be the most sensitive and vulnerable. To deal with international criticism, Israeli foreign policy combines both: the Russian defensive approach and the American strategy of prevention. On the one hand, Israeli foreign policy, shaped by the narrative of *Um Shmum* and the traditional reliance on American protection, easily cold-shoulders international criticism. Taking into consideration the scale of the criticism produced by international organizations, Israeli politicians seem to be much more effective in ignoring it compared with their Russian counterparts. The tendency of Israelis, as already observed in 1971 by Sir Ernest John Barnes, "to air their views so loudly and so frequently on anything that happens . . . [and] blow [the UN resolutions] up into international tragedies"[6] seems to be a useful strategy, because, after all, Israel is not as isolated as it is in its self-image.[7] When the Kremlin tried to fend off the international outrage regarding the Russian seizure of Crimea in 2014, it had to deal with economic sanctions detrimental to the Russian economy. In contrast, Israel has been, so far, relatively successful in withstanding the international criticism of its seizure of the West Bank—despite the fact that the European Union, one of the biggest critics of Israel, is also Israel's biggest trading partner.[8]

On the other hand, considering Israeli foreign policy in the context of international criticism, it is also important to point out the preventive tactics intended to minimize international outrage. While Washington "recruits" other states to support its actions by American political and economic power, Israel does it by self-restraint. To open a window for international political support for its actions, Israeli political leadership tolerates enemy attacks to an unprecedented scale, maneuvering between the danger of being accused of indecisiveness by the Israeli public and the attempt to minimize international accusations that Israel is bellicose.

In conclusion, it seems right to argue that international criticism of the American, Russian and Israeli military activities is not as effective in minimizing collateral damage and civilian casualties as one might think. The political leadership of these countries have very little reason to press their militaries to perform fundamental changes in their weaponry or concepts of operations because of enemy civilian casualties. It is important to highlight, however, that the general ineffectiveness of international criticism is rooted not only in successful foreign policy mechanisms developed by these countries to withstand this criticism, but also in the general framework of international humanitarian law, in context of which this criticism is articulated. And while the first was already intensively examined throughout this book, the latter will be discussed in more details in following sections of this chapter.

DOMESTIC POLITICS AND ENEMY CIVILIAN CASUALTIES

As was discussed throughout this book, each of the examined countries has experienced significant domestic political pressure to reduce the casualties among their own military personnel, e.g., the United States in Vietnam, Russia in Afghanistan and then in Chechnya, and Israel in the Second Lebanon War. While the presence of these domestic political pressures is undeniable, it seems that the societies of these countries have expressed much less (if any) political outrage when it comes to enemy civilian casualties. To understand this absence of domestic political pressure, each chapter addressed the cultural-historical narratives of each national character, finding that while the American, Russian and Israeli national heritages are very different, all of them show very little concern about the enemy, in general, and enemy civilian casualties, in particular. Throughout the analysis of the American, Russian and Israeli national characters, the examination focused on their respective position on the individualistic-versus-collectivistic continuum. While the American mind-set is characterized by its extreme individualism and the Russian one by its predisposition toward collectivism, the Israeli combination of these two places the Israeli character somewhere in the middle. Interestingly enough, all three have developed a strong differentiation between "us" and "others" that prevents them from projecting their appreciation of their own lives onto the lives of their enemies.

American, Russian and Israeli national characters are, undoubtedly, products of the unique historical experience of each one of these nations. Most interesting, however, is the fact that despite their differences, all three pay very little regard to foreigners' lives. The American mind-set, shaped by the violent experience of the frontier, perceives war as an act of vengeance or as a rightful act intended to "free" or "protect" local populations. Throughout their history, Americans considered wars as temporary and unfortunate events that disturb their "American Dream" and, therefore, have to be concluded as fast and decisively as possible. The combination of this approach toward war with the traditional inclination toward violence has left very little time and place for valuing enemy civilian populations. After the Vietnam War, Americans have become very sensitive to their own casualties; however, as their latest experiences in Iraq and Afghanistan proved, American society generally failed to project its rising appreciation of human life onto the enemy civilian population. From the early violent clashes on the frontier to this day and age, wars have been interpreted by the American mind-set as a fight of "brave American soldiers" against a "brutish enemy," who understands only violence and, therefore, has to be treated accordingly.

Unlike the American mind-set, the Russian national character perceives war as an integral part of the human condition. Russian history, rich with conflicts and suffering, cultivated the narratives of permanent struggle and readiness to sacrifice for the common good. Bred on the heritage of the Patriotic Wars, Russians perceive military operations as a matter of survival, rather than a way to make their lives better or safer. This narrative is so important in the Russian cultural context that leaders who failed to address it paid an enormous political price—Nicolas II during World War I, Mikhail Gorbachev during the war in Afghanistan, and Boris Yeltsin during the First Chechen campaign. It is not surprising then that President Putin, having learned these lessons, described Russian intervention in Syria as a matter of Russia's survival, because:

> More than 2,000 fighters from Russia and ex-Soviet Republics are in the territory of Syria [fighting for ISIS]. There is a threat of their return to us. So, instead of waiting for their return, we are better off helping Assad fight them on Syrian territory.[9]

When a war is a matter of survival, when people are ready to suffer to protect their homes and families, when they are led by a strong leader, Russian people demonstrate unprecedented levels of endurance and sacrifice. However, while this readiness to endure saved Russians from Napoleon and Hitler, and had helped them to survive extreme totalitarian regimes, it also shaped their minimal concern for enemy civilian casualties. In other words, the casualties of an enemy population during military operations are generally tolerated in the Russian cultural context, because a war, as Major General Alexander Vladimirov put it, "is an integral part of human existence,"[10] and Russians are ready to sacrifice themselves to win this war.

Interestingly enough, it seems right to argue that the Israeli understanding of the phenomenon of war combines, to a certain degree, the Russian traditional perception of permanent struggle with the American fast-cut and aggressive approach to war. On the one hand, the Israeli character, shaped by a combination of the political-military reality of the Middle East and the national cultivation of the myth of Masada, is predisposed to believe (similar to Russian culture that perceives Russia as a besieged fortress) that the whole world is against Israel. The very simple fact that Masada's defenders were arguably more like bandits and outlaws (who preferred to kill themselves, rather than be killed by the Romans) than "brave defenders" (who sacrificed themselves instead of being enslaved), is perceived by contemporary Israeli culture as an all-too detailed interpretation intended just for historians, rather than for schools.[11] The heritage of the long Jewish history of persecution shaped not only the Israeli cultural existential anxiety, but also the very clear distinction between Israel and everybody else, who ultimately must have an—either open or hidden—anti-Semitic agenda. Consequently, as Israelis see it, Israel must permanently and continuously defend itself against Palestinians, Arab countries, anti-Semitic Western countries, and the whole world, as Israel is the only safe place for Jews.

On the other hand, the economic and political constraints shaped the Israeli approach to war as something that has to be carried out as fast and decisively as possible. Having neither resources, nor international political approval for prolonged wars, the Israeli mind-set is predisposed toward aggressive peaks of fighting that will only provide relative peace for a certain period, rather than an overwhelming victory. As the Jews' struggle is perceived as an eternal phenomenon, and conclusive victory as an unachievable outcome, Israelis fight not to defeat their enemies irrefutably, but to deter them from future hostilities for a certain period. After each war, Israelis ask themselves: "Will deterrence be achieved and will there be quiet? For how long?"[12] If the meaning of victory is, as Amos Yadlin put it, "postponing the next conflict for as long as possible by depriving the enemy of capabilities and influencing its intentions,"[13] then the conduct of war has to be fast-cut and aggressive, and this approach leaves little space for the appreciation of enemy civilian lives. While Americans disregard the lives of their enemies because of American traditional imprudence and inclination toward violence, and the Russians because of their collectivistic characteristic that tends to keep societies as responsible for the action of individuals, Israelis combine both. Though American, Russian, and Israeli political leadership have been pressed by their people to minimize the casualties of their own military personnel, a similar pressure regarding enemy civilian casualties simply does not exist. And, while under domestic political pressure to minimize friendly casualties, politicians forced their militaries to develop and employ technologies accordingly, there has been no political reason to invest in the development of NLW.

To conclude the discussion about domestic politics and the absence of the RMA of NLW, it is important to discuss an additional factor that was not covered by this book, but is identified by the existing literature as influential in the domestic politics of military armament: the military-industrial complex. Since the end of World War II, and especially after President Eisenhower's famous farewell address in 1961, which warned the American people against an extension of the influence of the U.S. military-industrial complex on political-military decision-making,[14] scholars have continuously pointed to the significant role of the military-industrial complex in promoting novel military technologies.[15] While it is difficult to disagree with the fact that military organizations employ certain novel technologies due to the powerful political lobby of the military-industrial complex, this is not the case for NLW. As it was discussed throughout different chapters, American, Russian, and Israeli military-industrial complexes are able to produce effective NLW when they are specifically tasked by their military. Without this specific demand, however, it seems that big manufacturers are not very interested in promoting NLW, because, as one Israeli defense analyst put it:

> The defense industries have no economic motivation to develop enhanced NLW. On the one hand, the headache is enormous (for example, lawyers can spend months discussing whether the system answers the legal definition of non-lethality). On the other, the profits are diminutive. One lethal and sophisticated system, which after being sold to the IDF can also be sold to the rest of the world, will bring more profits than all NLW together.[16]

Moreover, as big industries are not interested in NLW, most of the American, Russian, and Israeli (and, in fact, all around the world) companies that produce them are small and medium sized manufacturers[17] that are short of the required economic and political power to support an effective lobbying for NLW. In other words, American, Russian, and Israeli political-military decision-makers do not press their military organizations to employ NLW, not only because domestic public opinion tolerates enemy civilian casualties, but also because their military-industrial complexes show very little interest in promoting these weapons.

MILITARY INTEREST IN NLW

The analysis of different political and cultural factors depicted the indifference of the American, Russian, and Israeli political leadership towards enemy civilian casualties and the subsequent lack of political pressure to employ less lethal armaments. The question is, however, why the military organizations of these countries still express an interest in NLW. Interestingly enough, despite the fact that each military has its own reasons to believe in the potential usefulness of NLW, their interests in these weapons are highly interconnected.

As was discussed in chapter 2, the idea of the military employment of NLW is the fruit of American military thought. After the end of the Cold War, the American military was struggling to find *a new great evil* to fight against. The political atmosphere in the United States at the beginning of the 1990s was best described by Francis Fukuyama's *The End of History and the Last Man*, where he argued for the unchallengeable triumph of liberal democracy that would eventually end the bloody wars of the twentieth century.[18] It is not surprising then that the U.S. military assessed the upcoming twenty-first century as "a century of shared understanding through information technology—a new era of relative peace"[19] and, therefore, it was assumed that the main future role of U.S. military forces would be to police the *Pax Americana*. While this understanding dominated and drove the American military attraction to NLW during the 1990s, the 9/11 attacks and the following declaration of a global War on Terror propelled the American military into interventions in Afghanistan and Iraq, which were anything but peacekeeping activities, as it had been assumed during the 1990s. On the one hand, without political support and no real military necessity, the interest in NLW ultimately dwindled. From 1997 to 2009, the overall investment in the research and development of NLW in the U.S. military was no more than \$386 million[20]—a diminutive fraction of the U.S. defense budget that did not leave a chance for the development of the revolutionary technologies promised during the 1990s. On the other, two different cultural factors prevented the American military from abandoning the idea of NLW completely. The first one is the American traditional techno-utopia in military affairs, and the fact that NLW provide a technological solution for something that the U.S. military had defined as a problem. The fact that this problem has not been seen as significant enough does not mean that the interest in the potential of this technology should be abandoned. The second factor is the inability of American military thought, shaped by the traditional rivalry between different services, to produce a coherent understanding of the strategic environment and future technologies that answer future military challenges. While different services within the American military are incapable of reaching an agreement on the future of NLW, they also are unable to disregard them completely.

In contrast to the American military, Russians have never considered technology as a "magic bullet" that solves problems on the battlefield. Moreover, the Russian military-industrial complex has traditionally been suffering from technological inferiority in comparison to the West, and rarely produced revolutionary novel military technologies. Consequently, Russian observation of Western technologies in general, and American in particular, has always been one of the main drivers behind not only Russian defense production, but also Russian military thought. While, as discussed in chapter 3, Russian military thinkers succeeded in developing more comprehensive concepts for possible employment of NLW, it was the American fascination with this technology that drove their interest. Moreover, as Western technological achievements frequently force Russian political-military decision-makers to urge their industries to develop similar capabilities, the American failure to develop operationally effective NLW reduced Russian interest and enthusiasm to press for further development and employment. Interestingly enough, it has not stopped Russian military thinkers (who have never been deterred by the gap between theory and practice) to continue developing concepts and theories for the possible employment of NLW. While American interest in NLW has been kept alive by the American traditional fascination with technologies and inability to coherently conceptualize their future implementations, Russians are still interested in NLW because of their traditional inclination toward abstract thinking and their belief that solutions lie not in technology, but rather in a sophisticated conceptual military theory.

As it was discussed in chapter 4, it seems right to argue that the IDF has more operational experience with NLW than both the American and Russian militaries together. Due to the routine deployment of military units in the Territories as law enforcement forces, the Israeli military has not only mastered typical NLW as tools for riot control, but also introduced several very innovative technologies. The success of NLW during *batash* missions, however, has also been a vital factor that denied the transition of NLW to large-scale military operations; as in the Israeli military tenet, what is good for *batash* is not suitable for war. The IDF's cultural predisposition to admire "doers" rather than "thinkers" explains not only the absence of a broader conceptual approach toward NLW, but also the technological setback between the peaks of civilian unrest in the Territories—the technologies of the First Intifada were forgotten before the Second, and technologies developed during the Second were abandoned after its end. While the IDF's interest in NLW has been driven solely due to the requirements of the *batash* realities, it is important to highlight the Israelis' increasingly American-style quest for technological solutions to military challenges. The Israeli military, lacking its own resources for technological adventures that are not supported by a clear military necessity, closely observes American technological achievements and, as the history of the latest information-technology RMA of precision-guided munitions teaches, Israelis are able to employ novel technologies developed in the United States, without having a conceptual framework for using it.[21] In other words, the IDF does not employ NLW during its military operations not only because the conceptual denial of NLW as *batash* weaponry, but also because the American military has not introduced NLW that are operationally employable on the battlefield.

The main reason why the American, Russian, and Israeli military organization still discuss the potential of NLW, despite the noticeable absence of political pressure to minimize enemy civilian casualties to a technologically possible minimum, is American unfading interest in these weapons. Since World War II, the U.S. military has been a technological leader in military affairs and NLW are not an exception. As long as Americans will show interest in NLW, it will inspire Russians to develop theories for their possible employment and Israelis to look at American manufacturers for possible acquisitions.

THE ABSENCE OF THE RMA OF NLW AND THE "DECLINING VIOLENCE" OF HUMAN NATURE

As it was demonstrated throughout this book, the problem of harming enemy civilian populations during military operations is an inherent part of the rhetoric of the American, Russian, and Israeli political leaderships. The need to avert enemy civilian casualties has indeed been defined as a military challenge that has to be met; it seems, however, that the tangible reactions to this challenge are not as significant as it is presented by the politicians' rhetoric. As the analyses of the American, Russian, and Israeli cases reveal, the political leaderships of these countries have very little external or domestic reasons to meet this challenge through fundamental revolutionary change in their conduct of war. In other words, the sociocultural and political contexts that have influenced the American, Russian, and Israeli political-military leadership in their decision-making processes regarding enemy civilian population, have necessitated giving *the impression* of minimizing enemy civilian casualties, rather than their *actual* minimization.

While the challenge of minimizing civilian casualties has been identified and defined as such, military organizations see it as of secondary importance. As the political leadership has not called for fundamental changes in the conduct of war, the military has been satisfied by the mere adaptation of existing capabilities accompanied by quite modest innovativeness. Due to their different cultural predispositions, the American, Russian, and Israeli military establishments have been engaging with the challenge of averting enemy civilian casualties differently—Americans by emphasizing the "humane" role of precision weapons; Russians by conceptualizing intangible military theories intended to minimize enemy civilian casualties; and the Israelis by emphasizing the "lawfare" and increased compliance with international humanitarian law as a way to avert collateral damage and unnecessary civilian casualties. In other words, none of the analyzed military organizations have defined *the minimization of enemy civilian casualties* as an unmet military challenge that requires a revolution in the form of innovative technological solutions combined with novel operational and organizational concepts. The RMA of NLW in the United States, Russia, and Israel has (so far) failed to materialize because the political leadership of these countries have asked their militaries to reduce the casualties to a politically acceptable level, rather than to the technologically possible minimum; in the political-sociocultural context of each country, support for the former appears to be significantly higher than for the latter.

The contemporary research that discusses the role and the place of violence in twenty-first century society can be generally divided into two schools of thought, and while both agree that violence is declining, their explanation of this trend is different. The most notable advocate of the first school is psychologist Steven Pinker. In his book *The Better Angels of Our Nature* he suggests that "today we may be living in the most peaceable era in our species' existence" because contemporary society has become more empathic, self-controlling, moral, and reasonable and has developed "a commitment that other living things, no matter how distant or dissimilar, be safe from harm and exploitation."[22] While Pinker generally argues that contemporary societies show more respect for human life than they used to in the past and, therefore, inter-human violence is declining, some scholars suggest an alternative explanation for this phenomenon. The main supporter of this second school of scholars is the U.S.-based British classicist and archaeologist Ian Morris. While in his *War—What Is It Good For?* he agrees with Pinker's observation on the trend, stating that "today's world [is] safer and richer than ever before,"[23] his explanation is entirely different. In contrast to Pinker's "tale of six trends, five inner demons, four better angels, and five historical forces,"[24] which was shaped by his background in psychology, Morris's analytical examination of "the whole ten-thousand-year-long story of war since the end of the last ice age" led him to conclude that, in fact, it is "war [that] has made the world safer" as "war made governments, and governments made peace."[25] While the failure of the RMA of NLW has been shaped by general public tolerance of enemy civilian casualties during military operations, it seems important to test these two explanations for the "declining violence" in twenty-first century in the light of the present research findings.

Pinker's main argument is that different historical social circumstances have forced humans to "reframe violence as a problem to be solved rather than a contest to be won." According to him, there are five historical developments that have led to this conclusion: (1) the development of a stronger state and judiciary with a monopoly on the legitimate use of force; (2) the international commerce that allows the exchange of goods and ideas over longer distances, making other people become more valuable alive than dead; (3) the feminization process that has empowered women and decreased cultural inclination toward violence that largely characterizes a male pastime; (4) the globalization and cosmopolitanism that have prompted societies to understand other people and embrace them, rather than fight them; (5) the increasing reasonability in human affairs that have been forcing people to recognize the futility of violence. These five historical developments, according to Pinker, have led people to favor their internal peaceable motives—"Better Angels"—Empathy, Self-Control, Moral Sense, and Reason, and to reject their inner drives toward aggression and violence—"Inner Demons"—Instrumental Violence, Dominance, Revenge, Sadism, and Ideology.[26] While it is not the aim of this book to directly criticize Pinker's theory, the American, Russian, and Israeli attitudes towards enemy civilian casualties, analyzed throughout this book, offers a much more selfish perception of human life that is based more on revenge and dominance, than on universal empathy, moral sense, and reason—and the resultant failure of military organizations to develop and employ weapons that would minimize unnecessary violence (i.e., NLW) is clear evidence of that. It is difficult to disagree with Pinker on the fact that the traceable decreasing violence within Western societies has been caused by "Better Angels," however, when it comes to the contemporary ways of war, it seems right to suggest that the only "angel" that restrains the violence is international humanitarian law. Though one may argue that this "angel" is a byproduct of human internal peaceable motives and liberal values, which have forced the international community to restrict the level of violence during military clashes, Chris Jochnick and Roger Normand, who critically analyzed the history of the laws of war, claim the opposite, stating that:

> [The] Examination of the historical development of these laws reveals that despite noble rhetoric to the contrary, the laws of war have been formulated deliberately to privilege military necessity at the cost of humanitarian values. As a result, the laws of war have facilitated rather than restrained wartime violence. Through law, violence has been legitimated.[27]

As it was discussed in chapter 1, international humanitarian law is a double-sided sword that intends to restrict the violence on the one hand, but on the other, create situations when, as Donald Rumsfeld stated, "our forces are allowed to shoot somebody and kill them, but they're not allowed to use a non-lethal riot control agent [i.e., NLW]."[28] International humanitarian law is a much-politicized topic that facilitates harming enemy civilians, rather than asks for their minimization. And NLW are not an exception, as, for example, analyzing the regulatory history of small arms munitions, Paul Cornish stated that:

> When a society feels under threat, from war, terrorism or violent crime, almost any effective weapon is likely to be seen as legitimate—regardless of the damage it can cause to the human body.[29]

His analysis points to the politicization of international arms control regimes and the cynical exploitation of international institutions for political reasons. His best example is the attempt by Swedish experts in 1974 to ban the American 5.56 mm cartridge under the auspices of the Red Cross due to "humanitarian concerns" and promote their own 7.62 mm caliber, which, in fact, was capable of delivering even worse wounds.[30] In other words, in his quite technical analysis, he points out the fact that the international legal system is controlled more by political self-interests covered by the rhetoric of "humanitarian concerns," rather by our "Better Angels"—moral sense and reason. Following Pinker's terminology, the RMA of NLW failed because, despite the fact that "Better Angels" overcame "Inner Demons" within internal human affairs of Western societies, these societies failed to do so in their relation with their enemies—and this, eventually, quite weakens the very argument of *The Better Angels of Our Nature*.

Unlike Pinker, who structures his arguments on the increasing morality of human beings, Ian Morris tries to produce a more pragmatic explanation of the "declining violence" in human affairs. While emphasizing only one of Pinker's historical social developments—a stronger state and judiciary with a monopoly on legitimate use of force—Morris argues that throughout history, violence has always been declining in big societies (ancient empires, early modern empires, or the contemporary *Pax Americana*) because:

> The only way to make these large societies work was for their rulers to develop stronger governments, and one of the first things these governments had to

> do, if they wanted to stay in power, was suppress violence within the society.[31]

In other words, it is not that humans have become "better souls" at the end of the twentieth century, but they repeatedly had been forced to disown violence by their own collective quest for a wealthier life. As Morris argues: "the men who ran these governments hardly ever perused policies of peacemaking purely out of the goodness of their hearts. They cracked down on killing because well-behaved subjects were easier to govern and tax." Analysing human history, Morris concludes that "governments have made us safer and the war is pretty much the only way we have discovered to make governments," and as "the winners of wars incorporate the loser into larger societies . . . with the passage of time—maybe decades, maybe centuries—the creation of bigger society tends to make everyone, the descendant of victors and vanquished alike, better off." In other words, wars create bigger societies, and bigger societies create peace within these societies. In his analysis of human history, Morris distinguished between two types of war. While the first one—"the productive way of war"—when a winner successfully incorporates the loser, creating bigger societies and making life safer and wealthier; the second one—"the counterproductive way of war"—when a stronger actor fails to incorporate weaker rivals, ultimately breaks down large societies, impoverishing people and making their lives more dangerous. He claims that after the "counterproductive" wars ruined ancient empires and engendered the dark ages, early modern Europe reinvented the "productive" way of war, discovering that:

> The wealth of nations could be increased most not by plundering or even taxing downtrodden subjects, but by using state power to make as many people, as possible, as free, as possible to trade in bigger and bigger markets.[32]

The declining violence and relative peace within the Western Hemisphere (in its political, rather than geographical, borders), according to Morris, is not an outcome of humans' better nature, but a result of 500 years of "productive" wars that were fought between the nations that inhabit this global, yet in many aspects limited, society. Interestingly enough, while Morris comes to a paradoxical conclusion that it is the process of waging ("productive") wars that make life safer, he does not claim that these wars must be accompanied by high levels of violence. In other words, according Morris, the most important characteristic of war is its "productiveness" (i.e., the inclusion of the loser's society into the society of the winner) and, as long as a war has "productive" ends, any level of violence (the highest possible or the lowest required) is acceptable. In the final chapter of his book, Morris analyzes the latest American global War on Terror, suggesting that the way the West has been fighting this war is, in fact, counterproductive, as the West has failed to incorporate rival societies (i.e., Afghani and Iraqi). Fighting terror, according to Morris, is reminiscent of Roman Emperor Domitian's dilemma, who in 85 A.D. drew the radical conclusion that "Rome no longer had much to gain from productive war."[33]

Morris states that "we kill because the grim logic of the game of death rewards it . . . [and] that is why we cannot just decide to end the war."[34] However, what if "killing" itself make wars "counterproductive"? After all, the "productive" aim of war in the twenty-first century is to make as many people, as possible, as free, as possible, winning their hearts and minds, rather than killing them. The proposition that "killing" makes wars "counterproductive" turns Morris's paradoxical conclusion that *wars create peace* even more puzzling, as it contradicts not only the logical perception of the consequences of war (i.e., destruction), but also shakes the very understanding of war as peak of human violence. While this seemingly controversial argument will be discussed in more detail in the last part of this chapter, it seems right to conclude that following Morris's theory, the RMA of NLW failed because 500 years of "productive" wars, which have made the contemporary world the safest since the Stone Age, were also always accompanied by extreme violence. The long-term history suggests that "the logic of the game of death" has been rewarding; however, as Morris himself claims: "all the running we have done in the last ten thousand years has transformed our society, changing the payoffs in the game."[35] Unfortunately, as discussed throughout this book, the social requirement to minimize enemy civilian casualties during this game has not been one of these transformations.

To conclude this discussion on the absence of the RMA of NLW and the "declining violence" it is important to answer the question in the title of this chapter—do we really care about enemy civilian casualties? On the one hand, Pinker's appeal for increasing empathy and morality that human beings express to one another proved itself incomplete in reflecting the way of war in the beginning of the twenty-first century. This chapter was written just several days after the November 13, 2015, Paris attacks, a series of coordinated terrorist assaults that occurred in the French capital, taking 130 lives. Two days after the attacks, French President Francois Hollande stated that "France will be merciless toward these barbarians," ordering the biggest French air raid on ISIS targets that has taken place in Syria so far, a "massive" attack as described by the French defense ministry spokesman.[36] On November 21, 2015, eight days after the attacks, the UN Security Council adopted resolution 2249 (2015) that called upon Member States to take "all necessary measures" to prevent and suppress terrorist acts on territory under the control of ISIS in Syria and Iraq,[37] proving that our "Inner Demons" are very much alive and kicking. While in its call, the UN Security Council insisted that Member States have to act "in compliance with international law, in particular with the United Nations Charter, as well as international human rights, refugee and humanitarian law"[38]; it is important to remember, that, as noted above, none of these "laws" explicitly demands the minimizing of enemy civilian casualties. When it comes to waging wars against foreign societies, it is difficult not to concur with Ralph Peters's already cited observation that:

> The citizenry of the United States, in fact, will tolerate the killing of enormous numbers of foreigners, so long as that killing does not take too long, victory is clear-cut, friendly casualties are comparatively low, and the enemy dead do not have names, faces, and families.[39]

While Peters focused primarily on American society, his conclusion seems to be generalizable and it depicts well the minor part played by "Better Angels" in minimizing the harming of enemy civilians. In other words, from a moral perspective, due to the historical cultural predispositions shaped by the long history of bloody wars and violence, we just do not care about the civilian population of a society that raised people who have dared to aim a blow at us, as long as the war is fast, decisive, with no painful costs (human, as well as financial), and the terrible consequences of our fighting are hidden from our eyes and minds.

On the other hand, according to Ian Morris, there is a more pragmatic explanation of the relationship between war and violence, and, in fact, violence during war is just a tool, rather than a goal, of war. The success of a war, according to Morris, is measured by its level of "productiveness" (i.e., the winner's ability to absorb the loser, thus, creating a larger society), rather than the level of applied violence. Following this logic, it is possible to assume that the level of violence in contemporary conflicts is no more than a cultural relic of the`long-term tradition of making war. According to this theory, we do not care about enemy civilian casualties because our history tells us not only that war makes us safer and wealthier, but also that war ultimately means unprecedented (thought regulated) violence, as General William Tecumseh Sherman put it in 1864: "War is cruelty and you cannot refine it."[40] The question is, however, if this much enrooted historical cultural narrative makes wars in the twenty-first century "counterproductive"—would we care more for enemy civilians?—if not for moral reasons, then, at least, for pragmatic selfishness, because, if Morris is right, fighting "counterproductive" wars ultimately leads to the collapse of the society. Interestingly enough, the analysis of America's failure to fight "productive" wars in Iraq and Afghanistan offers a small ray of optimism.

A RAY OF OPTIMISM?

Analysing the historical examples of technologies that failed to ignite an RMA, Lieutenant Colonel Gregory Wilmoth differentiate between two different results to which the military employment of initially failing technologies might lead. The first one is an absolute rejection due to technological unfeasibility. According to Wilmoth, the best example of such technology was the glider, which after a very short operational success in World War II became obsolete due to its technological inferiority and more successful alternative technologies (i.e., air-assault aircraft and helicopter). The second pathway describes a technology, which initially fails due to different technological disadvantages and lacking conceptualization, but which, after a period of time, makes a come-back when the operational needs require it. The best example, according to Wilmoth, is the technology of in-flight refueling. Pioneered in the 1920s to extend the range of wooden biplanes, the idea of in-flight refueling was generally abandoned due to many technological obstacles and general lack of operational need, as aircraft range and endurance improved during the 1930s. In the late 1940s, however, in-flight refueling got a second chance to extend the range of strategic bombers in the context of the Cold War. During the following decades, the technology quickly spread throughout different services to extend the range of tactical strike aircrafts. As Wilmoth concluded: "What was once a stun and then a niche technology blossomed into a widespread innovation."[41]

Keeping in mind these two possible results of a failed RMA, and the analysis of the military employment of NLW discussed throughout this book, it seems right to argue that the story of NLW remains more the story of in-flight refuelling, than that of gliders. NLW have not been rejected by the military for more than a quarter of century and, so far, mainly due to lacking conceptualization of operational needs, these weapons have been a niche technology intended for law enforcement capabilities. Therefore, it could well be that, similar to the technology for in-flight refueling, a clearly defined military necessity would turn NLW into a widespread innovation. The history of past RMAs suggests that the military necessity for an RMA comes only when there is a military challenge that requires development of novel organizational and operational concepts for the employment of innovative technologies. While it impossible to foresee the future of NLW, the analysis of the military experience during in the first decade of the twenty-first century offers a ray of optimism.

One of the most distinctive characteristics of warfare in the first decade of the twenty-first century was the great confidence that military organizations felt regarding the information-technology RMA that occurred during the 1990s. The advocates of this RMA believed that a combination of intelligence, surveillance, and reconnaissance; advanced command, control, communications, computers, and information; and precision-strike weapons offers a rapid and decisive victory with low friendly casualties and limited collateral damage.[42] Focused on the prevalence of air power, the proponents of this RMA believed that it could save "enemy lives through the use of precision to minimize non-combatant fatalities, and friendly lives by the substitution of technology for manpower and the creation of battlefield conditions in which land elements . . . can do their job without significant resistance."[43] While, in theory, this RMA was promised to provide an overwhelming victory, the practice proved the opposite. The American experience in Iraq and Afghanistan,[44] as well as the IDF's encounter with Hezbollah in Lebanon in 2006,[45] proved that this RMA has very little to offer, as Colin Gray put it: "it is not unreasonable to ask why it has been that our ever-improving battle space knowledge has been compatible with so troubled a course of events in the 2000s in Iraq and Afghanistan."[46]

The interesting fact is that the failure of this RMA can be easily explained by an examination of its roots—the information-technology RMA was developed in the context of the Cold War, when the U.S. military was looking for new operational concepts and new technologies to meet the challenge of the deep-echeloned Soviet rear.[47] In other words, the unmet military challenge that had been bridged by this RMA was "the orderly advance of their [Soviet] follow-on echelons"[48] during symmetrical conflict against a traditional military. It is not surprising then that this RMA was very helpful in defeating the Iraqi military (twice, in 1991 and in 2003) but failed to provide satisfactory results against insurgencies and terror organizations. This failure, however, is rooted not in technological sophistication of the enemy, but in the different nature of the conflict. While the information-technology RMA was intended to defeat conventional military, "insurgencies are defeated not by killing insurgents, but by winning the support of the population and thus denying the insurgents both refuge and recruits."[49] In other words, as many scholars already pointed out, counter-insurgency and counter-terrorism are about winning "hearts and mind," rather than a bold fighting against an "invisible" enemy.[50] And, from that perspective, the information-technology RMA has had, indeed, very little to offer.

It seems, however, that the American military has been learning lessons from its failure in Iraq and Afghanistan, and the understanding that in fighting counter-insurgency, "priority should be given to securing the civilian population, not hunting down insurgents"[51] has begun to spread through the circles of the U.S. military decision-makers. In 2015, commenting on the American air strikes on ISIL in Syria, Lieutenant General Bob Otto, the U.S. Air Force's deputy chief of staff for intelligence, surveillance, and reconnaissance, stated that:

> We believe if you inadvertently kill innocent men, women and children, then there's a backlash from that . . . we might kill three and create 10 terrorists. It really goes back to the question of are we killing more than we are making?[52]

In other words, the American military has begun to understand that causing civilian casualties is pragmatically counterproductive in fighting insurgencies and terrorist organizations. Changing military doctrine, however, as Alvin and Heidi Toffler claim, is "like trying to stop a tank armor by throwing marshmallows at it. The military, like any huge modern bureaucracy, resists innovation—especially if the change implies the downgrading of certain units and the need to learn new skills and to transcend service rivalries."[53] And, therefore, it is not surprising that the U.S. military intuitively looks for a solution in an existing technology, as Colonel Steve Warren, a spokesman for the anti-ISIL campaign, stated: "the amount of care that we have taken to preserve civilian life, to preserve civilian infrastructure, is unprecedented" because the air-strikes of the U.S. and its allies "are the most precise in the history of warfare."[54] The NATO-led Libya campaign in 2011, however, already showed that neither precision weapons, nor the strict rules of engagement, make war "productive." Despite the fact that "advanced munitions, particularly low-collateral, inert and fused weapons such as Hellfire, Dual Mode Seeker Brimstone and Paveway IV were highly prized" and the campaign "was conducted throughout with very restrictive rules of engagement, with a mandate to protect the population and minimize collateral damage to the infrastructure in order to help the country back on its feet after the cessation of violence"[55]—the violence in Libya has not ceased, and it seems that a more fundamental transformation has to occur to make contemporary counter-insurgency wars more "productive."

Christopher Coker, in his *Humane Warfare*, *The Future of War*, and most recently, *Warrior Geeks*, proposes an idea that by exploiting different novel technologies, the killing in war may become redundant.[56] While he explores the radical idea that war might be "technicized" to a level that "we may be able to pass beyond the 'death barrier,'"[57] this military transformation will require not only technological maturity, but also the socio-political environment that will facilitate these fundamental changes. In other words, before Coker's technologies (e.g., genetic engineering of warriors, robots controlled by artificial intelligence, cyborgs) will fundamentally change the conduct of war, they have to pass the test of political-military necessity, i.e., there has to be a clearly defined political-military challenge that these technologies are intended to bridge, and the political leadership has to have a socio-political reason to cope with this challenge.

The RMA of NLW, as any fundamental military transformation, is no different. However, as discussed above, the future of NLW will be determined by the U.S. military, and there are two reasons why the observation given by Lieutenant General Bob Otto offers a ray of optimism. The first is the fact that U.S. political-military decision-makers have come to the understanding that counter-insurgency presents an unmet challenge that has to be bridged by a combination of novel operational and organizational concepts and innovative technologies.[58] Only time will show whether NLW will be a part of the required military transformation, but, as Theo Farrell and Terry Terriff put it, "whereas previously the focus had been on destroying targets," today "it may be sufficient or indeed preferable to disable targets,"[59] and NLW are intended to do exactly that. The second reason for optimism regarding the future of NLW is, according to Ian Morris's theory, the "counterproductiveness" (i.e., the inability of winners to incorporate losers into a larger society) of contemporary ways of war against insurgencies and terrorism. The U.S.-led West failed to make Iraqi and Afghani (as well as Libyan, Yemeni, Syrian, etc.) people as free as possible to trade in bigger and bigger markets. To make counter-insurgency "productive," its success or failure has to be measured not by body counts or footage of "liberated" territory, but according to the public opinion of the local population. As the lowest possible level of enemy civilian casualties is essential in achieving this, NLW are, again, intended to do exactly that.

The opponents of military employment of NLW might claim that these weapons are suitable for police officers, rather than war-fighters. According to Ian Morris, however, the United States is the "globocop" of the early twenty-first century, and it seems that it is beginning to understand that in conducting wars against insurgencies, "policing" (i.e., disabling, incapacitating, minimizing collateral damage, etc.) may be more effective than "warfighting" (i.e., destroying and killing). The RMA of NLW may eventually occur, not because of the moral or political costs of causing enemy civilian casualties, but due to the potential effectiveness of these weapons in counterinsurgency. As the history of military innovations suggests, however, without a clear and decisive political directive and support, this can be a very long process.

NOTES

1. Tim Craig, Missy Ryan, and Thomas Gibbons-Neff, "By Evening, a Hospital. By Morning, a War Zone," *Washington Post*, October 10, 2015, https://www.washingtonpost.com/world/national-security/story-of-how-a-kunduz-hospital-was-shelled-by-us-gunship-in-question/2015/10/10/1c8affe2-6ebc-11e5-b31c-d80d62b53e28_story.html [accessed May 10, 2018].

2. "Obama Apologises to MSF President for Kunduz Bombing," *BBCNews*, October 7, 2015, http://www.bbc.co.uk/news/world-us-canada-34467631 [accessed May 10, 2018].

3. Vladimir Putin, *Speech at the UN General Assembly*, September 28, 2015, https://russian.rt.com/article/119712 [accessed May 10, 2018], (Russian).

4. "Russia's Unguided Weapons in Syria Could Ricochet on Moscow," *Time*, October 2, 2015, http://time.com/4059442/russia-bombing-syria-isis-assad/ [accessed May 10, 2018].

5. Elizabeth Foht, "Rossiyskiye korabli v Kaspiyskom more nanesli udary po pozitsiyam IG v Sirii" ["Russian Ships in the Caspian Sea Attacked the Positions of ISIS in Syria"], *RBC*, October 7, 2015, http://www.rbc.ru/politics/07/10/2015/561502039a7947add1998f52 [accessed May 10, 2018].

6. John Barnes, "The Intransigent Israeli?," British Embassy, Tel Aviv, November 18, 1971, republished in Bar, Shmuel, "The Intransigent Israeli?—A Document from the Archives of the British Foreign Office (FCO/17 1554)," *Israel Journal of Foreign Affairs* 4, no. 1 (2010): 103–13.

7. Efraim Inbar, *Israel eina mevudedet [Israel Is Not Isolated]* (Tel Aviv: Begin-Sadat Center for Strategic Studies, 2013).

8. Israel State Bureau of Statistics, *Israel's Foreign Trade in Goods, by Country—September 2015*, Jerusalem, October 25, 2015, http://www.cbs.gov.il/reader/newhodaot/hodaa_template_eng.html?hodaa=201516283, [accessed October 30, 2015].

9. *Vladimir Putin in an interview to CBS News*, September 27, 2015, http://www.cbsnews.com/news/vladimir-putin-russian-president-60-minutes-charlie-rose/, [accessed May 10, 2018].

10. Alexander Vladimirov, *Osnovy Obshchey Teorii Voyny [Fundamentals of the General Theory of War]*, Volume 1 (Moscow: Moskovskiy Finansovo-Promyshlennyy Universitet "Sinergiya," 2013), 160.

11. Nachman Ben-Yehuda, "The Masada Mythical Narrative and the Israeli Army," in Edna Lomsky-Feder and Eyal Ben-Ari (ed.), *The Military and Militarism in Israeli Society* (New York: State University of New York Press, 1999); Yael Zerubavel, "The Multivocality of a National Myth: Memory and Counter-Memories of Masada," in Robert Wistrich and David Ohana (ed.), *The Shaping of Israeli Identity: Myth, Memory and Trauma* (London: Fank Cass, 1995).

12. Amos Yadlin, "The Strategic Balance of Operation Protective Edge: Achieving the Strategic Goal Better, Faster, and at a Lower Cost," in Anat Kurz and Shlomo Brom (ed.), *The Lessons of Operation Protective Edge* (Tel Aviv: Institute for National Security Studies, 2014), 205.

13. Yadlin, "The Strategic Balance," 211.

14. Dwight Eisenhower, *Farewell Radio and Television Address to the American People*, January 17, 1961, http://www.presidency.ucsb.edu/ws/index.php?pid=12086&st=farewell&st1, [accessed May 10, 2018].

15. For example, Michael Armacost, *The Politics of Weapons Innovations: The Thor-Jupiter Controversy* (New York: Columbia University Press, 1969); Carroll Pursell, *The Military-Industrial Complex* (New York: Harper & Row Publishers, 1972); Paul Koistinen, *The Military-Industrial Complex* (New York: Praeger Publishers, 1980); Helen Caldicott, *The New Nuclear Danger. George W. Bush's Military-Industrial Complex* (New York: The New Press, 2002); Peter Dombrowski and Eugene Gholz, *Buying Military Transformation: Technological Innovation and the Defense Industry* (New York: Columbia University Press, 2006).

16. Amir Rapaport, "Metach gavoha [High Tension]," *Israel Defense*, April 26, 2011, http://www.israeldefense.com/en/node/9371 [accessed May 10, 2018].

17. Homeland Security Research Cooperation (HSRC), *Non-Lethal Weapons: Technologies and Global Market—2012–2020*, 2011, http://www.homelandsecurityresearch.com/2011/10/non-lethal-weapons-technologies-global-market-2012-2020/ [accessed May 10, 2018].

18. Francis Fukuyama, *The End of History and the Last Man* (Penguin Books, 1992).

19. Department of the Army Headquarters, United States Army, *Force XXI Operations: A Concept for the Evolution of Full-Dimensional Operations for the Strategic Army of the Early Twenty-First Century TRADOC Pamphlet 525–5*, Fort Monroe, VA, August 1, 1994, 2.

20. United States Government Accountability Office (GAO), *DOD Needs to Improve Program Management, Policy, and Testing to Enhance Ability to Field Operationally Useful Non-Lethal Weapons: Report to Congressional Requesters*, Washington, DC, April 2009, 18.

21. See Dima Adamsky, *The Culture of Military Innovation: The Impact of Cultural Factors on the Revolution in Military Affairs in Russia, the US and Israel* (Stanford: Stanford University Press, 2010).

22. Steven Pinker, *The Better Angels of Our Nature: A History of Violence and Humanity* (London: Pinguin Books, 2012), xix, 714.

23. Ian Morris, *War—What Is It Good For? The Role of Conflict in Civilisation, from Primates to Robots* (London: Profile Books, 2014), 10.

24. Pinker, *The Better Angels of Our Nature*, xxii.

25. Morris, *War—What Is It Good For?*, 10, 8.

26. Pinker, *The Better Angels of Our Nature*.

27. Chris Jochnik and Roger Normand, "The Legitimation of Violence: A Critical History of Laws of War," *Harvard International Law Journal* 35, no. 1 (1994): 50.

28. Quoted in Brad Knickerbocker, "The Fuzzy Ethics of Nonlethal Weapons," *Christian Science Monitor*, February 14, 2003, http://www.csmonitor.com/2003/0214/p02s01-usmi.html [accessed May 10, 2018].

29. Paul Cornish, "Unlawful Wounding: Codifying Interaction between Bullets and Bodies," in Paul Cornish and Nicholas Saunders (ed.), *Bodies in Conflict: Corporeality, Materiality and Transformation* (London: Routledge, 2014), 20.

30. Cornish, "Unlawful Wounding," 14, 19.

31. Morris, *War–What Is It Good For?*, 8.

32. Ibid., 8, 9, 337–38.

33. Ibid., 113.

34. Ibid., 377.

35. Ibid., 378.

36. Doug Bolton, "French Warplanes Bombard Isis Raqqa Strongholds Two Days after Paris Attacks," *Independent*, November 18, 2015, http://www.independent.co.uk/news/world/middle-east/french-warplanes-bombard-isis-strongholds-in-raqqa-two-days-after-paris-attacks-which-killed-132-a6735636.html [accessed May 10, 2018].

37. UN Security Council, *UN Security Council Resolution 2249 (2015) on ISIL*, New York, November 21, 2015.

38. Ibid.

39. Ralph Peters, *Fighting for the Future: Will America Triumph?* (Mechanicsburg: Stackpole Books, 2001), 105.

40. General William Tecumseh Sherman in a letter to the Mayor of Atlanta, September 12, 1864.

41. Gregory Wilmoth, "False-Failed Innovations," *Joint Force Quarterly* 22 (1999): 51–57.

42. Elizabeth Stanley-Mitchell, "The Digital Battlefield: What Army Transformation Efforts Say about its Professional Jurisdiction," in Don Snider and Gayle Watkins (ed.), *The Future of the Army Profession* (New York: McGraw Hill, 2003), 155–78.

43. Benjamin Lambeth, *The Transformation of American Air Power* (Ithaca, NY: Cornell University Press, 2000), 303.

44. Stephanie Carvin and Michael Williams, *Law, Science, Liberalism and the American Way of Warfare* (Cambridge: Cambridge University Press, 2015), chapter 5; Richard Lock-Pullan, "U.S. Military Strategy, Strategic Culture, and the 'War on Terror,'" in John Owens and John Dumbrell (ed.), *America's "War on Terrorism": New Dimensions in U.S. Government and National Security* (Lanham, MD: Lexington Books, 2008).

45. Avi Kober, "The Israel Defence Forces in the Second Lebanon War: Why the Poor Performance?," *The Journal of Strategic Studies* 30, no. 1 (2008): 3–40.

46. Colin Gray, *Making Strategic Sense of Cyber Power: Why the Sky Is Not Falling* (Carlisle: Strategic Studies Institute, 2013), 51.

47. For example, Alvin Toffler and Heidi Toffler, *War and Anti-War: Survival at the Dawn of the 21st Century* (Boston: Little, Brown and Company, 1993), chapter 7; Diego Ruiz Palmer, "The NATO-Warsaw Pact Competition in the 1970s and 1980s: A Revolution in Military Affairs in the Making or the End of a Strategic Age," *Cold War History* 14, no. 4 (2014): 533–74.

48. Toffler and Toffler, *War and Anti-War*, 50.

49. James Dobbins, "Iraq: Winning the Unwinnable War," *Foreign Affairs* 84, no. 1 (2005): 19.

50. For example, Michael Howard, "What's in a Name? How to Fight Terrorism," *Foreign Affairs* 81, no. 1 (2002): 8–13; Thomas Mockaitis, "Winning Hearts and Minds in the 'War on Terrorism,'" *Small Wars & Insurgencies* 14, no.1 (2003): 21–38.

51. Dobbins, "Iraq: Winning," 19.
52. Lieutenant General Bob Otto quoted in "Russia's Unguided Weapons in Syria Could Ricochet on Moscow."
53. Toffler and Toffler, *War and Anti-War*, 52.
54. Colonel Steve Warren quoted in "Russia's Unguided Weapons in Syria Could Ricochet on Moscow."
55. Elizabeth Quintana, "The War from the Air," in Adrian Johnson and Saqeb Mueen (ed.), *Short War, Long Shadow: The Political and Military Legacies of the 2011 Libya Campaign*, Whitehall Report 1–12, (London: Royal United Services Institute for Defence and Security Studies, 2012), 34, 31.
56. Christopher Coker, *Humane Warfare* (London: Routledge, 2002); Christopher Coker, *The Future of War: The Re-Enchantment of War in the Twenty-First Century* (Oxford: Black-Well, 2004); Christopher Coker, *Warrior Geeks: How 21st Century Technology is Changing the Way We Fight and Think About War* (London: Hurst & Co., 2013).
57. Coker, *The Future of War*, 141.
58. Theo Farrell and Terry Terriff, "Military Transformation in NATO: A Framework for Analysis," in Theo Farrell, Terry Terriff, and Frans Osinga (ed.), *Transformation Gap?: American Innovations and European Military Change* (Stanford: Stanford University Press, 2010)
59. Farrell and Terriff, "Military Transformation in NATO," 5.

Bibliography

1977 Additional Protocol to the Geneva Conventions of August 12, 1949, and relating to the Protection of Victims of International Armed Conflicts (Protocol I) (adopted June 8, 1977, entered into force December 7, 1978), 16 ILM 1391.

"'36 killed' in raids on Baghdad." *The Guardian*, March 27, 2003, http://www.theguardian.com/world/2003/mar/27/iraq.sallybolton [accessed May 10, 2018].

Abman, Oren. "Ha'ma'avar le'lehima m'shigra ve'mi'batash [The Transition to Fighting from Routine and from Batash]." *Ma'aruchot* 441 (2012): 21.

Ackroyd, Carol, et al. *The Technology of Political Control*, Second Edition. London: Pluto Press Limited, 1980.

Adams, James. *The Epic of America*. London: George Routledge & Sons Ltd., 1940.

Adamsky, Dima. "Through the Looking Glass: The Soviet Military-Technical Revolution and the American Revolution in Military Affairs." *The Journal of Strategic Studies* 31, no. 2 (2008): 257–94.

Aftergood, Steven. "The Soft-Kill Fallacy." *The Bulletin of the Atomic Scientists* 50, Issue 5 (1994): 40–45.

Alabin, Peter. *Trekhvekovaya godovshchina goroda Samary [Three-Century-Old Anniversary of the City of Samara]*. Samara: Gubernskaya Zemskaya Tipografiya, 1887.

Albek, Yehuda. "Pizur hafganot seder: ha'etgar ve'hama'ane [Riot Control: The Challenge and the Solution]." *Ma'arachot* 449 (2013): 24–29.

Alekseyev, Veniamin. "Ekaterinburg." *Rodina* 8 (2003): 100–145.

Alexander, John. "Shoot, But Not to Kill—Non-Lethal Weapons Have Yet to Establish a Military Niche." *Jane's International Defense Review*, June 1, 1996.

Alexander, John. *Future War: Non-Lethal Weapons in 21st Century Warfare*. New-York: St. Martin's Press, 1999.

Alkhuzai, Amir, et al. "Violence-Related Mortality in Iraq from 2002 to 2006." *New England Journal of Medicine* 358, no. 2 (2008): 484–93.

Allison, Graham, Paul Kelley, and Richard Garwin. *Nonlethal Weapons and Capabilities: Report of an Independent Task Force*. New York: Council on Foreign Relations Press, 2004.

Amnesty International. *Amnesty International Annual Report 2005*, p. ii, https://www.amnesty.org/en/documents/pol10/0001/2005/en/ [accessed May 10, 2018].

Anders, Åslund. "An Assessment of Putin's Economic Policy." *CESifo Forum*, no. 9(2) (2008): 16–21.

Antipov, V., and S. Novichkov. "K voprosu o razrabotke i primenenii neletal'nykh sredstv porazheniya na khimicheskoy osnove [Regarding the Question of Development and Employment of Non-Lethal Means Based on Chemicals]." *Voenaya Misl'*, no. 9 (2009): 54–61.

Arbatov, A., and V. Dvorkin. "Voyennaya reforma: proschety i puti ikh ispravleniya [Military Reform: Miscalculations and the Ways to Their Correction]." In M. Lipman (ed.), *Povestka dnya dlya novogo prezidentstva [Agenda for the New Presidency]*. Moscow: Moscow Carnegie Centre, Rossiyskaya Politicheskaya Entsiklopediya, 2012.

Arkin, William. *Divining Victory Airpower in the 2006 Israel-Hezbollah War*. Maxwell: Air University Press, 2007.

———. "Smart Bombs, Dump Targeting?" *The Bulletin of the Atomic Scientists* 56, no. 3 (2000): 46–53.

Armacost, Michael. *The Politics of Weapons Innovations: The Thor-Jupiter Controversy*. New York: Columbia University Press, 1969.

Arreguin-Toft, Ivan. *How the Weak Win Wars: A Theory of Asymmetric Conflict*. Cambridge: Cambridge University Press, 2005.

Asada, Sadao. "The Mushroom Cloud and National Psyches." In Laura Hein and Mark Selden (eds.), *Living with the Bomb: American and Japanese Cultural Conflicts in the Nuclear Age*. Armonk: M. E. Sharpe, 1997.

Averkov, V. "Prinyatiye vneshnepoliticheskikh resheniy v Rossii [Foreign Policy Decision-Making in Russia]," *Mezhdunarodnyye Protsessy* 10, no. 2(29) (2012): 59–70.

Avidor, Gidon (ed.). *Hamashmauyot haben'leumiyot, hamishpatiyot vehamusriyot belehima bekerev uh'lusiya ezrahit [The International, Legal and Moral Consequences of Warfare in Civil Environment]*. Latrun: Zvi Meitar Institute for Land Warfare Studies, 2010.

Avnery, Shlomo. *The Making of Modern Zionism*. New York: Basic Books, 1981.

Azulai, Yuval. "Technologiya hal-hereg: haim hayah nitan limnoa hasa'ara be 'marmara'? [Non-Lethal Technology: Whether Was It Possible to Prevent the 'Marmara' Storm?]." *Globes*, September 20, 2011, http://www.globes.co.il/news/article.aspx?did=1000683960 [accessed May 10, 2018].

Bacevich, Andrew (ed.). *The Imperial Tense: Prospects and Problems of American Empire*. Chicago: Ivan R. Dee, 2003.

Bacevich, Andrew, and Eliot Cohen (eds.). *War over Kosovo*. New York: Columbia University Press, 2001.

Baev, P. "Trayektoriya voyennoy reformy v Rossii [The Trajectory of the Military Reform in Russia]," *Pro et Contra* 16, Issue 4–5 (2012): 171–84.

Bagnall, Nigel. *The Israeli Experience: A Study of Quality*. Surrey: Tactical Doctrine Retrieval Cell, Staff College, 1974.

Bar-Joseph, Uri (ed.). *Israel National Security Towards the 21st Century*. London: Frank Cass, 2001.

Bar-Tal, Daniel. *Lechiot im ha'sichsuch: nituah psichologi-havrati shel' he'hevra ha'yehudit be'israel [Living with the Conflict: Socio-Psychological Analysis of the Jewish Society in Israel]*. Jerusalem: Karmel, 2007.

Baranovskii, V. "Osnovnyye parametry sovremennoy sistemy mezhdunarodnykh otnosheniy – chast' III [The Major Parameters of the Modern System of International Relations – Part III]." *Polis: Politicheskii Issledovaniya*, no. 5 (2012): 148–56.

Barnes, John. "The Intransigent Israeli?," British Embassy, Tel Aviv, 18 November 1971, republished in Shmuel Bar, "The Intransigent Israeli? – A document from the Archives of the British Foreign Office (FCO/17 1554)." *Israel Journal of Foreign Affairs* 4, no. 1 (2010): 103–13.

Barnett, Roger. *Asymmetric Warfare*. Dulles: Brassey's, 2003.

Barrett, Laurence. *Gambling with History: Ronald Reagan in the White House*. Garden City: Doubleday, 1983.

Barry, John, Michael Everett, and Allen Peck. "Nonlethal Military Means: New Leverage for a New Era." *Harvard University, John F. Kennedy School of Government, National Security Program Policy Analysis Paper*, no. 94–101, 1994.

Baruch, Pnina. "Operation Protective Edge: Legality and Legitimacy." *INSS Insight*, no. 577, 2014.

Bazak, Yuval. "Responding to the Need for International Legitimacy: Strengthening the IDF Strike Force." *Military and Strategic Affairs* 3, no.2 (2011): 3–17.

Bein, Yohanan. "Israel, ha'um ve'ha'ma'araha le'shalom [Israel, the U.N. and the Struggle for Peace]." In Moshe Igar, Yosef Guvrin, and Rei Oded (eds.), *Misrad ha'hutz—50 ha-shanim ha'rishonot [The Foreign Affairs Office—the 50 First Years].* Jerusalem, Keter, 2002.

Ben-Gurion, David. "Israel ben ha'umot [Israel between the Nations]." *Israel Government Year Book*, 1952.

Ben-Horin, Roi. "Neshek Al-Hereg: Ha'alaha, Hama'ase ve-mashebenehem." *Idkun Astrategi* 4, no.1 (2001): 18–24.

———. "Non-Lethal Weapons: Theory, Practice, and what Lies Between." *Strategic Assessment* 2, no. 4 (2001, January): 21–27.

Ben-Israel, Yitzhak. *Milchemet tilim Harishona: Israel – Hizballa (kaitz 2006) [The First Missile War: Israel – Hezbollah (2006)].* Tel Aviv: Tel Aviv University Press, 2007.

———. "Neshel al-heger [Non Lethal Weapon]." *Ma'arachot* 363 (1999): 20–24.

———. "Torat ha'yahasut shel bniyat ha'koah [The Theory of Relativity of the Military Build-up]." *Ma'arachot* 352–353 (1997): 33–42.

Ben-Meir, Dov. *Mediniyut Hutz [Foreign Policy].* Tel Aviv: Yedioth Ahronot and Chemed Books, 2011.

Ben-Meir, Yehuda. "Israeli Public Opinion and the Second Lebanon War." In Shlomo Brom and Meir Elran (ed.), *The Second Lebanon War: Strategic Perspectives.* Tel Aviv: Institute for National Security Studies, 2007.

———. "Operation Cast Lead: Political Dimensions and Public Opinion." *Strategic Assessment* 11, no. 4 (2009): 29–34.

Ben-Shalom, Uzi. "'Faser lematzav ham'em': mah'shevot al neshek pahot katlani be'sde hakrav ['Faser Switched to Shock': Thoughts on Less Lethal Weapons on the Battlefield]." *Be'yabasha*, no. 15 (2010, September).

Ben-Simhon, Kobi. "Hatargil hamasriah; hamafginim mi'Na'alin tzrihim liknot deodorant [A Stinky Hoax; The Rioters from Na'alin Have to Buy Deodorant]." *Ha'aretz Online*, September 4, 2008, http://www.haaretz.co.il/misc/1.1347475 [accessed May 10, 2018].

Ben-Yehuda, Nachman. "The Masada Mythical Narrative and the Israeli Army." in Edna Lomsky-Feder and Eyal Ben-Ari (eds.), *The Military and Militarism in Israeli Society.* New York: State University of New York Press, 1999.

Ben-Yosef, Moriya. "Muetzet Hait'tztaidut Beneshek Pahot Katlani [The Employment of Non-Lethal Weapons is Speeding Up]." *IsraelDefense,* June 5, 2011.

Benbow, Tim. "Talking 'Bout Our Generation? Assessing the Concept of 'Fourth-Generation Warfare.'" *Comparative Strategy* 27, no. 2 (2008): 148–63.

Bender, Bryan. "USAF Seeks Air-Dropped Non-Lethal Weapons." *Jane's Defence Weekly*, March 24, 1999.

Berdyaev, Nikolai. *Novoye Srednevekov'ye - Razmyshleniya o sud'be Rossii i Yevropy [New Middle Ages: Reflections on the Destiny of Russia and Europe].* Moscow: Feniks-HDS Press, 1991.

———. *Sud'ba Rossii [The Destiny of Russia]*, reprint edition from 1918 publication. Moscow: Filosofskoye Obshchestvo SSSR, 1990.

Berlin, Isaiah. *The Soviet Mind: Russian Culture under Communism.* Washington, DC: Brookings Institution Press, 2004.

Berman, Lazar. "Capturing Contemporary Innovation: Studying IDF Innovation against Hama and Hezbollah." *Journal of Strategic Studies* 35, no.1 (2012): 121–47.

Berzi, Tamas. "European Reactions to Israel's Gaza Operation." Jerusalem Center for Public Affairs, *Articles on Europe & Israel* 8, no. 20 (2009).

Betts, Richard. *American Force: Dangers, Delusions, and Dilemmas in National Security.* New York: Columbia University Press, 2013.

Bialer, Uri. "Ben-Gurion ve'she'elat ha'orientatziya ha'beinleumit shel' Israel 1948–1956 [Ben-Gurion and the Question of the International Orientation of Israel 1948–1956]." *Cathedra*, no. 43 (1987).

———. *"Our Place in the World" – Mapai and Israel's Foreign Policy Orientation 1947–1952.* Jerusalem: Magnes Press, Hebrew University, 1981.

———. "Top Hat, Tuxedo and Cannons: Israeli Foreign Policy from 1948 to 1956 as a Field of Study." *Israel Studies* 7, no. 1 (2002): 1–80.

Billington, Ray (ed.). *The Frontier Thesis: Valid Interpretation of American History?* New York: Holt, Rinehart and Winston, 1966.

Black, Jeremy. *War and Technology*. Bloomington: Indiana University Press, 2013.

———. *War and the Cultural Turn*. Cambridge: Polity Press, 2012.

Bobb, Justin. *Non-Lethal Weaponry: Applications to AC-130 Gunships*. Maxwell Air Force Base, Alabama: Air University, 2002.

Bogaturov, A. "Tri pokoleniya vneshnepoliticheskikh doktrin Rossii [The Three Generations of the Russian Foreign Policy Doctrines]." *Mezhdunarodnyye Protsessy* 5, no. 1(3) (2007): 54–69.

Boggs, Carl (ed.). *Masters of War: Militarism and Blowback in the Era of American Empire*. New York: Routledge, 2003.

———. "Outlaw Nation: The Legacy of U.S. War Crimes." In Carl Boggs (ed.), *Masters of War: Militarism and Blowback in the Era of American Empire*. New York: Routledge, 2003.

Boher, Ziv, and Mark Osiel. "Proportionality in Military Force at War's Multiple Levels: Averting Civilian Casualties vs. Safeguarding Soldiers." *Vanderbilt Journal of Transnational Law* 46, no. 747 (2013): 747–822.

Bolger, Daniel. *Savage Peace: Americans at War in the 1990s*. New York, Presidio, 1995.

Bolton, Doug. "French Warplanes Bombard Isis Raqqa Strongholds Two Days after Paris Attacks." *Independent*, November 18, 2015, http://www.independent.co.uk/news/world/middle-east/french-warplanes-bombard-isis-strongholds-in-raqqa-two-days-after-paris-attacks-which-killed-132–a6735636.html [accessed May 10, 2018].

Boorstin, Daniel. *The Americans: The Colonial Experience*. New York: Vintage Books, 1958.

———. *The Americans: The Democratic Experience*. New York: Vintage Books, 1973.

———. *The Americans: The National Experience*. New York: Vintage Books, 1965.

Boot, Max. "The New American Way of War." *Foreign Affairs* 82, no. 4 (2003): 41–58.

———. *War Made New: Technology, Warfare and the Course of History, 1500 to Today*. New York: Gotham Books, 2006.

Booth, William. "Israeli Government to Refugees: Go Back to Africa or Go to Prison." *The Washington Post*, May 14, 2015, https://www.washingtonpost.com/world/middle_east/toughening-its-stance-toward-migrants-israel-pushes-africans-to-leave/2015/05/14/e1637bce-f350–11e4–bca5–21b51bbdf93e_story.html [accessed May 10, 2018].

"Bowe Bergdahl release: US Parties Clash on Afghan Deal." *BBC New*, June 2, 2014, http://www.bbc.co.uk/news/world-asia-27658821 [accessed May 10, 2018].

Boyd, David. "The Search for Low Hanging Fruit: Recent Developments in Non-Lethal Technologies." In Malcolm Dando (ed.), *Non-Lethal Weapons: Technological and Operational Prospects*, Jane's Special Report, November 2000.

Bradford Non-Lethal Weapons Research Project (BNLWRP). https://bradscholars.brad.ac.uk/handle/10454/3960 [accessed May 10, 2018].

Bregman, Ahron, and Jihan El-Tahri. *The Fifty Years War: Israel and Arabs*. London: Penguin Books & BBC Worldwide, 1998.

Brockhaus and Efron Encyclopedic Dictionary. Petersburg, 1907, Vol. XVI (Russian).

Brom, Shlomo (ed.). *In the Aftermath of Operation Pillar of Defense The Gaza Strip, November 2012*. Tel-Aviv: Institute for National Security Studies, 2012.

Brom, Shlomo, and Meir Elran (ed.). *The Second Lebanon War: Strategic Perspectives*. Tel-Aviv: Institute for National Security Studies, 2007.

Builder, Carl. *The Masks of War: American Military Styles in Strategy and Analysis*. Baltimore: John Hopkins University Press, 1989.

Buley, Benjamin. *The New American Way of War: Military Culture and the Political Utility of Force*. New York: Routledge, 2008.

Burenok, V., A. Leonov, and A. Pronin. *Voyenno-Ekonomicheskiye i Innovatsionnyye Aspekty Integratsii Netraditsionnykh Vidov Oruzhiya v Sostav Sistemy Vooruzheniya [Military-Economic and Innovative Aspects of the Integration of Non-traditional Weapons]*. Moscow: Izdatel'skaya Gruppa Granitsa, 2014.

Burenok, V., V. Korchak, and S. Smirnov. "Oruzhiye neletal'nogo deystviya—sostavnaya chast' sistemy vooruzheniya budushchego [Weapons with Non-Lethal Action—The Constitutive Part of the Future Weaponry System].” *Vestnik Akademii Voyennykh Nauk*, no. 4 (2007): 117–27.

Burnham, Gilbert, et al. *The Human Cost of the War in Iraq: A Mortality Study, 2002–2006*. Baltimore: Johns Hopkins University, 2006.

Burutin, A., et al. “Sistematizatsiya tipovykh skhem primeneniya oruzhiya neletal’nogo deystviya [The Systematization of Typical Scenarios of the Implementation of Weapons with Non-Lethal Action].” *Strategicheskaya Stabil’nost’*, no. 2(51) (2010): 18–26.

Bush, George. *A Distinctly American Internationalism*. Simi Valley, November 19, 1999, https://www.cnsnews.com/news/article/distinctly-american-internationalism [accessed May 10, 2018].

———. *A Period of Consequences*. Charleston, September 23, 1999, http://fas.org/spp/starwars/program/news99/92399_defense.htm [accessed May 10, 2018].

———. *Address Before a Joint Session of the Congress on the United States Response to the Terrorist Attacks of September 11*. Washington, September 20, 2001, http://www.presidency.ucsb.edu/ws/?pid=64731 [accessed May 10, 2018].

———. *Address to the Nation on Iraq From Cincinnati, Ohio*. Cincinnati, October 7, 2002, http://www.presidency.ucsb.edu/ws/?pid=73139 [accessed May 10, 2018].

———. *Address to the Nation on the Terrorist Attacks*. Washington, September 11, 2001, http://www.presidency.ucsb.edu/ws/?pid=58057 [accessed May 10, 2018].

———. *Letter to Congressional Leaders Reporting on the National Emergency With Respect to Iraq*, Washington, DC. February 11, 1991, http://www.presidency.ucsb.edu/ws/?pid=19295 [accessed May 10, 2018].

———. *Radio Address to the Nation on the Persian Gulf Crisis*. Washington, January 5, 1991, http://www.presidency.ucsb.edu/ws/?pid=19193 [accessed May 10, 2018].

———. *Remarks Accepting the Presidential Nomination at the Republican National Convention in New York City*. New York, September 2, 2004, http://www.presidency.ucsb.edu/ws/index.php?pid=72727& [accessed May 10, 2018].

———. *Remarks at the United States Military Academy in West Point*. West Point, New York, January 5, 1993, http://www.presidency.ucsb.edu/ws/?pid=20414 [accessed May 10, 2018].

———. *Remarks Following a Meeting with the National Security Team*, Washington, 12 September 2001, http://www.presidency.ucsb.edu/ws/?pid=58058 [accessed May 10, 2018].

———. *Remarks in a Victory Celebration*. Washington, November 4, 2004, http://www.presidency.ucsb.edu/ws/index.php?pid=62664 [accessed May 10, 2018].

———. *Remarks on the War on Terror*. Washington, September 6, 2006, http://www.presidency.ucsb.edu/ws/index.php?pid=779 [accessed May 10, 2018].

———. *Remarks to the National Endowment for Democracy*, New York, October 6, 2005, http://www.presidency.ucsb.edu/ws/?pid=73821 [accessed May 10, 2018].

———. *The President’s News Conference Following the North Atlantic Treaty Organization Summit Meeting in Brussels*. Brussels, May 30, 1989, http://www.presidency.ucsb.edu/ws/?pid=17077 [accessed May 10, 2018].

———. *The President’s Radio Address*. Washington, October 6, 2001, http://www.presidency.ucsb.edu/ws/?pid=24998 [accessed May 10, 2018].

Buyanov, V. “Rossiya v globaliziruyushchemsya mire: politicheskiy aspekt [Russia in the Globalizing World: Political Aspect].” in S. Smul’skii, O. Abramova, and V. Buyanov (eds.), *Vneshnyaya politika Rossii: teoriya i praktika [The Foreign Policy of Russia: Theory and Practice]*. Moscow: Kniga i Biznes, 2013.

Caldicott, Helen. *The New Nuclear Danger: George W. Bush’s Military-Industrial Complex*. New York: The New Press, 2002.

Campbell, John. “The United States and the Arab-Israeli Conflict.” In Robert Freedman (ed.), *World Politics and the Arab-Israeli Conflict*. New York: Pergamon Press, 1979.

Carafano, James. “The Future of Anti-Terrorism Technologies.” *Heritage Lectures*, no. 885 (2005).

Carter, Jimmy. *The President’s News Conference*. Chicago, May 25, 1978, http://www.presidency.ucsb.edu/ws/?pid=30852 [accessed May 10, 2018].

Carvin, Stephanie, and Michael Williams. *Law, Science, Liberalism and the American Way of Warfare*. Cambridge: Cambridge University Press, 2015.

Center for Army Lessons Learned (CALL). *Afghanistan Civilian Casualties Prevention,* Handbook 12–16, June 2012.

Chang, Gordon. *Friends and Enemies: The United States, China, and the Soviet Union, 1948–1972.* Stanford: Stanford University Press, 1990.

Chernichenko, S. "Dvoynyye Standarty [Double Standartds]." *Mezhdunarodnaya Zhizn,'* no. 12 (2008): 63–71.

Chomsky, Noam. *Interventions*. London: Penguin Group, 2007.

Chuvatkin, N. "Irritanty v sredstvakh neletal'nogo deystviya [Irritant in Non-Lethal Means]." *Proceeding to the International Conference 'Special Means of Non-Lethal Action,'* Moscow, 2012.

Cimbala, Stephen (ed.). *The George W. Bush Defense Program: Policy, Strategy, and War.* Washington, DC: Potomac Books, Inc., 2010.

Citino, Robert. *The Quest for Decisive Victory: From Stalemate to Blitzkrieg in Europe 1899–1940.* Lawrence: University Press of Kansas, 2002.

Clarke, Michael. "The Diplomacy that Led to War in Iraq." In Paul Cornish (ed.), *The Conflict in Iraq, 2003.* New York: Palgrave Macmillan, 2004.

Clinton, William. *Address to the Nation on Airstrikes Against Serbian Targets in the Federal Republic of Yugoslavia (Serbia and Montenegro).* Washington, March 24, 1999, http://www.presidency.ucsb.edu/ws/?pid=57305 [accessed May 10, 2018].

———. *Interview with Dan Rather of CBS News.* Washington, March 31, 1999, http://www.presidency.ucsb.edu/ws/?pid=57340 [accessed May 10, 2018].

———. *Message to the Congress Transmitting a Report on Somalia.* Washington, October 13, 1993. http://www.presidency.ucsb.edu/ws/?pid=47197 [accessed May 10, 2018].

Cohen, Amichai, and Stuart Cohen. "Israel and International Humanitarian Law: Between the Neo-Realism of State Security and the 'Soft Power' of Legal Acceptability." *Israel Studies* 16, no. 2 (2011): 1–23.

Cohen, Eliot. "Kosovo and the New American Way of War." In Andrew Bacevich and Eliot Cohen (ed.), *War over Kosovo.* New York: Columbia University Press, 2001.

Cohen, Eliot, Michael Eisenstadt, and Andrew Bacevich. *Knives, Tanks & Missiles: Israel's Security Revolution.* Washington: The Washington Institute for Near East Policy, 1998.

Cohen, Raymond. *Culture and Conflict in Egyptian-Israeli Relations: A Dialogue of the Deaf.* Bloomington: Indiana University Press, 1990.

Cohen, Stuart (ed.). *Democratic Societies and Their Armed Forces: Israel in Comparative Context.* London: Frank Cass, 2000.

Coker, Christopher. *Humane Warfare*. London: Routledge, 2002.

———. *The Future of War: The Re-Enchantment of War in the Twenty-First Century.* Oxford: BlackWell, 2004.

———. *Warrior Geeks: How 21st Century Technology is Changing the Way We Fight and Think About War.* London: Hurst & Co, 2013.

Coleman, Stephen. "Discrimination and Non-Lethal Weapons: Issues for the Future Military." In David Lovell and Igor Primoratz (ed.), *Protecting Civilians During Violent Conflict: Theoretical and Practical Issues for the 21st Century.* Surrey: Ashgate Publishing Limited, 2012.

Collective Security Treaty Organization (CSTO). *Soglasheniye o Kollektivnykh silakh operativnogo reagirovaniya Organizatsii Dogovora o kollektivnoy bezopasnosti [Agreement on the Collective Rapid Response Forces of the Collective Security Treaty],* Moscow, June 14, 2009.

Commager, Henry. *The American Mind: An Interpretation of American Thought and Character Since the 1880's.* New York: Yale University Press, 1950.

———. *The Empire of Reason: How Europe Imagined and America Realized the Enlightenment.* Garden City, NJ: Anchor Press, 1977.

The Command and General Staff School, *Principles of Strategy for an Independent Corps or Army in a Theater Of Operations.* Fort Leavenworth: The Command and General Staff School Press, 1936.

Convention on Prohibitions or Restrictions on the Use of Certain Conventional Weapons Which May be Deemed to be Excessively Injurious or to Have Indiscriminate Effects (adopted October 10, 1980, and entered into force December 2, 1983), 19 ILM 1523.

Convention on the Prohibition of the Development, Production, and Stockpiling of Bacteriological (Biological) and Toxin Weapons and on Their Destruction (adopted April 10, 1972, and entered into force on March 26, 1975), 11 ILM 309.

Convention on the Prohibition of the Development, Production, Stockpiling and Use of Chemical Weapons and on Their Destruction (adopted January 13, 1993, entered into force April 29, 1997), 32 ILM 800.

Convention on the Prohibition of the Use, Stockpiling, Production and Transfer of Anti-Personnel Mines and on Their Destruction (adopted September 18, 1997, entered into force March 1, 1999) 36 I.L.M 1507.

Cordesman, Anthony, and Arleigh Burke. *The "GAZA WAR": A Strategic Analysis*. Washington D.C.: Center for Strategic & International Studies (CSIS), 2009.

Cornish, Paul. "Unlawful Wounding: Codifying Interaction between Bullets and Bodies." In Paul Cornish and Nicholas Saunders (ed.), *Bodies in Conflict: Corporeality, Materiality and Transformation*. London: Routledge, 2014.

——— (ed.). *The Conflict in Iraq, 2003*. New York: Palgrave Macmillan, 2004.

Cornish, Paul, and Nicholas Saunders (ed.). *Bodies in Conflict: Corporeality, Materiality and Transformation*. London: Routledge, 2014.

Council of European Union. *EU Statement Following Israel's Operation Cast Lead in Gaza Calling for an Immediate and Permanent Ceasefire*, December 30, 2008.

Coupland, Robin, and Dominique Loye. "Legal and Health Issues: International Humanitarian Law and the Lethality or Non-Lethality of Weapons." In Malcolm Dando (ed.), *Non-Lethal Weapons: Technological and Operational Prospects*, Jane's Special Report, November 2000.

Cox, Michael. *US Foreign Policy after the Cold War: Superpower Without a Mission?* London: Royal Institute of Foreign Affairs, 1995.

Craig, Tim, Missy Ryan, and Thomas Gibbons-Neff. "By Evening, a Hospital. By Morning, a War Zone." *Washington Post*, October 10, 2015, https://www.washingtonpost.com/world/national-security/story-of-how-a-kunduz-hospital-was-shelled-by-us-gunship-in-question/2015/10/10/1c8affe2–6ebc-11e5–b31c-d80d62b53e28_story.html [accessed May 10, 2018].

Cullen, Timothy. *Lethality, Legality, and Reality: Non-Lethal Weapons for Offensive Air Support*. Maxwell Air Force Base: Air Command and Staff College, 2008.

Dando, Malcolm (ed.). *Non-Lethal Weapons: Technological and Operational Prospects*, Jane's Special Report, November 2000.

———. *A New Form of Warfare: The Rise of Non-Lethal Weapons*. London: Brassey's Ltd, 1996.

Danieli, Michal. "'Noal' Khanibal: adif hayal harug ni'asher hayal hatuf? ['Khanibal' Procedure: Is a Dead Soldier Preferable to a Kidnapped Soldier?]," *Mako*, October 17, 2011, http://mobile.mako.co.il/pzm-magazine/Article-d4a692186511331006.htm&sCh=3d385dd2dd5d4110&pId=1093150966 [accessed May 10, 2018].

Davis, Paul. "Military Transformation? Which Transformation, and What Lies Ahead?" In Stephen Cimbala (ed.), *The George W. Bush Defense Program: Policy, Strategy, and War*. Washington: Potomac Books, Inc., 2010.

Davison, Neil. *"Non-Lethal" Weapons*. Eastbourne: Palgrave Macmillan, 2009.

Dayan, Moshe. "Peulot Tzvayiot be'yemei shalom [Military Operations in Peacetime]." *Ma'arachot* 119 (1959): 55–63.

Defense Science Board, Department of Defense, *Challenges to Military Operations in Support of U.S. Interests*, Volume II, Main Report, Washington DC, December 2008.

Department of the Army Headquarters, United States Army, *AirLand Operations: A concept for the Evolution of AirLand Battle for the Strategic Army of the 1990s and Beyond: TRADOC Pamphlet 525–5*, Fort Monroe, August 4, 1991.

Department of the Army Headquarters, United States Army, *Concept for Nonlethal Capabilities in Army Operations: TRADOC Pamphlet 525–73*, Fort Monroe VA, December 1, 1996.

Department of the Army Headquarters, United States Army, *Force Operating Capabilities: TRADOC Pamphlet 525–66*, Fort Monroe, March 7, 2008.

Department of the Army Headquarters, United States Army, *Force XXI Operations: A Concept for the Evolution of Full-Dimensional Operations for the Strategic Army of the Early Twenty-First Century TRADOC Pamphlet 525–5*, Fort Monroe VA, August 1, 1994.

Dobbins, James. "Iraq: Winning the Unwinnable War." *Foreign Affairs* 84, no. 1(2005): 16–25.

Dombrowski, Peter, and Eugene Gholz. *Buying Military Transformation: Technological Innovation and the Defense Industry*. New York: Columbia University Press, 2006.

Donnelly, Christopher. *Red Banner: The Soviet Military System in Peace and War*. London: Jane's Information Group, 1988.

Dostoyevsky, Fyodor. *Dnevnik pisatelya [*A *Writer's Diary]*. Kaliningrad: Yantarnii Rasskaz, 2005.

———. *Polnoye sobraniye sochineniy v 30 tomakh. Tom 21– Dnevnik pisatelya 1873 god [The Omnibus in Thirty Volumes. Volume 21 – Writer's Diary 1873]*. Leningrad: Nauka, 1980.

Dotan, Amos. "Technologiyat meil' ruach: yoter mi'mahapecha technologit be'lechimat shir'yon [The Technology of Meil' Ruach : More than a Revolution in Armour Warfare]." *Ma'arachot* 448 (2013): 68–74.

Duncan, James. "A Primer on the Employment of Non-Lethal Weapons." *Naval Law Review* 45 (1998): 1–55.

Dunlap, Charles. "Law and Military Interventions: Preserving Humanitarian Values in 21st Conflicts." *The Proceedings to the Humanitarian Challenges in Military Intervention Conference*, Carr Center for Human Rights Policy Kennedy School of Government, Harvard University, Washington, November 29, 2001.

Dzyaloshinskiy, I. "Kul'tura, zhurnalistika, tolerantnost' [Culture, Journalism, Tolerance]." In I. Dzyaloshinskiy and M. Dzyaloshinskaya (ed.), *Rossiyskiye SMI: kak sozdayetsya obraz vraga [Russian Mass Media: How the image of enemy is created]*. Moscow: Moskovskoe Byuro po Pravam Cheloveka "Academia," 2008.

Dzyaloshinskiy, I., and M. Dzyaloshinskaya (ed.). *Rossiyskiye SMI: kak sozdayetsya obraz vraga [Russian Mass Media: How the Image of Enemy is Created]*. Moscow: Moskovskoe Byuro po Pravam Cheloveka "Academia," 2008.

Echevarria, Antulio. *Reconsidering the American Way of War: US Military Practice from the Revolution to Afghanistan*. Washington: Georgetown University Press, 2014.

———. *Towards an American Way of War*. Carlisle: Strategic Studies Institute, 2004.

"Eduto shel' nitzav Alik Ron bephnei va'adat Or, 3 september 2001, helek shlishi [The Testimony of Major General Alik Ron to the Or Commission, 3 September 2001, Part 6]." *Ha'aretz Online*, September 4, 2001, http://www.haaretz.co.il/misc/1.1555776 [accessed May 10, 2018].

Efremov, E. "Matematicheskoye modelirovaniye primeneniya oruzhiya neletal'nogo deystviya v mirotvorcheskikh operatsiyakh [The Mathematical Simulation of the Employment of Weapons with Non-Lethal Action during Peacemaking Operations]." *Voennaia Misl,'* no. 4 (2012): 39–43.

Eisenhower, Dwight. *Farewell Radio and Television Address to the American People*. January 17, 1961, http://www.presidency.ucsb.edu/ws/index.php?pid=12086&st=farewell&st1 [accessed May 10, 2018].

———. *Radio and Television Address to the American People on the Need for Mutual Security in Waging the Peace*. Washington, May 21, 1957, http://www.presidency.ucsb.edu/ws/index.php?pid=11042&st=&st1 [accessed May 10, 2018].

Eran, Oded. "International Aspects of Operation Pillar of Defence." *INSS Insight*, no. 386 (2012).

Eugene Jarecki. *Why We Fight*. Axiom Films, 2005.

European Working Group Non-Lethal Weapons (EWG-NLW), http://www.non-lethal-weapons.com/index.html [accessed May 10, 2018].

Evron, Yair. "Deterrence and its Limitations." In Shlomo Brom and Meir Elran (ed.), *The Second Lebanon War: Strategic Perspectives*. Tel Aviv: Institute for National Security Studies, 2007.

Farrell, Theo. *The Norms of War: Cultural Beliefs and Modern Conflict*. London: Lynne Rienner Publishers, 2005.

Farrell, Theo, and Terry Terriff. "Military Transformation in NATO: A Framework for Analysis." In Theo Farrell, Terry Terriff, and Frans Osinga (ed.), *Transformation Gap?: American Innovations and European Military Change*. Stanford: Stanford University Press, 2010.

——— (ed.). *The Sources of Military Change: Culture, Politics, Technology*. London: Lynne Rienner, 2002.

Farrell, Theo, Terry Terriff, and Frans Osinga (ed.). *Transformation Gap?: American Innovations and European Military Change*. Stanford: Stanford University Press, 2010.

The Federal Law of the Russian Federation № 3–ФЗ, *O Politsii [About Police]*, Moscow, March 1, 2011, Articles 21–22.

Fedotov, Georgii. *Sud'ba i grekhi Rossii [The Fate and the Sins of Russia]*, Volume II. Saint-Petersburg: Sofiya, 1991.

Ferling, John. *A Wilderness of Miseries: War and Warriors in Early America*. Westport: Greenwood Press, 1980.

Fidler, David. "Non-Lethal Weapons and International Law—Three Perspectives on the Future." *Medicine, Conflict and Survival* 17, no. 3 (2001): 194–206.

———. "The International Legal Implications of 'Non-Lethal' Weapons." *Michigan Journal of International Law* 21, no. 51 (1999): 51–100.

———. "The Meaning of Moscow: 'Non-Lethal' Weapons and International Law in the Early 21st Century." *International Review of the Red Cross* 87, no. 857 (2005): 525–52.

Finkelstein, Norman. *Beyond Chutzpah: On the Misuse of Anti-Semitism and the Abuse of History*, second edition. London: Verso, 2008.

Florovskii, Georgii. "Yevraziyskiy soblazn [Eurasian Temptation]." In L. Novikova and I. Sizemskaya (ed.), *Mir Rossii - Yevraziya: Antologiya [The World of Russia - Eurasia: An Anthology]*. Moscow: Vysshaya shkola, 1995.

Foht, Elizabeth. "Rossiyskiye korabli v Kaspiyskom more nanesli udary po pozitsiyam IG v Sirii," ["Russian Ships in the Caspian Sea Attacked the Positions of ISIS in Syria"]. *RBC*, October 7, 2015, http://www.rbc.ru/politics/07/10/2015/561502039a7947add1998f52 [accessed May 10, 2018].

Ford, Gerald. *Address Before a Joint Session of the Congress Reporting on the State of the Union*, Washington, January 15, 1975, http://www.presidency.ucsb.edu/ws/?pid=4938 [accessed May 10, 2018].

Fossum, Robert, and John Roth. *The American Dream*. South Shields: Peterson Printers, 1981.

Freedman, Robert (ed.). *World Politics and the Arab-Israeli Conflict*. New York: Pergamon Press, 1979.

Fridman, Ofer. "'Are We Ready for the Revolution of Nonlethal Weapons?': Using a Comprehensive RMA Model to Examine the Current Strategic Situation." *Comparative Strategy* 32, no. 3 (2013): 192–206.

Fridman, Ofer. "Revolutions in Military Affairs that Did not Occur: A Framework for Analysis." *Comparative Strategy* 35, no. 5 (2016): 388–406.

Fukuyama, Francis. *The End of History and the Last Man*. Penguin Books, 1992.

"Full text: Obama's victory speech." *BBC News*, November 5, 2008, http://news.bbc.co.uk/1/hi/world/americas/us_elections_2008/7710038.stm [accessed May 10, 2018].

Gad, Yair. "Israeli Existential Anxiety: Cultural Trauma and the constitution of National Character." *Social Identities: Journal of the Study of Race, Nation and Culture* 20, no. 4–5 (2014): 346–62.

———. *Tzofen ha'isareliyut: aseret ha'dibrot shel' shnot he'al'paiim [The Code of Israeliness: The Ten Commandments for the 21 st century]*. Jerusalem: Keter Books, 2011.

Galdi, Theodor. *Revolution in Military Affairs?—Competing Concepts, Organizational Responses, Outstanding Issues*. Congressional Research Service Report to the Congress (CRS 951170F, December 11, 1995).

"Gaza EU Statement a Misunderstanding—Czech Minister." *Reuters*, January 4, 2009, https://www.reuters.com/article/idUSL4734544 [accessed May 10, 2018].

Gazit, Shlomo. *Hamakel vehagezer: hamimshal haisraeli be'yehuda veshomron [The Stick and the Carrot: The Israeli Government in Judea and Samaria]*. Tel Aviv: Zmora-Bitan, 1985.

Giglio, James. *The Presidency of John F. Kennedy*. Lawrence: University Press of Kansas, 1991.

Girgus, Sam (ed.). *The American Self: Myth, Ideology, and Popular Culture*. Albuquerque: University of New Mexico Press, 1981.

"The Goldstone Report." *Journal of Palestine Studies* 39, no. 2 (2010, Winter): 109.

Golov, Avner. "Rethinking the Deterrence of Hamas." In Anat Kurz and Shlomo Brom (ed.), *The Lessons of Operation Protective Edge*. Tel Aviv: Institute for National Security Studies, 2014.

Gordon, Philip, and Jeremy Shapiro. *Allies at War: America Europe and the Crisis over Iraq*. New York: McGraw-Hill, 2004.

Graebner, Norman, Richard Burns, and Joseph Siracusa. *Reagan, Bush, Gorbachev: Revisiting the End of the Cold War*. Westport: Greenwood Publishing Group, 2008.

Granger, Dewey. *Integration of Lethal and Nonlethal Fires: The Future of the Joint Fires Cell*. Fort Leavenworth: Command and General Staff College, 2009.

Gray, Colin. *Irregular Enemies and the Essence of Strategy: Can the American Way of War Adapt?* Carlisle: Strategic Studies Institute, 2006.

———. *Making Strategic Sense of Cyber Power: Why the Sky Is Not Falling*. Carlisle: Strategic Studies Institute, 2013.

———. *Modern Strategy*. Oxford: Oxford University Press, 1999.

———. "National Style in Strategy: The American Experience." *International Security* 6, no. 2 (1981): 21–47.

———. *Out of the Wilderness: Prime Time for Strategic Culture*. Fort Belvoir: Defence Threat Reduction Agency, 2006.

———. *Perspectives on Strategy*. Oxford: Oxford University Press, 2013.

———. *Strategy for Chaos: Revolutions in Military Affairs and The Evidence of History*. London: Frank Cass Publishers, 2002.

Grenier, John. *The First Way of War: American War Making on the Frontier, 1607–1814*. Cambridge: Cambridge University Press, 2005.

Grinberg, Hanan. "Ha'til she'kofef et egrof ha'plada [The Missile that Bended the Iron Fist]." *Ynet*, http://www.ynet.co.il/articles/0,7340,L-3297389,00.html [accessed May 10, 2018].

Grinfeld, Gabi. "Lean halchu ha'lekhachim kulam? [Where Did All Lessons Go To?]." *Ma'arachot* 380–381 (2001): 83.

Gronik, Arie. "Yahasei Israel ve'um – shalosh tkufot ve'shalosh she'elot [The Israel-U.N. Relations – Three Periods and Three Questions]." In Benjamin Noiberger and Arie Gronik (ed.), *Mediniut Hutz bein imut le'esderim – Israel 1948–2008 [The Foreign Policy between Conflict and Agreement – Israel 1948–2008]*, Vol. B. Raanana: The Open University of Israel, 2008.

Gross, Michael. *Moral Dilemmas of Modern War: Torture, Assassination, and Blackmail in an Age of Asymmetrical Conflict*. New York: Cambridge Press, 2010.

———. "The Second Lebanon War: The Question of Proportionality and the Prospect of Non-Lethal Warfare." *Journal of Military Ethics* 7, no. 1 (2008): 1–22.

Gubanov, V. *Russkiy natsional'nyy kharakter v kontekste politicheskoy zhizni Rossiii [Russian National Character in the Context of the Russian Political Life]*. Sent-Petersburg: Izdatel'skiy Tsentr SPbGMGU, 1999.

Gudkov, Lev. "God chechenskoy voyny v obshchestvennom mnenii Rossii [A Year of Chechen Way in the Russian Public Opinion]." *Druzhba Narodov*, no. 2 (1996).

———. "Pamyat' o voyne i massovaya identichnost' rossiyan [Memory of War and the Russian Identity]." *Neprikosnovennyy Zapas*, no. 2–3(40–41) (2005).

———. "Tsinizm 'neperekhodnogo' obshchestva [Cynicism of 'Non-transitional' Society]." *Vestnik obshchestvennogo mneniya. Dannyye. Analiz. Diskussii* 2(76) (2005): 43–82.

Ha'rel, Amos. "Tzal inase lefazer hafganot hamoniyot bepagazei tankim im k'liey plastik sh'osim rak ra'ash [The IDF Will Try to Disperse Massive Riots with Tanks' Shells with Plastic Warheads that Make Noise Only]." *Ha'aretz Online*, February 28, 2005, http://www.haaretz.co.il/misc/1.1505848 [accessed May 10, 2018].

Handel, Michael. *Israel's Political Military Doctrine*. Cambridge, MA: Harvard University Press, 1973.

Harkabi, Yehoshafat. "Hir'urim al torat ha'bitachon ha'leumi [Thoughts about the National Security Doctrine]." *Ma'arachot*, no. 270–271 (1979): 42–45.

———. *The Bar Kokhba Syndrome: Risk and Realism in International Politics*. Chappaqua: Rossel Books, 1983.

Harkov, Lahav. "Former British Commander in Afghanistan: No Army Acts with as Much Discretion as IDF Does." *Jerusalem Post*, April 9, 2014.

Harrison, Richard. *The Russian Way of War: Operational Art*, 1904–1940. Lawrence, KS: University Press of Kansas, 2001.

Harvardt, A., et al. "Perspektivy primeneniya oruzhiya neletal'nogo deystviya v urbanizirovannom rayone [The Perspectives of the Employment of Weapons with Non-Lethal Action in Urban Area]." *Strategicheskaya Stabil'nost'*, no. 1(61) (2014): 64–71.

Heal, Charles. "The Quest for the Magic Bullet." In Malcolm Dando (ed.), *Non-Lethal Weapons: Technological and Operational Prospects*, Jane's Special Report, November 2000.

Hein, Laura, and Mark Selden (ed.). *Living with the Bomb: American and Japanese Cultural Conflicts in the Nuclear Age*. Armonk: M.E. Sharpe, 1997.

Heller, Mark. "Israeli Deterrence in the Aftermath of Protective Edge." In Anat Kurz and Shlomo Brom (ed.), *The Lessons of Operation Protective Edge*. Tel Aviv: Institute for National Security Studies, 2014.

———. "The International Dimension: Why So Few Constrains on Israel?" In Shlomo Brom and Meir Elran (ed.), *The Second Lebanon War: Strategic Perspectives*. Tel Aviv: Institute for National Security Studies, 2007.

Herring, G. K. "The War in Afghanistan: A Strategic Analysis." In Williamson Murray (ed.), *National Security Challenges for the 21st Century*. Carlisle: US Army War College, 2003.

Herzog, Chaim. *The Arab-Israeli Wars: War and Peace in the Middle East*, 3rd edition, revised and updated by Shlomo Gazit. Barnsley: Frontline Books, 2010.

Heuser, Beatrice. *Nuclear Mentalities? Strategies and Beliefs in Britain, France and the FRG*. London: MacMillan Press, 1998.

———. *The Bomb: Nuclear Weapons in their Historical, Strategic and Ethical Context*. London: Longman, 2000.

Hirsch, Adam. "The Collision of Military Cultures in Seventeenth-Century New England." *Journal of American History* 74, no. 4 (1988): 1187–212.

Homeland Security Research Cooperation (HSRC). *Non-Lethal Weapons: Technologies and Global Market—2012–2020*, 2011, http://www.homelandsecurityresearch.com/2011/10/non-lethal-weapons-technologies-global-market-2012–2020/ [accessed May 10, 2018].

Howard, Michael. "What's in a Name? How to Fight Terrorism." *Foreign Affairs* 81, no. 1 (2002): 8–13.

Hundley, Richard. *Past Revolutions Future Transformation: What Can History of Revolutions in Military Affairs Tell Us about Transforming the US Military?* Santa Monica: RAND, 1999.

Hunt, Michael. *Ideology and the U.S. Foreign Policy*. New Haven: Yale University Press, 1987.

Huntington, Samuel. "The Erosion of American National Interests." *Foreign Affairs* 76, no. 5 (1997).

The IDF General Officer Commanding Army Headquarters. *Puma*, http://mazi.idf.il/4986–5648–HE/IGF.aspx [accessed May 10, 2018] (Hebrew).

Igar, Moshe, Yosef Guvrin, and Rei Oded (ed.). *Misrad ha'hutz – 50 ha-shanim ha'rishonot [The Foreign Affairs Office—The 50 First Years]*. Jerusalem: Keter, 2002.

Ikenberry, John. "America's Imperial Ambitions." *Foreign Affairs* 81, no. 5 (2002): 44–60.

Ilyin, Ivan. *Natsional'naya Rossiya, nashi zadachi, [National Russia, Our Goals]*. Moscow: Eskimo Algoritm, 2011.

———. *Puti Rossii [Russian Ways]*. Moscow: Vagrius, 2007.

———. *Sushchnost' i svoyeobraziye Russkoy kul'tury [The Essence and Peculiarity of Russian Culture]*. Moscow: Russkaya Kniga-XXI Vek, 2007.

Ilyin, L., and V. Rikhlin. "Inkapasitanty kak oruzhiye neletal'nogo deystviya [Incapacitants as Weapons with Non-Lethal Action]." *Voennaia Misl'*, no. 9 (2014): 37–42.

Inbar, Efraim. *Israel eina mevudedet [Israel Is Not Isolated]*. Tel-Aviv: Begin-Sadat Center for Strategic Studies, 2013.

The Independent International Fact-Finding Mission on the Conflict in Georgia. *Report*, Volume II, 2009.

"Infinite Justice, out—Enduring Freedom, in." *BBCNews*, September 25, 2001, http://news.bbc.co.uk/1/hi/world/americas/1563722.stm [accessed May 10, 2018].

Inosemzev, Vladislav. *Poteryannoye desyatiletiye [The Lost Decade]*. Moscow: Moskovskaya shkola politicheskih issledovanii, 2013.

The Institute for Non-Lethal Defense Technologies (INLDF), https://www.arl.psu.edu/content/institute-non-lethal-defense-technologies-0 [accessed May 10, 2018].

The Intelligence and Terrorism Information Center (ITIC). *Estimate of the Number of Hamas Operatives Killed in Operation Protective Edge (Part Eleven)*, Tel-Aviv, July 20, 2015.

"Interview with A. Starunskii, Former Deputy Commander of the Joint Group of Forces in the North Caucasus (2002–2003)," published in *Zashchita i Bezopasnost'*, no. 4(47) (2008): 6–8 (Russian).

"Interview with Defence Minister Yitzhak Rabin," published in *Jerusalem Post*, February 16, 1989.

"Interview with K. Kosachev, the Chairman of the State Duma Committee on Foreign Affairs," published in *Mezhdunarodnaya Zhizn,'* no. 4 (2010): 4–9 (Russian).

"Interview with S. Ryabkov, the Deputy Foreign Minister of the Russian Federation," published in *Mezhdunarodnaya Zhizn,'* no. 1 (2014): 4–16 (Russian).

"Interview with S. Ryabkov, The Deputy Minister of Foreign Affairs of the Russian Federation," published in *Mezhdunarodnaya Zhizn,'* no. 8 (2014): 2–12 (Russian).

"Interview with Sergey Lavrov, the Foreign Minister of the Russian Federation," published in *Mezhdunarodnaya Zhizn,'* no. 9 (2012): 20 (Russian).

The Israel Democracy Institute. *The Peace Index: August 2014*, August 12, 2014, http://www.peaceindex.org/indexMonthEng.aspx?num=283&monthname=August#.VkIiwdLhDcs [accessed May 10, 2018].

The Israel Democracy Institute. *The Peace Index: October 2015*, October 5, 2015, http://www.peaceindex.org/indexMonthEng.aspx?num=298 [accessed May 10, 2018].

Israel Ministry of Foreign Affairs. *FAQ: The Operation in Gaza—Factual and Legal Aspects*, August 16, 2009, http://mfa.gov.il/MFA/ForeignPolicy/FAQ/Pages/FAQ-Operation_in_Gaza-Legal_Aspects.aspx#13 [accessed May 10, 2018].

———. *Goldstone Report: Interviews with PM Netanyahu on Israeli Television* (communicated by the Prime Minister's Media Adviser), September 17, 2009, http://mfa.gov.il/MFA/PressRoom/2009/Pages/Goldstone_Report_Interviews_PM_Netanyahu_17–Sep-2009.aspx [accessed May 10, 2018].

———. *Israel Rejects General Assembly Resolution*, November 5, 2009,http://mfa.gov.il/MFA/PressRoom/2009/Pages/Israel_rejects_General_Assembly_resolution_5–Nov-2009.aspx [accessed May 10, 2018].

———. *Israel's Analysis and Comments on the Gaza Fact-Finding Mission Report* (communicated by the Foreign Ministry Spokesperson), September 15, 2009, http://mfa.gov.il/MFA/PressRoom/2009/Pages/Israel_initial_reaction_Goldstone_Mission_15–Sep-2009.aspx [accessed May 10, 2018].

———. *The 2014 Gaza Conflict, 7 July – 26 August 2014: Factual and Legal Aspects*, Jerusalem, May 2015.

The Israel State Bureau of Statistics. *Israel's Foreign Trade in Goods, by Country - September 2015*, Jerusalem, October 25, 2015, http://www.cbs.gov.il/reader/newhodaot/hodaa_template_eng.html?hodaa=201516283 [accessed May 10, 2018].

Jentleson, Bruce. *American Foreign Policy: The Dynamics of Choice in the 21st Century*, 3rd edition. New York: W.W. Norton & Company, 2007.

Jervis, Robert. *American Foreign Policy in a New Era*. New York: Routledge, 2005.

Jochnik, Chris, and Roger Normand. "The Legitimation of Violence: A Critical History of Laws of War." *Harvard International Law Journal* 35, no. 1 (1994): 49–95.

Johnson, Adrian, and Saqeb Mueen (ed.). *Short War, Long Shadow: The Political and Military Legacies of the 2011 Libya Campaign*, Whitehall Report 1–12. London: Royal United Services Institute for Defense and Security Studies, 2012.

Johnson, Lyndon. *Remarks at Syracuse University on the Communist Challenge in Southeast Asia*, New York, August 5, 1964, http://www.presidency.ucsb.edu/ws/?pid=26419 [accessed May 10, 2018].

———. *Remarks to Committee Members on the Need for Additional Appropriations for Military Purposes in Viet-Nam and the Dominican Republic*, Washington, May 4, 1965, http://www.presidency.ucsb.edu/ws/?pid=26938 [accessed May 10, 2018].

Joint Non-Lethal Weapons Directorate, *Annual Report 2012: Non-Lethal Weapons for Complex Environments*, 2012.

———. *Current Non-Lethal Weapons*, http://jnlwp.defense.gov/CurrentNonLethalWeapons.aspx [accessed May 10, 2018].

———. *Developed Non-Lethal Weapons: Acoustic Hailing Devices,* http://jnlwp.defense.gov/CurrentNonLethalWeapons/AcousticHailingDevices.aspx [accessed May 10, 2018].

———. *Developing Non-Lethal Weapons*, http://jnlwp.defense.gov/DevelopingNonLethalWeapons.aspx [accessed May 10, 2018].

———. *Developing Non-Lethal Weapons: The Pre-Emplaced Electric Vehicle Stopper*, http://jnlwp.defense.gov/DevelopingNonLethalWeapons/PreEmplacedElectricVehicleStopper.aspx [accessed May 10, 2018].

———. *Future Non-Lethal Weapons,* http://jnlwp.defense.gov/FutureNonLethalWeapons.aspx [accessed May 10, 2018].

———. *Mission Payload Module – Non-Lethal Weapons System (MPM-NLWS) Fact Sheet,* October 2011.

———. *Non-Lethal Weapons (NLW): Reference Book*, June 30, 2011.

Joint Non-Lethal Weapons Program, *Annual Report*, 2012, 7.

———. *Non-Lethal Weapons Requirement Fact Sheet*, http://jnlwp.defense.gov/pdf/pressroom/NLW%20Requirements%20Fact%20Sheet%20Oct%202011.pdf [accessed: 24 June 2015].

———. *Organization*, http://jnlwp.defense.gov/About/Organization.aspx [accessed: 24 June 2015].

———. *Purpose*, https://jnlwp.defense.gov/About/Planning-Guidance-Vision-Mission-Definition/ [accessed May 10, 2018].

Kagan, Robert. *Paradise & Power: America and Europe in the New World Order*, Revised and Expanded Edition. London: Atlantic Books, 2004.

Kagan, Robert, and William Kristol (ed.). *Present Dangers: Crisis and Opportunity in American Foreign and Defense Policy*. San Francisco: Encounter Books, 2000.

Kane, Thomas. *Military Logistics and Strategic Performance*. London: Frank Cass, 2001.

Kaplan, Robert. *Imperial Grunts: The Ameriran Military on the Ground*. New York: Random House, 2005.

Kaplinsky, Moshe. "The IDF in the Years before the Second Lebanon War." *Military and Strategic Affairs* 1, no. 2 (2009): 25–37.

Karr, Ronald. "'Why Should You Be So Furious?': The Violence of the Pequot War." *Journal of American History* 85, no. 2 (1998): 876–909.

Kaspi, Dan. "Shuv yesh lanu oiev meshutaf [We Have a Mutual Enemy Again]." *Ynet*, July 31, 2009, http://www.ynet.co.il/articles/0,7340,L-3753712,00.html [accessed May 10, 2018].

Kennedy, John. *Inaugural Address*, Washington, January 20, 1961, http://www.presidency.ucsb.edu/ws/?pid=8032 [accessed May 10, 2018].

———. *Radio and Television Report to the American People on the Soviet Arms Buildup in Cuba*, Washington, October 22, 1962, http://www.jfklibrary.org/Research/Research-Aids/JFK-Speeches/Cuba-Radio-and-Television-Report_19621022.aspx [accessed May 10, 2018].

Kharkhordin, Oleg. *The Collective and the Individual in Russia*. Berkeley: University of California Press, 1999.

Khlobustov, Oleg. "Chechnya, Terrorizm i Kontrterroristicheskiye Operatsii v Zerkale Obshchestvennogo Mneniya [Chechnya, Terrorism and Counter-Terrorism Operations in the Mirror of Public Opinion]." *Vlast,'* no. 5 (2003): 54–59.

Khoury, Jack, et al. "Israel Girds for Diplomatic War Over 'Biased' UN Gaza Report." *Ha'aretz*, September 15, 2009, http://www.haaretz.com/israel-girds-for-diplomatic-war-over-biased-un-gaza-report-1.7831 [accessed May 10, 2018].

Kier, Elizabeth. *Imagining War: French and British Military Doctrine between the Wars*. Princeton: Princeton University Press, 1997.

Kimmerling, Baruch. "Making Conflict a Routine: Cumulative Effects of the Arab-Jewish Conflict Upon Israeli Society." In Moshe Lissak (ed.), *Israeli Society and Its Defense Establishment: The Social Political Impact pf a Protracted Violent Conflict*. London: Frank Cass, 1984.

———. *The Invention and Decline of Israeliness: State, Society, and the Military*. Berkley: University of California Press, 2001.

Klar, Yechiel, Noa Schori-Eyal, and Yonat Klar. "The 'Never Again' State of Israel: The Emergence of the Holocaust as a Core Feature of Israeli Identity and Its Four Incongruent Voices." *Journal of Social Issues* 69, no. 1 (2013): 125–43.

Klieman, Aaron. *Israel & The World after 40 Years*. McLean: Pergamon-Brassey's International Defense Publishers, 1990.

Knickerbocker, Brad. "The Fuzzy Ethics of Nonlethal Weapons." *Christian Science Monitor*, February 14, 2003, http://www.csmonitor.com/2003/0214/p02s01–usmi.html [accessed May 10, 2018].

Kober, Avi. "The IDF in the Second Lebanon War: Why the Poor Performance?" *Journal of Strategic Studies* 31, no. 1: 3–40.

———. "The Intellectual and Modern Focus in Israeli Military Thinking as Reflected in *Ma'arachot* Articles, 1948–2000." *Armed Forces & Society* 30, no. 2 (2003): 141–60.

———. "The Rise and Fall of Israeli Operational Art, 1948–2008." In John Olsen and Martin Creveld Van (ed.), *The Evolution of Operational Art: From Napoleon to the Present*. Oxford, Oxford University Press, 2010.

———. "What Happened to Israeli Military Thought?" *Journal of Strategic Studies* 34, no. 5 (2011): 707–32.

Koistinen, Paul. *The Military-Industrial Complex*. New York: Praeger Publishers, 1980.

Kokoshin, Andrei. *Soviet Strategic Thought, 1917–91*. Cambridge, MA: MIT Press, 1998.

Kondrashev, Andrey. *Krim. Put' na Rodinu [Crimea . The Way Home]*. Vserossiyskaya Gosudarstvennaya Televizionnaya i Radioveshchatel'naya Kompaniya, 2015 (Russian).

Koplow, David. *Death by Moderation: The U.S. Military's Quest for Useable Weapons*. New York: Cambridge University Press, 2010.

———. *Non-Lethal Weapons: The Law and Policy of Revolutionary Technologies for the Military and Law Enforcement*. New York: Cambridge Press, 2006.

———. "Red-Teaming NLW: A Top Ten List of Criticisms about Non-Lethal Weapons." *Case Western Reserve Journal of International Law* 47, Issue 1 (2015): 229–38.

Korchak, V., and S. Smirnov. "Regulirovaniye normativno-pravovykh otnosheniy pri sozdanii oruzhiya neletal'nogo deystviya [The Legislative Regulation Regarding the Development of Weapons with Non-Lethal Action]." *Kompetentnost',* no. 4–5(45–46) (2007): 3–10.

Kotlyar, V. "'Otvetstvennost' pri zashchite' i 'Arabskaya Vesna' ['The Responsibility to Protect' and the 'Arab Spring']." *Mezhdunarodnaya Zhizn,'* no. 9 (2012): 99–114.

Krepinevich, Andrew. "Cavalry to Computer: The Pattern of Military Revolutions." *National Interest*, no. 37 (1994): 30–42.

Kuperman, Raanan. "The Impact of Internal Politics on Israel's Reprisal Policy During the 1950s." *Journal of Strategic Studies* 24, no.1: 1–28.

Kurz, Anat, and Shlomo Brom (ed.). *The Lessons of Operation Protective Edge*. Tel-Aviv: Institute for National Security Studies, 2014.

Lamb, Timothy. *Emerging Non-Lethal Weapons Technology and Strategic Policy Implications for 21 st Century Warfare*. Carlisle Barracks: U.S. Army War College, 1998.

Lambeth, Benjamin. *The Transformation of American Air Power*. Ithaca: Cornell University Press, 2000.

Landau, Emily, and Azriel Bermant. "Iron Dome Protection: Missile Defense in Israel's Security Concept." In Anat Kurz and Shlomo Brom (ed.), *The Lessons of Operation Protective Edge*. Tel-Aviv: Institute for National Security Studies, 2014.

Lapid, Efraim. "Haintifada harishona, ha'avanim she'banu et heskem Oslo [The First Intifada, the Stones that Built the Oslo Agreements]." *IsraelDefense*, December 8, 2013.

Lapkin, V., and V. Pantin. "Dinamika obraza Rossii i tsikly reform-kontrreform [The Dynamics of Russia and the Reform/Counter-Reform Cycles]." *Rossiya i musul'manskiy mir*, no. 5 (2008): 15–22.

Larson, Eric, and Bogdan Savych. *American Public Support for U.S. Military Operations from Mogadishu to Baghdad*. Santa Monica: The RAND Corporation, 2005.

———. *Misfortunes of War: Press and Public Reaction to Civilian Deaths in Wartime*. Santa Monica: The RAND Corporation, 2006.

Larson, Robert. *The British Army and the Theory of Armoured Warfare, 1918–1940*. London: Associated University Press, 1984.

Lavrov, Sergey. "Kak okonchatel'no zavershit' kholodnuyu voynu? [How to finally end the Cold War?]." *Mezhdunarodnaya Zhizn'*, no. 5 (2009): 3–12.

———. *Speech at the First Forum of Young Diplomats*, April 25, 2014, published in *Mezhdunarodnaya Zhizn,'* no. 5 (2014): 5–20 (Russian).

———. "Vneshnepoliticheskaya filosofiya Rossii [The Foreign Policy Philosophy of Russia]." *Mezhdunarodnaya Zhizn,'* no. 4 (2013): 2.

Leffler, Melvyn, and Jeoffrey Lergo (ed.). *To Lead the World: American Strategy after the Bush Doctrine*. Oxford: Oxford University Press, 2008.

Lenin, Vladimir. *Sotsializm i voyna [Socialism and War]*. Moscow: Partizdat TsK VKPb, 1937.

Leon, Dan, and Yehuda Adlin (ed.). *Chaim Weizmann: Statesman of the Jewish Renaissance*. Jerusalem: Alpha Press, 1974.

Levinson, Aleksey. "Voyna, voyny, voyne... [War, Wars, to War. . .]." *Neprikosnovennyy Zapas*, no. 2–3(40–41) (2005).

Levy, Shai. "Ason ha'nagmashim: lama tzal meshtamesh be'kli lo memugan mi-1971? [The Disaster of APCs: Why does the IDF uses vehicles with no armour from 1971?]." *Mako*, July 22, 2014, http://mobile.mako.co.il/pzm-israel-wars/operation-protective-edge/Article-2fab4a8f83e5741006.htm [accessed May 10, 2018].

———. "Neshek (Hal) Hereg [(Non) Lethal Weapons]." *Mako*, http://www.mako.co.il/pzm-magazine/Article-c4b5b1a6475a531006.htm [accessed May 10, 2018].

Lewer, Nick (ed.). *The Future of Non-Lethal Weapons: Technologies, Operations' Ethics and Law*. London: Frank Cass Publishers, 2002.

Lewer, Nick, and Steven Schofield. *Non Lethal Weapons: A Fatal Attraction? Military Strategies and Technologies for 21st Century Conflict*. London: Zed Books, 1997.

Lewis, Adrian. *The American Culture of War: The History of U.S. Military Force from World War II to Operation Iraqi Freedom*. New York: Routledge, 2007.

Lieutenant Colonel Y. and Colonel Y. "Mischakei milchama memuchshavim be tzal [Computerised War-games in the IDF]." *Ma'arachot* 343 (1995): 28–31.

Lieven, Anatol. *Chechnya: Tombstone of Russian Power*. New Haven, CT: Yale University Press, 1999.

Lindsay, James. "George W. Bush, Barack Obama and the Future of US Global Leadership." *International Affairs* 87, no. 4 (2011): 765–79.

Lipman, M. (ed.). *Povestka dnya dlya novogo prezidentstva [Agenda for the New Presidency]*. Moscow: Moscow Carnegie Centre, Rossiyskaya Politicheskaya Entsiklopediya, 2012.

Lissak, Moshe (ed.). *Israeli Society and Its Defense Establishment: The Social Political Impact of a Protracted Violent Conflict*. London: Frank Cass, 1984.

Littlewood, Jez. "NLWs and Urban Warfare: Aspects of US Thinking." In Malcolm Dando (ed.), *Non-Lethal Weapons: Technological and Operational Prospects*, Jane's Special Report, November 2000.

Lock-Pullan, Richard. *US Intervention Policy and Army Innovation: From Vietnam to Iraq*. London: Routledge, 2006.

———. "U.S. Military Strategy, Strategic Culture, and the 'War on Terror.'" In John Owens and John Dumbrell (ed.), *America's "War on Terrorism": New Dimensions in U.S. Government and National Security*. Lanham: Lexington Books, 2008.

Lomsky-Feder, Edna, and Eyal Ben-Ari (ed.). *The Military and Militarism in Israeli Society*. New York: State University of New York Press, 1999.

Lorell, Mark, and Charles Kelly. *Casualties, Public Opinion, and Presidential Policy during Vietnam War*. Santa Monica: The RAND Corporation, 1985.

Lorenz, F. M. "Non-Lethal Force: The Slippery Slope to War?" *Parameters* XXVI, no. 3 (1996): 52–62.

Lossky, Nikolay. *Kharakter Russkogo naroda – Kniga 1 [The Character of the Russian People – Book 1]*. Frankfurt Main: Possev-Velrlag, 1957.

Lovelace, Douglas, and Steven Metz. *Nonlethality and American Land Power: Strategic Context and Operational Concepts*. Carlisle: Strategic Studies Institute, 1998.

Lovell, David, and Igor Primoratz (ed.). *Protecting Civilians During Violent Conflict: Theoretical and Practical Issues for the 21st Century*. Surrey: Ashgate Publishing Limited, 2012.

Lowenheim, Oded, and Gadi Heimann. "Revenge in International Politics." *Security Studies* 17, no. 4: 685–724.

Lupsha, Peter. "American Values and Organized Crime: Suckers and Wiseguyes." In Sam Girgus (ed.), *The American Self: Myth, Ideology, and Popular Culture*. Albuquerque: University of New Mexico Press, 1981.

Luz, Ehud. "'Jewish Ethics' as an Argument in the Public Debate Over the Israeli Reaction to Palestinian Terror." *Israeli Studies* 7, no. 3 (2002): 134–56.

Lyaschenko, V. *Torgovlya Oruzhiyem: oruzhiye neletal'nogo deystviya, sozdaniye, proizvodstvo, primeneniye, operatsii na mirovom rynke [Trade in Weapons: Weapons with Non-Lethal Action, Development, Production, Employment, Operations on the World Market]*. Moscow: Economika, 2011.

Mack, Andrew. "Why Big Nations Lose Small Wars: The Politics of Asymmetric Conflict." *World Politics* 27, no. 2 (1975): 175–200.

Mahnken, Thomas. *Technology and the American Way of War Since 1945*. New York: Columbia University Press, 2010.

———. *United States Strategic Culture*. Washington: Science Applications International Corp, 2006.

Malcolm Dando (ed.). *Non-Lethal Weapons: Technological and Operational Prospects*, Jane's Special Report, November 2000.

Malone, David. *The International Struggle Over Iraq: Politics and the UN Security Council 1980–2005*. Oxford: Oxford University Press, 2006.

Mandel, Robert. *The Meaning of Military Victory*. London: Lynne Rienner Publishers, 2006.

Manor, Jonatan. "Kishalon ha'asbara or kishalon ha'mediniyut? [A Failure of the Hasbara or a Failure of the Policy?]." In Benjamin Noiberger and Arie Gronik (ed.), *Mediniut Hutz bein imut le'esderim – Israel 1948–2008 [The Foreign Policy between Conflict and Agreement – Israel 1948–2008]*, Vol. B. Raanana: The Open University of Israel, 2008.

Mantzur-Karmel, Tamir. "Likrat hel hagana mirchavit [Towards a Territorial Defence Force]." *Ma'arachot* 459 (2015): 20–27.

MARCH GROUP LLC, http://www.shoker.ru/ (Russian) [accessed May 10, 2018].

Massingham, Eve. "Conflict without Casualties . . . A Note of Caution: Non-Lethal Weapons and International Humanitarian Law." *International Review of the Red Cross* 94, no. 886 (2012): 673–85.

Mazarr, Michael. *The Military Technical Revolution: A Structural Framework. Final Report of the CSIS Study Group on the MTR*. Washington, DC: Center for Strategic and International Studies, 1993.

McMaster, H. R. "On War: Lessons to be Learned." *Survival* 50, no. 1 (2008): 19–30.

McNab, Robert, and Scott Richard. "Non-Lethal Weapons and the Long Tail of Warfare." *Small Wars & Insurgencies* 20, no. 1 (2009): 141–59.

Mearsheimer, John, and Stephen Walt. *The Israel Lobby and US Foreign Policy*. New York: Penguin Books, 2007.

Medvedev, Dmitry. *Press conference after a meeting with French President Nicolas Sarkozy*, 8 September 2008, http://kremlin.ru/events/president/transcripts/1330 [accessed May 10, 2018] (Russian).

———. *Russia Forward!*, 2009, http://www.kremlin.ru/transcripts/5413 [accessed: 08 August 2014] (Russian).

———. *Statement in Relation to the Recognition of the Independence of Abkhazia and South Ossetia*, Moscow, August 26, 2008, http://news.kremlin.ru/news/1223 [accessed May 10, 2018] (Russian).

———. "Why I had to Recognise Georgia's Breakaway Regions." *Financial Times*, August 26, 2008, https://www.ft.com/content/9c7ad792–7395–11dd-8a66–0000779fd18c [Accessed May 10, 2018].

Meernik, James. *The Political Use of Military Force in US Foreign Policy*. Aldershot: Ashgate Publishing Limited, 2004.

Merget, Kenneth. "Non-Lethal Weapons and the Possibility of Radical New Horizons for the Laws of War: Why Kill, Wound and Hurt (Combatants) at All?" *Social Science Research Network*, July 1, 2008, http://papers.ssrn.com/sol3/papers.cfm?abstract_id=1295348 [accessed May 10, 2018].

Merom, Dror. "Hamasokim neged hafganot hamoniyot [Helicopters against Massive Demonstrations]." *Bit'aon Hel' Avir*, no. 67(168) (1989): 14–15.

Merom, Gil. *How Democracies Lose Small Wars*. Cambridge: Cambridge University Press, 2003.

Mesloh, Charlie, Mark Henych, and Ross Wolf. *Less Lethal Weapon Effectiveness, Use of Force, and Suspect & Officer Injuries: A Five-Year Analysis. Report to the National Institute of Justice*. Florida Gulf Coast University, 2008.

Metz, Steven. "Non-Lethal Weapons: A Progress Report." *Joint Force Quarterly* 28 (2001): 18–22.

———. "Non-Lethality and the Revolution in Military Affairs." In Malcolm Dando (ed.), *Non-Lethal Weapons: Technological and Operational Prospects*, Jane's Special Report, November 2000.

Metz, Steven, and James Kievit. *Strategy and the Revolution in Military Affairs: From Theory to Policy*. Carlyle: US Army Strategic Studies Institute, 1995.

Michaeli, Sarit. *Crowd Control: Israel's Use of Crowd Control Weapons in the West Bank*. Jerusalem: B'tselem, 2013.

Milbank, Dana. "At UN, Bush Is Criticised over Iraq." *The Washington Post*, 24 September 2003, http://www.washingtonpost.com/archive/politics/2003/09/24/at-un-bush-is-criticized-over-iraq/a8b7d7eb-15be-4b47–a197–6275867a26e7/ [accessed May 10, 2018].

The Ministry of Defense of the Russian Federation. *Kontseptsiya sozdaniya i primeneniya oruzhiya neletal'nogo deystviya v Vooruzhennykh Silakh Rossiyskoy Federatsii [The Conception to the Development and Employment of Weapons with Non-Lethal Action in the Armed Forces of the Russian Federation]*, Moscow, 2009.

The Ministry of Defense of the Russian Federation. The Online Version of the *Encyclopedia of the Strategic Missile Troops* (*Entsiklopediya Raketnykh Voysk Strategicheskogo Naznacheniya*), http://encyclopedia.mil.ru/encyclopedia/dictionary/details_rvsn.htm?id=13770@morfDictionary [accessed May 10, 2018].

———. The Online Version of the *Military Encyclopaedic Dictionary* (*Voyennyy Entsiklopedicheskiy Slovar'*), http://encyclopedia.mil.ru/encyclopedia/dictionary/details.htm?id=7781@morfDictionary [accessed May 10, 2018].

Ministry of Foreign Affairs of the Russian Federation. *The Foreign Policy Concept of the Russian Federation-2000*, Moscow, 2000 (Russian).

———. *The Foreign Policy Concept of the Russian Federation-2008*, Moscow, 2008 (Russian).

———. *The Foreign Policy Concept of the Russian Federation-2013*, Moscow, 2013 (Russian).

The Ministry of Interior of the USSR. *Ob utverzhdenii Instruktsii po primeneniyu spetsial'nykh sredstv individual'noy zashchity Lichnogo sostava i drugikh mer bezopasnosti v organakh vnutrennikh del i Podrazdeleniya vnutrennikh voysk SSSR [Decree to Approve the Instruc-*

tions for the Use of Special Means for the Individual Self-Defence of the Personnel and Other Security Measures by the Law Enforcement Agencies and Internal Forces of the Ministry of Interior of the U.S.S.R.], Moscow, December 26, 1978.

———. *Ob utverzhdenii Instruktsii o poryadke primeneniya rabotnikami militsii khimichesko-go Sredstva spetsial'nogo naznacheniya "Cheremukha-10" [Decree to Approve the Instructions for the Use of Chemical Special Means "Cheremuha-10" by Police Officers]*, Moscow, November 19, 1981.

———. *Ob utverzhdenii Nastavleniya po primeneniyu spetsial'nykh sredstv "Cheremukha" [Decree to Approve the Manual for the Use of Special Means "Cheremuha"]* , Moscow, September 24, 1970.

"Minoborony: Na Forume 'Armiya-2015' Pokazhut Netraditsionnyye Vidy Oruzhiya [Ministry of Defense: Non-Traditional Types of Weapons Will be Demonstrated at 'Army-2020' Show']." *RiaNovosti*, April 21, 2015, http://ria.ru/defense_safety/20150421/1059915256.html [accessed May 10, 2018].

Mintz, Alex. "The Military-Industrial Complex: The Israeli Case." In Moshe Lissak (ed.), *Israeli Society and Its Defense Establishment: The Social Political Impact pf a Protracted Violent Conflict*. London: Frank Cass, 1984.

Mockaitis, Thomas. "Winning Hearts and Minds in the 'War on Terrorism.'" *Small Wars & Insurgencies* 14, no.1 (2003): 21–38;

Moiseev, A. "O nekotorykh mezhdunarodno-pravovykh pozitsiyakh po ukrainskomu voprosu [About Some International Law Positions on the Ukrainian Issue]." *Mezhdunarodnaya Zhizn'*, no. 5 (2014): 53–65.

Moiseev, V., and V. Orlyanskii. "Oruzhiye neletal'nogo deystviya i printsipy taktiki [Weapons with Non-Lethal Action and the Principles of Tactics]." *Voennaia Misl'*, no. 6 (2011): 26–33.

"More than 1,000,000 Iraqis Murdered." *Opinion Research Business*, September 2007.

Morris, Ian. *War—What Is It Good For? The Role of Conflict in Civilisation, from Primates to Robots*. London: Profile Books, 2014.

Muciano, Gil. "A Matter of Honour: A Review of Israeli Decision Making during the Second Lebanon War." *The Atkin Paper Series*, The International Centre for the Study of Radicalization and Political Violence, March 2011.

Mueller, John. "Public Opinion as a Constrain on U.S. Foreign Policy: Assessing the Perceived Value of American and Foreign Lives." Unpublished text prepared for presentation at National Convention of International Studies Association at Los-Angeles, California, March 14–18, 2000.

Muranov, Artem. "Administrativno-pravovoy rezhim oruzhiya neletal'nogo deystviya: problemy formirovaniya i realizatsii [The Regulatory System of Weapons with Non-Lethal Action: Problems of Formulation and Implementation]." PhD thesis, The All-Russian Research Institute of the Ministry of Internal Affairs of the Russian Federation, 2008.

Murray, Donette. *US Foreign Policy and Iran: American-Iranian Relations Since the Islamic Revolution*. New York: Routledge, 2009.

Murray, Williamson. "Contingency and Fragility of the German RMA." In Williamson Murray and MacGregor Knox (ed.), *The Dynamics of Military Revolution 1300–2050*. Cambridge: Cambridge University Press, 2001.

——— (ed.). *National Security Challenges for the 21st Century*. Carlisle: US Army War College, 2003.

——— (ed.). *The Emerging Strategic Environment: Challenges of the Twenty-first Century*. Westport: Greenwood Publishing Group, 1999.

———. "The Future of American Military Culture: Does Military Culture Matter?" *Orbis* 43, Issue 1 (1999): 27–42.

———. "Thinking about Revolution in Military Affairs." *Joint Force Quarterly* 16 (1997): 69–76.

———. "Transformation and Professional Military Education: Past as Prologue to the Future." In Williamson Murray (ed.), *National Security Challenges for the 21st Century*. Carlisle: US Army War College, 2003.

Murray, Williamson and MacGregor Knox. "Future behind Us." In Williamson Murray and MacGregor Knox (ed.), *The Dynamics of Military Revolution 1300–2050*. Cambridge: Cambridge University Press, 2001.

——— (ed.). *The Dynamics of Military Revolution 1300–2050*. Cambridge: Cambridge University Press, 2001.

———. "Thinking about revolutions in Warfare." In Williamson Murray and MacGregor Knox (ed.), *The Dynamics of Military Revolution 1300–2050*. Cambridge: Cambridge University Press, 2001.

Nagovitsyn, A. "Ispol'zovaniye oruzhiya neletal'nogo deystviya kollektivnymi silami ODKB [The Employment of Weapons with Non-Lethal Action by CSTO Forces]." *Zashchita i Bezopasnost'*, no. 3(58) (2011): 2–3.

Nagovitsyn, A., A. Grudzinskii', and A. Sporykhin. "Oruzhiye neletal'nogo deystviya i perspektivy yego ispol'zovaniya v interesakh sil Organizatsii Dogovora o kollektivnoy bezopasnosti [Weapons with Non-Lethal Action and the Perspectives of Their Employment in the Interest of Collective Security Treaty Organization]." *Voenaya Misl,'* no.3 (2011): 51–59.

National Research Council. *An Assessment of Non-Lethal Weapons: Science and Technology*. Washington, 2003.

NATO. *NATO Policy on Non-Lethal Weapons*, Brussels, October13, 1999.

"Netanyahu Holocaust Remarks: Israeli PM Criticized." *BBCNews*, October 21, 2015, http://www.bbc.co.uk/news/world-middle-east-34594563 [accessed May 10, 2018].

Netanyahu, Benjamin. *Speech at the Holocaust Martyrs' and Heroes' Remembrance Day Ceremony*, April 11, 2010, http://www.pmo.gov.il/English/MediaCenter/Speeches/Pages/speechshoa1104010.aspx [accessed May 10, 2018].

———. *Speech to US Congress,* July 10, 1996, http://mfa.gov.il/MFA/MFA-Archive/1996/Pages/PM%20Netanyahu-%20Speech%20to%20US%20Congress-%20July%2010–%201996.aspx [accessed May 10, 2018].

Nevins, Allan, and Henry Commager. *America: The Story of Free People*. Oxford: Oxford University Press, 1977.

Nicgorski, Walter, and Ronald Weber (ed.). *An Almost Chosen People: The Moral Aspirations of Americans*. Notre Dame: The University of Notre Dame Press, 1976.

Nixon, Richard. *Second Annual Report to the Congress on United States Foreign Policy*, Washington, 25 February 1971, http://www.presidency.ucsb.edu/ws/?pid=3324 [accessed May 10, 2018].

Noiberger, Benjamin, and Arie Gronik (ed.). *Mediniut Hutz bein imut le'esderim – Israel 1948–2008 [The Foreign Policy between Conflict and Agreement – Israel 1948–2008]*, Vol. B. Raanana: The Open University of Israel, 2008.

Noiman, Noam. "Lohamat mishpat: iumim ve'hizdamnuyiot [Lawfare: Threats and Opportunities]." *Ma'arachot*, 449 (2013): 16–23.

Norbut, Gerald. *Non-Lethal Weapons: Force Enabler for the Operational Commander Conducting Peace Operations*. Newport, RI: Naval War College, 2001.

Novikova, L., and I. Sizemskaya (ed.). *Mir Rossii - Yevraziya: Antologiya [The World of Russia - Eurasia: An Anthology]*. Moscow: Vysshaya shkola, 1995.

Nutley, Erik. *Non-Lethal Weapons: Setting Our Phasers on Stun? Potential Strategic Blessings and Curses of Non-Lethal Weapons on the Battlefield*, Center for Strategy and Technology Occasional Paper No. 34. Maxwell Air Force Base, Alabama: Air University, 2003.

"Obama Apologises to MSF President for Kunduz Bombing." *BBCNews*, October 7, 2015, http://www.bbc.co.uk/news/world-us-canada-34467631 [accessed May 10, 2018].

Obama, Barack. *Remarks at the National Archives and Records Administration*, Washington, May 21, 2009, http://www.presidency.ucsb.edu/ws/?pid=86166 [accessed May 10, 2018].

———. *Remarks to the United Nations General Assembly in New York City*, New York, September 24, 2014, http://www.presidency.ucsb.edu/ws/index.php?pid=107615 [accessed May 10, 2018].

Ofer, Dalia. "We Israelis Remember, but How?: The Memory of the Holocaust and the Israeli Experience." *Israel Studies* 18, no. 2 (2013): 70–85.

Ofer, Zvi, and Avi Kober (ed.). *Ichut ve'kamut: dilemot be'bniyat ha'koah ha'tzva [Quality and Quantity: Dilemmas of Building a Military Force]*. Tel Aviv: Ma'arachot, 1985.

Ogawa, James. *Evaluating the U.S. Military's Development of Strategic And Operational Doctrine for Non-Lethal Weapons in a Complex Security Environment*. Fort Leavenworth: Command and General Staff College, 2003.

Olajos, Eugene, and Woodhall Stopford (ed.). *Riot Control Agents: Issues in Toxicology, Safety, and Help*. Boca Raton: CRC Press LLC, 2004.

Olsen, John, and Martin Creveld Van (ed.). *The Evolution of Operational Art: From Napoleon to the Present*. Oxford, Oxford University Press, 2010.

Orbons, Sjef. "Non-Lethal Weapons: Peace Enables or Troublesome Force? Assessing the Role of CS and Baton Rounds in the Northern Ireland Conflict." *Small Wars & Insurgencies* 22, no.3 (2011): 467–94.

———. "Non-Lethality in Reality: A Defense Technology Assessment of its Political and Military Potential." PhD Dissertation, The University of Amsterdam, 2013.

Oren, Amir. "Nishkal shigur "Bo'esh" leknesiyat hamolad [It Was Considered to Deploy "Skunk" to the Church of the Nativity]." *Ha'aretz*, April 12, 2002, http://news.walla.co.il/item/208321 [accessed May 10, 2018].

Orgen, Kenneth. "Humanitarian Law in the Articles of War Decreed in 1621 by King Gustavus II Adolphus of Sweeden." *International Review of the Red Cross*, no. 313 (1996): 438–42.

Ornstein, Norman. "Foreign Policy and the 1992 Elections." *Foreign Affairs* 72, no. 3 (1992): 1–16.

Osgood, Robert. *Alliances and American Foreign Policy*. Baltimore: The John Hopkins Press, 1968.

Ovchinnikov, V. "Lokal'nyye vooruzhennyye konflikty i bezopasnost' Rossii [Local Armed Conflicts and the Security of Russia]." *Zashchita i Bezopasnost,'* no. 4(47) (2008): 9–11.

Owens, William. "The American Revolution in Military Affairs." *Joint Force Quarterly* 10 (1996): 37–38.

Owens, John, and John Dumbrell (ed.). *America's "War on Terrorism": New Dimensions in U.S. Government and National Security*. Lanham: Lexington Books, 2008.

Panarin, Alexander. *Rossiya v tsivilizatsionnom protsesse: mezhdu atlantizmom i yevraziystvom [Russia in the civilisation process: between Atlanticism and Eurasianism]*. Moscow: Istitut filosofii RAN, 1995.

Panteleev, S. "Attraktory vrazhduyushchikh storon v lokal'nykh konfliktakh s primeneniyem sredstv neletal'nogo vozdeystviya [Attractors of the Fighting Sides in Local Conflicts with the Employment of Weapons with Non-Lethal Action]." *Voennaia Misl'*, no. 4 (2012): 11–16.

Pantin, Vladimir. "Istoricheskaya rol' rossii mezhdu zapadom i vostokom [The Historical Role of Russia between the West and the East]." *Istoriya i sovremennost'*, no. 2 (2012): 93–112.

Parsons, Dan. "Army, Marine Corps Succeed in Rapidly Fielding Specialized Individual Weapons (UPDATED)." *National Defence*, January 2013.

Patterson, Eric. *Just War Thinking: Morality and Pragmatism in the Struggle against Contemporary Threats*. Lanham, MD: Lexington Books, 2007.

Pedro, Sharon. *Utzma raka ke'nehes leumi [Soft Power as a National Asset]*. Jerusalem: The Jewish People Policy Planning Institute, 2004.

Peled, Tal. "Am bone tzava bone am – ha'umnam? [Nation Builds Military That Builds Nation – Is That So?]." *Ma'arachot* 442 (2012): 68–72.

Person, Stow. *American Minds: A History of Ideas*. New York: Henry Holt and Company, 1958.

Peters, Ralph. *Fighting for the Future: Will America Triumph?* Mechanicsburg: Stackpole Books, 2001.

Peterson, John, and Mark Pollack (ed.). *Europe, America, Bush: Transatlantic Relations in the Twenty-First Century*. London: Routledge, 2003.

Petrelli, Niccolo. "Israel's Struggle Against Hamas." PhD Dissertation, Roma Tre University, 2014.

Petrosyan, Levon. "Konflikt mentalitetov kak osnovaniye protivostoyaniya politicheskikh kul'tur: Rossiya i severoatlanticheskiy soyuz v mire posle kholodnoy voyny [The Conflict of Mentalities as a Basis for the Political Cultures Stand-Off: Russia and The North Atlantic Treaty Organization in the Post-Cold War World].” PhD thesis, Moscow State Pedagogical University, 2003.

Petrov, I. “Oruzhiye nesmertel’nogo deystviya [Weapons with Non-Lethal Action].” *Vecherniaia Moskva*, no. 190 (2004 October 8).

Pinker, Steven. *The Better Angels of Our Nature: A History of Violence and Humanity*. London: Penguin Group, 2011.

Pinkney, Alphonso. *The American Way of Violence*. New York: Random House, 1972.

Power, Samantha. “Legitimacy and Competence.” In Melvyn Leffler and Jeoffrey Lergo (ed.), *To Lead the World: American Strategy after the Bush Doctrine*. Oxford: Oxford University Press, 2008.

Press-conference with Sergey Lavrov, Russian Foreign Minister, published in *Mezhdunarodnaya Zhizn,'* no. 2 (2013), 11 (Russian).

Pressman, Jeremy. “Israeli Unilateralism and Israeli-Palestinian Relations 2001–2006.” *International Studies Perspectives*, no. 7 (2006): 360–76.

Primakov, E. “Vneshnyaya politika vo vse bol’shey stepeni napravlyayetsya na vosstanovleniye Rossii v kachestve derzhavy mirovogo klassa [Foreign Policy is More Directed to the Restoration of Russia as a World-class Power].” *Mezhdunarodnaya Zhizn,'* no. 1–2 (2008): 48–51.

Pronin, A., A. Leonov, and L. Kaplyarchuk. “Osnovnyye kriterii voyenno-ekonomicheskoy otsenki oruzhiya neletal’nogo deystviya [The Major Criteria of Military-Economic Evaluation of Weapons with Non-Lethal Action].” *Voennaia Misl',* no.10 (2012): 43–50.

Protocol for the Prohibition of the Use in War Asphyxiating, Poisonous or Other Gases, and of Bacteriological Methods of Warfare (adopted June 17, 1925, entered into force February 8, 1928), 94 L.N.T.S. 65.

Protocol on Blinding Laser Weapons (Protocol IV) (adopted October 13, 1995, entered into force July 30, 1998), 35 ILM 1218.

Protocol on Non-Detectable Fragments (Protocol I) (adopted October 10, 1980, entered into force December 2, 1983), 19 ILM 1529.

Protocol on Prohibitions or Restrictions on the Use of Incendiary Weapons (Protocol III) (adopted October 10, 1980, entered into force December 2, 1983), 19 ILM 1535.

Protocol on Prohibitions or Restrictions on the Use of Mines, Booby-Traps and Other Devices (Protocol II) (initially adopted October 10, 1980, entered into force December 2, 1983; review adopted in May 3, 1996, entered into force December 3, 1998), 35 ILM 1206.

Public Committee against Torture in Israel and Palestinian Society for the Protection of Human Rights and the Environment vs. The Government of Israel (High Court of Justice 769/02, December 11, 2005), Paragraphs 34–37 (Hebrew).

Pursell, Carroll. *The Military-Industrial Complex*. New York: Harper & Row Publishers, 1972.

Putin, Vladimir. “Byt’ sil’nymi: garantii natsional’noy bezopasnosti dlya Rossii [Being strong: National Security Guarantees for Russia].” *Rossiyskaya gazeta*, no. 5708, February 17, 2012.

———. *Speech at the UN General Assembly*, September 28, 2015, https://russian.rt.com/article/119712 [accessed May 10, 2018] (Russian).

———. *Speech of the President of the Russian Federation*, March 18, 2014, http://kremlin.ru/events/president/news/20603/videos [accessed May 10, 2018] (Russian).

———. *2005 Address to the Federal Assembly*, http://kremlin.ru/events/president/transcripts/22931 [accessed: 10 Мaы 2018] (Russian).

———. *2006 Speech at Meeting with the Ambassadors and Permanent Representatives of the Russian Federation*, http://kremlin.ru/events/president/transcripts/23669 [accessed May 10, 2018] (Russian).

———. *2007 Speech at the Munich Security Conference*, http://www.stringer-news.com/publication.mhtml?Part=50&PubID=7070 [accessed May 10, 2018] (Russian).

———. *2012 Address to the Federal Assembly*, http://kremlin.ru/events/president/news/17118 [accessed May 10, 2018] (Russian).

———. *2012 Speech at Meeting of Ambassadors and Permanent Representatives of the Russian Federation to the International Organizations*, published in *Mezhdunarodnaya Zhizn,'* no. 7 (2012): 4 (Russian).

——— *2014 Speech at Meeting the Ambassadors and Permanent Representatives of the Russian Federation*, published in *Mezhdunarodnaya Zhizn,'* no. 7 (2014): 3–12 (Russian).

Quille, Gerrard. "The Revolution in Military Affairs Debate and Non-Lethal Weapons." In Nick Lewer (ed.), *The Future of Non-Lethal Weapons: Technologies, Operations, Ethics and Law*. London: Frank Cass Publishers, 2002.

Quintana, Elizabeth. "The War from the Air." In Adrian Johnson and Saqeb Mueen (ed.), *Short War, Long Shadow: The Political and Military Legacies of the 2011 Libya Campaign*, Whitehall Report 1–12. London: Royal United Services Institute for Defence and Security Studies, 2012.

Rabin, Yitzhak. *Speech in the Knesset Regarding the U.N. General Assembly Resolution from 10 November 1975 about Zionism*, November 11, 1975 (Hebrew).

Rabotyazhev, N., and E. Solov'yev. "Ot 'Vozvrashcheniya v Mirovuyu Tsivilizatsiyu' k 'Suverennoy Demokratii': Evolyutsiya vneshnepoliticheskikh kontseptsiy Rossiyskikh 'Partiy Vlasti' [From the 'Comeback to the World Civilization' to the 'Sovereign Democracy': An Evolution of the Foreign Policy Conceptions of the Russian 'Parties in Power']." *Politiya*, no. 1(68) (2013): 36–54.

Rapaport, Amir. "Metach gavoha [High Tension]." *IsraelDefense*, April 26, 2011, http://www.israeldefense.com/en/node/9371 [accessed May 10, 2018].

Reagan, Ronald. *Election Eve Address "A Vision for America*," Washington, November 3, 1980, http://www.presidency.ucsb.edu/ws/?pid=85199 [accessed May 10, 2018].

———. *Message to the Congress Transmitting the Fiscal Year 1984 Budget*, Washington, January 31, 1983, http://www.presidency.ucsb.edu/ws/?pid=40539 [accessed May 10, 2018].

———. *New Year's Messages of President Reagan and President Mikhail Gorbachev of the Soviet Union*, Washington, January 1, 1989, http://www.presidency.ucsb.edu/ws/?pid=35313 [accessed May 10, 2018].

Record, Jeffrey. *Beating Goliath*. Washington: Potomac Books, 2009.

Regev, Motti. "To Have a Culture of Our Own: on Israeliness and Its Variants." *Ethnic and Racial Studies* 23, no. 2 (2000): 223–47.

Rejali, Darius. *Torture and Democracy*. Princeton, NJ: Princeton University Press, 2008.

Richmond, Yale. *From Nyet to Da: Understanding the Russians*. Yarmouth, ME: Intercultural Press, 1996.

Ricks, Thomas. "Rotation Policy: Iraq & Afghanistan Were Really the Wars We Fought 1 Year at a Time." *Foreign Policy*, May 27, 2015, http://foreignpolicy.com/2015/05/27/rotation-policy-iraq-afghanistan-were-really-the-wars-we-fought-1–year-at-a-time/ [accessed May 10, 2018].

Roberts, Adam, and Richard Guelff (ed.). *Documents on the Laws of War*, 3rd edition. New York: Oxford University Press, 2002.

Roberts, Michael. "The Military Revolution, 1560–1660." In Clifford Rogers (ed.), *The Military Revolution Debate: Readings on the Military Transformation of Early Modern Europe*. Oxford: Westview Press, 1995.

Robinson, Julian. "Non Lethal Warfare and the Chemical Weapons Convention." *A Further HSP submission to the OPCW Open-Ended Working Group on Preparations for the Second CWC Review Conference*, October 24, 2007, http://www.sussex.ac.uk/Units/spru/hsp/Papers/421rev3.pdf [accessed May 10, 2018].

Rodnik, Rafael. "Haderch le'hasagat legitimatziya [The Way to Achieve a Legitimacy]." *Ma'arachot* 449 (2013): 9–15.

Rogers, Clifford. "'As if a New Sun Had Arisen': England's Fourteenth-Century RMA." In Williamson Murray and MacGregor Knox (ed.). *The Dynamics of Military Revolution 1300–2050*. Cambridge: Cambridge University Press, 2001.

——— (ed.). *The Military Revolution Debate: Readings on the Military Transformation of Early Modern Europe*. Oxford: Westview Press, 1995.

Rogov, K. "Sverkhbol'shinstvo dlya sverkhprezidenstva [Supermajority for Super-presidency]." *Pro et Contra* 17, Issue 3–4 (2013).

Roosevelt, Franklin. *Christmas Eve Message to the Nation*, Washington, December 24, 1941, http://www.presidency.ucsb.edu/ws/?pid=16073 [accessed May 10, 2018].

Rosen, Stephen. "Military Effectiveness: Why Society Matters." *International Security* 19, no. 4 (1995): 5–31.

———. "The Impact of the Office of Net Assessment on the American Military in the Matter of the Revolution in Military Affairs." *Journal of Strategic Studies* 33, no. 4 (2010): 569–482.

Rosenberg, Barbara. "'Non-Lethal' Weapons May Violate Treaties." *The Bulletin of the Atomic Scientists* 50, no. 5 (1994): 44–45.

Rosner, Shmuel, and Michael Herzog. *Jewish Values and Israel's Use of Force in Armed Conflict: Perspectives from World Jewry*. Jerusalem: The Jewish People Policy Institute, 2015.

Rubin, Alissa. "The Conflict in Iraq: 40 Killed in U.S. Attack on Iraqi Border Village." *Los Angeles Times*, May 20, 2004. http://articles.latimes.com/2004/may/20/world/fg-iraq20 [accessed May 10, 2018].

Rudnitskaya, E. (ed.). *Mirotvorchestvo v Rossii—Tserkov,' Politika, Mysliteli, ot rannego srednevekov'ya do rubezha XIX-XX stoletiy [Peacemaking in Russia: The Church, Politics, Thinkers: from the early Middle Ages to the Boundary between the 19th and 20th Centuries]*. Moscow: Nauka, 2003.

Rudnitskaya, E., and N. Lisovoi. "Mirotvorcheskaya paradigma russkoy mysli [The Peacemaking Paradigm of the Russian Thought]." In E. Rudnitskaya (ed.), *Mirotvorchestvo v Rossii—Tserkov,' Politika, Mysliteli, ot rannego srednevekov'ya do rubezha XIX-XX stoletiy [Peacemaking in Russia: The Church, Politics, Thinkers: From the Early Middle Ages to the Boundary between the 19th and 20th Centuries]*. Moscow: Nauka, 2003.

Ruiz Palmer, Diego. "The NATO-Warsaw Pact Competition in the 1970s and 1980s: A Revolution in Military Affairs in the Making or the End of a Strategic Age." *Cold War History* 14, no. 4 (2014): 533–74.

Rumsfeld, Donald. *DoD News Briefing with Secretary Rumsfeld and General Pace*, Washington, June 14, 2005, http://archive.defense.gov/Transcripts/Transcript.aspx?TranscriptID=3854 [accessed May 10, 2018].

"Russia's Unguided Weapons in Syria Could Ricochet on Moscow." *Time*, October 2, 2015, http://time.com/4059442/russia-bombing-syria-isis-assad/ [accessed May 10, 2018].

Sade, Sharon. "Israel's Defence Industry in the 21st Century: Challenges and Opportunities." *Strategic Assessment* 7, no. 3 (2004): 31–40.

Sagi, Raz. "Mi'tzava ha'hagana le'tzava ha'hagana ha'mishpatit [From the Defence Force to the Legal Defence Force]." *Ma'arachot* 462 (2015): 56–59.

Saletan, William. "Israel's Unprecedented Steps to Avoid Civilian Casualties." *The National Post*, July 16, 2014.

Savrasov, S. "Kontseptsiya primeneniya oruzhiya neletal'nogo deystviya v boyevykh operatsiyakh sukhoputnykh voysk VS SSHA [The NLW Deployment Policy of the U.S. Army's in Combat Operations]." *Zarubezhnoe voennoe obozrenie*, no. 10 (2009): 37–44.

Schwartz, Barry, et al. "The Recovery of Masada: A Study in Collective Memory." *The Sociological Quarterly* 27, no. 2 (1986): 147–64.

Scott, Richard. *Conflict Without Casualties: Non-Lethal Weapons in Irregular Warfare*. Monterey: Naval Postgraduate School, 2007.

———. "Nonlethal Weapons and the Common Operating Environment." *Army Magazine* 50, no. 4 (2010): 21–26.

Security Planning Corporation. *Non-Lethal Weapons for Law Enforcement: Research Needs and Priorities: A Report to the National Science Foundation*. Washington, 1972.

Seger, Idan. "Ei'n makom letauiyot: kah tzal' metkonen lemehumot [There Is No Place for Mistakes: This Is the Way in Which the IDF Prepares itself for the Riots]." *Mako*, September 20, 2011, http://www.mako.co.il/pzm-magazine/Article-c7ddcfef4e18231006.htm [accessed May 10, 2018].

Seliktar, Ofira. *New Zionism and the Foreign Policy System of Israel*. Carbondale: Southern Illinois University Press, 1986.

Selivanov, V. "Oruzhiye neletal'nogo deystviya kak sredstvo bor'by s terrorizmom, obespecheniya mirotvorcheskikh operatsiy i operatsiy po pravoprinuzhdeniyu [Non-Lethal Weapons as Means for Struggle against Terrorism, Provision on Peacekeeping and Law Enforcement Operations]." *Vestnik Rossii'skoi' Akademii Estestvennih Nauk*, no. 4 (2005): 1–11.

Selivanov, V., D. Levin, and I. Ilyin. "Metodicheskiye osnovy programmno-tselevogo upravleniya razvitiyem oruzhiya neletal'nogo deystviya [Methodological Basics of Management by Objectives of Non-Lethal Weapons Development]." *Voprosy Oboronnoy Tekhniki, Seriya 16*, no. 3–4(69–70) (2014): 36–50.

Selivanov, V., D. Levin, and S. Lyuwin. "Arkhitektura i realizatsiya bazy dannykh 'Oruzhiye Neletal'nogo Deystviya' [The Architecture and Realization of the Data-base 'Weapons with Non-Lethal Action']." *Voprosi Oboronoi Tehniki – Seriya 16*, no. 7–8 (2012): 63–76.

Selivanov, V., V. Leonov, and N. Bagdasaryan. "Social Aspects of NLW Development." In *Proceeding to the 3 rd European Symposium on Non-Lethal Weapons*. Ettlingen: Fraunhofer-Institut für Chemische Technologie ICT, 2005.

Seybolt, Taylor, Jay Aronson, and Baruch Fischhoff (ed.). *Counting Civilian Casualties: An Introduction to Recording and Estimating Nonmilitary Deaths in Conflict*. Oxford: Oxford University Press, 2013.

Shachar, Amihud. "Ha'baya eina be'tzal [The Problem is Not in the IDF]." *Ma'arachot*, vol. 380–381 (2001): 88–89.

Shain, Yossi, and Neil Rogachevsky. "Between JDate and J Street: US Foreign Policy and the Liberal Jewish Dilemma in America." *Israel Journal of Foreign Affairs* 3 (2011): 37–48.

Shakleina, T. "Obshchnost' i razlichiya v strategiyakh Rossii i SSHA [Common Ground and Divergences in the Strategies of Russia and the U.S.A.]." *Mezhdunarodnyye Protsessy* 11, no. 2(33) (2013): 6–19.

Shakleina, T. *Rossiya i SSHA v mirovoy politike [Russia and the USA in the World's Politics]*. Moscow: Aspekt Press, 2012.

Shalev, Aryeh. *The Intifada: Causes and Effects*. Jerusalem: Jerusalem Post, 1991.

Shalom, Zaki. "Is the IDF Prepared to Face a Regular War against the Arab States?" *Military and Strategic Affairs* 1, no. 1 (2009, April): 17–23.

Shapir, Yiftah. "Iron Dome: The Queen of Battle." In Shlomo Brom (ed.), *In the Aftermath of Operation Pillar of Defense The Gaza Strip, November 2012*. Tel Aviv: Institute for National Security Studies, 2012.

Sharp, Jane. "The US-UK 'Special Relationship' after Iraq." In Paul Cornish (ed.), *The Conflict in Iraq, 2003*. New York: Palgrave Macmillan, 2004.

Sharvit, Moshe. "Mekoma shel ha'technologia be'doktrina shel tzal [The Place of Technology in the IDF's Doctrine]." *Ma'arachot* 393 (2004): 81–83.

Shaw, Martin. *The New Western Way of War*. Cambridge: Polity Press, 2005.

Shelah, Ofer. "Mekavim letov, mitkonenim le r'a' [Hoping for Good, Preparing for Bad]." *NRG*, September 17, 2011, http://www.nrg.co.il/online/1/ART2/286/345.html [accessed 10 May 2018].

Sher, Yonatan. "Hafsad lo ba' be'hashbon [A Riot Is not an Option]." *Be'yabasha*, No. 24, 2010.

Shetopal, E. "Tsennostnyye kharakteristiki rossiyskogo politicheskogo protsessa i strategiya razvitiya strany [The Characteristics of Values of the Russian Political Process and the Country's Development Strategy]." *Polis: Politicheskii Issledovaniya*, no. 2 (2014): 61–71.

Shishkina, O. "Post-Soviet Foreign Policy Foundations: State Interests vs. Identities." *Moscow State Institute of International Relations (MGIMO) Blog - Govoryat Eksperty MGIMO*, 9 June 2014, https://mgimo.ru/about/news/experts/253273/ [accessed May 10, 2018].

Shmelev, N. (ed.). *Rossiya v mnogoobrazii tsivilizatsiy [Russia in the Diversity of Civilizations]*. Moscow: Ves' Mir, 2011.

Shtultz, Guy-Zahar. "Haf'alat koah medureget be'imutim besviva ezrahit [The Employment of Graduate Force in Conflicts within Civilian Environment]." *Ma'arachot* 438 (2011): 8–17.

Siboni, Gabriel. "War and Victory." *Military and Strategic Affairs* 1, no. 3 (2009): 39–49.

Sikorskii, Ivan. *Cherty iz psikhologii slavyan - rech' na torzhestvennom zasedanii Slavyansko-go blagotvoritel'nogo obshchestva 14 maya 1895 goda [Lines from the Psychology of Slavs-A Speech at the Ceremonial Meeting of the Slavic Charitable Society on 14 May 1895].* Kiev: Tipografiya imeni Chokolova, 1895.

Silliman, Stephen. "The 'Old West' in the Middle East: U.S. Military Metaphors in Real and Imagined Indian Country." *American Anthropologist* 110, Issue 2 (2008): 237–47.

Silnikov, A. "Vidy spetsial'nykh sredstv sostoyashchikh na vooruzhenii OVD, i ikh klassifikatsiya [Types and Classification of Special Equipment Adopted by the Russian Interior Ministry Units]." *Vestnik Sankt-Peterburgskogo universiteta MVD Rossii,* no. 3(51) (2011): 66–70.

Siltman, Frank, and John Frisbie. "Fire Support Just Got Harder: Adding Nonlethal Fires as a Core Competency." *Fires Bulletin*, July–September 2008, 6–8.

Simchayoff, Elad, and Udi Segal. "Doh ha'um al mivtza 'tzuk eitan' ushar, rak aratzot habrit hitnagda [The UN Report about Operation Protective Edge Was Approved, only the U.S. Was Against]." *Mako,* http://www.mako.co.il/news-military/israel-q3_2015/Article-268b8dec0445e41004.htm?sCh=31750a2610f26110&pId=786102762 [accessed May 10, 2018].

Siniscalchi, Joseph. "Non-Lethal Technologies: Implications for Military Strategy." *Center for Strategy and Technology Occasional Paper*, no. 3 (1998).

Sloan, Stanley. *Permanent Alliance? – NATO and the Transatlantic Bargain from Truman to Obama*. London: Bloomsbury Publishing, 2010.

Smelser, Marshall. "An Understanding of the American Revolution." in Walter Nicgorski and Ronald Weber (ed.), *An Almost Chosen People: The Moral Aspirations of Americans*. Notre Dame, IN: The University of Notre Dame Press, 1976.

Smirnov, A. "Kontseptsii vneshney politiki Rossiyskoy Federatsii: sravnitel'nyy analiz [The Foreign Policy Concept of the Russian Federation – Comparative Analysis]." In S. Smul'skii, O. Abramova, and V. Buyanov (ed.), *Vneshnyaya politika Rossii: teoriya i praktika [The Foreign Policy of Russia: Theory and Practice]*. Moscow: Kniga i Biznes, 2013.

Smul'skii, S., O. Abramova, and V. Buyanov (ed.). *Vneshnyaya politika Rossii: teoriya i praktika [The Foreign Policy of Russia: Theory and Practice]*. Moscow: Kniga i Biznes, 2013.

Snider, Don, and Gayle Watkins (ed.). *The Future of the Army Profession*. New York: McGraw Hill, 2003.

Snigur, V. "Ovladet' gorodom bez zhertv i razrusheniy [To Capture a City without Casualties or Damage]." *Armei'skii' Sbornik*, no. 12 (2006): 50–52.

Snyder, Jack. *The Soviet Strategic Culture: Implications for Limited Nuclear Operations*. Santa Monica, CA: The RAND Corporation, 1977.

Sofer, Roni. "Lieberman: Goldstone Report Meant to Destroy Israel's Image." *Yedioth Ahronoth*, September 17, 2009, http://www.ynetnews.com/articles/0,7340,L-3778143,00.html [accessed May 10, 2018].

———. "Peres: Goldstone Report Makes Mockery of History." *Yedioth Ahronoth*, September 16, 2009, http://www.ynetnews.com/articles/0,7340,L-3777814,00.html [accessed May 10, 2018].

Sofer, Sason. *Lidata shel ha'mah'shava hamedinit be'israel [The birth of the Political Thought in Israel]*. Jerusalem: Shoken, 2001.

———. "Towards Distant Frontiers: The Course of Israeli Diplomacy." *Israel Affairs* 10, no. 1–2 (2004): 1–9.

Solonevich, Ivan. "Doklad o mezhdunarodnom polozhenii [Report about the International Situation]." *Nasha Strana*, no. 60 (1950): 1–2.

———. *Narodnaya Monarkhiya [People's Monarchy]*. Moscow: Algoritm, 2011.

———. "Tak chto zhe yest' demokratiya? [So, What is Democracy?]." *Nasha Strana*, no. 35 (p. 1–2), no. 36 (p.1–2), no. 38 (p. 1–2), 1950.

Solov'yev, Vladimir. *President*. Masterskaya Movie Company, 2015 (Russian).

Sondhaus, Lawrence. *Strategic Culture and Ways of War*. New York: Routledge, 2006.

Spanier, John, and Steven Hook. *American Foreign Policy since World War II*, 13th edition, Washington, DC: Congressional Quarterly Inc, 1995.

"SPIEGEL Interview: Merkel: Guantanamo Mustn't Exist in Long Term." *Spiegel International Online*, January 9, 2006, http://www.spiegel.de/international/spiegel-interview-merkel-guantanamo-mustn-t-exist-in-long-term-a-394180.html [accessed May 10, 2018].
Spillius, Alex. "Barack Obama 'proposes to move terrorists suspects from Guantanamo Bay.'" *The Telegraph*, November 10, 2008, http://www.telegraph.co.uk/news/worldnews/northamerica/usa/3417913/Barack-Obama-proposes-to-move-terrorists-suspects-from-Guantanamo-Bay.html [accessed May 10, 2018].
Stanley-Mitchell, Elizabeth. "The Digital Battlefield: What Army Transformation Efforts Say about its Professional Jurisdiction." In Don Snider and Gayle Watkins (ed.), *The Future of the Army Profession*. New York: McGraw Hill, 2003.
Stanton, Martin. "What Price Sticky Foam?" *Parameters* XXVI, no. 3 (1996): 63–68.
The State Comptroller of Israel. *Annual Report 54A for 2003 Year*, Jerusalem, 2003, 89–90 (Hebrew).
The State of Israel. *The Full Report of the Commission of Inquiry into the Events of Military Engagement in Lebanon 2006*, January 2008 (Hebrew), 425.
———. *The Partial Report of the Commission of Inquiry into the Events of Military Engagement in Lebanon 2006*, April 2007 (Hebrew).
Steele, Ian. *Warpaths: Invasions of North America*. New York: Oxford University Press, 1994.
Steinberg, Gerald. "Lawfare." *Hamas, the Gaza War and Accountability under International Law, Updated Proceedings to an International Conference, 18 June 2009*, Jerusalem Center for Public Affairs and the Konrad Adenauer Stiftung, Jerusalem, 2013.
Stockholm International Peace Research Institute (SIPRI). *The Problem of Chemical and Biological Warfare. Volume I: The Rise of CB Weapons*, 1971.
Stokes, Bruce. "70 Years after Hiroshima, Opinions Have Shifted on Use of Atomic Bomb." *Pew Research Center*, August 4, 2015, http://www.pewresearch.org/fact-tank/2015/08/04/70–years-after-hiroshima-opinions-have-shifted-on-use-of-atomic-bomb/ [accessed May 10, 2018].
Stone, John. "The British Army and the Tank." In Theo Farrell and Terry Terriff (ed.), *The Sources of Military Change: Culture, Politics, Technology*. London: Lynne Rienner, 2002.
Stupitskiy, E., and L. Kuznetsova. "Issledovaniye vozmozhnosti sozdaniya sredstv elektroshokovogo neletal'nogo vozdeystviya na bioob"yekty na bol'shikh rasstoyaniyakh [The Research of the Possibility to Create Means for Non-Lethal Electroshock Impact on Biological Objects from Significant Distance]." *Voenaya Misl'*, no. 4 (2013): 38–46.
Sutherland, Ronald. "Chemical and Biochemical Non-Lethal Weapons." *Stockholm International Peace Research Institute (SIPRI) Policy Paper*, no. 23 (2008).
Swett, Charles. *Strategic Assessment: Non-Lethal Weapons*, Office of the Assistant Secretary of Defense for Special Operations and Low-Intensity Office, November 9, 1993.
Tafolla, Tracy, David Trachtenberg, and John Aho. "From Niche to Necessity: Integrating Non-Lethal Weapons into Essential Enabling Capabilities." *Joint Force Quarterly* 66 (2012): 71–79.
Tal, Israel. *Bitachon leumi: meatim mul rabim [National Security: Few against Many]*. Tel Aviv: Dvir, 1996.
Taser International Inc.: http://www.taser.com [accessed May 10, 2018].
Taylor, George (ed.). *The Turner Thesis: Concerning the Role of the Frontier in American History*. Lexington: D.C. Heath and Company, 1972.
Team Bravo/Cohort 19. *Viable Short-Term Directed Energy Weapon Naval Solutions: A Systems Analysis of Current Prototypes*. Monterey, CA: Naval Postgraduate School, 2013.
Telem, Inbal, Shmuel Tzvaog, and Benjamin Noiberger (ed.). *Mediniyut ha'hutz shel Israel – kovetz mismachim [The Foreign Policy of Israel – Documents Collection]*, Vol. A. Raanana: The Open University of Israel, 2004.
Teveth, Shabtai. *Moshe Dayan: The Soldier, the Myth, the Legend*. Steimatzky's Agency with Weidenfeld & Nicolson, 1972.
"Text of George Bush's speech." *TheGuardian*, http://www.theguardian.com/world/2001/sep/21/september11.usa13 [accessed May 10, 2018].
Thomas, Mark. *Non-Lethal Weaponry: A Framework for Future Integration*. Maxwell Air Force Base: Air Command and Staff College, 1998.

"Ties between Israel and US 'worst in 35 years.'" *BBCNews*, March 15, 2010, http://news.bbc.co.uk/1/hi/8567706.stm [accessed May 10, 2018].

Tilford, Earl. *The Revolution in Military Affairs: Problems and Cautions*. Carlisle, PA: Strategic Studies Institute, US Army War College, 1995.

Tirman, John. *The Deaths of Others: The Fate of Civilians in America's Wars*. Oxford: Oxford University Press, 2011.

"To Israel with Love." *Economist*, August 3, 2006, http://www.economist.com/node/7255198 [accessed May 10, 2018].

Toffler, Alvin, and Heidi Toffler. *War and Anti-War: Survival at the Dawn of the 21st Century*. Boston: Little, Brown and Company, 1993.

Trenin, Dmitri. "Putin's Biggest Challenge Is Public Support." *Carnegie Moscow Center*, January 15, 2015, http://carnegie.ru/2015/01/15/putin-s-biggest-challenge-is-public-support/hzl5 [accessed May 10, 2018].

Trenin, Dmitri, and Aleksei Malashenko. *Russia's Restless Frontier: The Chechnya Factor in Post-Soviet Russia*. Washington: Carnegie Endowment for International Peace, 2004.

Truman, Harry. *Special Message to the Congress on Greece and Turkey*, Washington, March 12, 1947, http://www.presidency.ucsb.edu/ws/index.php?pid=12846&st=truman&st1=special+message [accessed: 10 May 2015].

Turner, Frederic. "Later Explanations and Developments." In Ray Billington (ed.), *The Frontier Thesis: Valid Interpretation of American History?* New York: Holt, Rinehart and Winston, 1966.

———. "The Significance of the Frontier in American History." In George Taylor (ed.), *The Turner Thesis: Concerning the Role of the Frontier in American History*. Lexington: D.C. Heath and Company, 1972.

Tzur, Yaakov. "Haim hiskil tzal lehafik et lekhei milhamotav ve'le'yasmam? [Whether Did the IDF Succeed to Generate the Lessons from Its Wars and Learn from Them?]." *Ma'arachot* 378–379: 12 19.

Umbrashko, K. "Istoricheskiy obzor: 'novonikolayevskaya guberniya – novosibirskaya oblast': lyudi, sobytiya, fakty [Historical Overview: 'Novonikolayevsk Province - Novosibirsk Region: People, Events, Facts'']," published in the proceedings to the *2nd Russian Scientific-Practical Conference Dedicated to the 90th Anniversary of the Novonikolayevsk Province*, Part 1, Novosibirsk, October 25, 2011, 6–28.

UN General Assembly. *Resolution 3379: Elimination of All Forms of Racial Discrimination*, November 10, 1975, New York.

UN Human Rights Council. *Human Rights in Palestine and Other Occupied Arab Territories: Report of the United Nations Fact-Finding Mission on the Gaza Conflict*, September 25, 2009.

———. *Report of the Independent Commission of Inquiry Established Pursuant to Human Rights Council Resolution*, June 24, 2015.

UN Security Council. *UN Security Council Resolution 2249 (2015) on ISIL*, New York, November 21, 2015.

UN Security Council. *Veto List*, http://research.un.org/en/docs/sc/quick [accessed: October 27, 2015].

United States Marine Corps. *Non-Lethal Weapons (NLW) Training and Readiness (T&R) Manual*, Washington, 2012.

United States Government Accountability Office (GAO). *DOD Needs to Improve Program Management, Policy, and Testing to Enhance Ability to Field Operationally Useful Non-Lethal Weapons: Report to Congressional Requesters*, Washington, April 2009.

U.S. Department of Defense. *Policy for Non-Lethal Weapons, Directive 3000.3E*, Washington DC, September 27, 2017.

The U.S. Army Program Executive Office Soldier. *Equipment Portfolio*, 2014.

U.S. Department of State, Policy Planning Staff. *Non-Lethal and Discriminate Weapons and Technologies*, Washington, DC, 1991.

U.S. Joint Chiefs of Staff. *Capstone Concept for Joint Operations: Joint Force 2020*, Washington, September 10, 2012.

U.S. Joint Chiefs of Staff. *Joint Publication 1–02, Department of Defense Dictionary of Military and Associated Terms*, 8 November 2010 (as Amended Through October 15, 2016).

Vais, Efrat. "Lerishona: Magav Pizer Mafginei Smol' betotah sirahon [For the First Time: The Border Police Dispersed Leftist Demonstrators with a Stinky Cannon]." *Yedioth Ahronoth*, 10 August 2008, http://www.ynet.co.il/articles/0,7340,L-3580288,00.html [accessed May 10, 2018].

Van Aelst, Peter, and Stefaan Walgrave. "New Media, New Movements? The Role of the Internet in Shaping the 'Anti-Globalization' Movement." *Information, Communication & Society* 5, no. 4 (2002): 465–93.

Van Creveld, Martin. *The Age of Airpower*. New York: Public Affairs, 2011.

Vickers, Rhiannon. "Harold Wilson, the British Labour Party, and the War in Vietnam." *Journal of Cold War Studies* 10, no. 2 (2008): 41–70.

Vilke, Gary, and Theodore Chan. "Less Lethal Technology: Medical Issues." *Policing: An International Journal of Police Strategies & Management* 30, no. 3 (2007): 341–57.

Vitman, Noam. "4 shanim le'oferet yatzuka: ktzinei ha'uchlusiya hafchu le'helek integrali mi'hakoach ha'lochem [4 Years after the Cast Lead: Population-during-Warfare Officers Have Become an Integral Part of the Fighting Forces]." *IDF Website*, http://www.idf.il/1133–17987–he/Dover.aspx [accessed May 10, 2018].

Vladimir Putin in an interview to CBS News. September 27, 2015, http://www.cbsnews.com/news/vladimir-putin-russian-president-60–minutes-charlie-rose/ [accessed May 10, 2018].

Vladimirov, Alexander. *Osnovy Obshchey Teorii Voyny, [Fundamentals of General Theory of War]*, Volume 1. Moscow: Moskovskiy Finansovo-Promyshlennyy Universitet "Sinergiya," 2013.

Vladimirov, V., and G. Chernikh. "Sostoyaniye i osnovnyye napravleniya razvitiya oruzhiya neletal'nogo deystviya, sredstv i sposobov zashchity ot nego [The Current Situation and Main Directions for the Development of NLW, the Means and Ways of Protection against Them]." *Strategiya Grazhdanskoy Zashchity': Problemy i Issledovaniya* 2, no. 1(2) (2012): 13–22.

Vladimirov, V. "Sovremennaya voyna i grazhdanskaya oborona [Contemporary War and Civilian Defence]." *Strategiya Grazhdanskoy Zashchity: Problemy i Issledovaniya* 2, no. 1 (2012): 471–81.

Voetberg, Jeffrey. *Non-Lethal Weapons: Considerations for the Joint Force Commander*. Newport, RI: Naval War College, 2007.

Von Clausewitz, Carl. *On War*, edited and translated by Michael Howard and Peter Paret. Princeton: Princeton University Press, 1989.

Voronin, E. "Problema legitimnosti vooruzhennogo vmeshatel'stva. Liviyskiy Casus Belli [The Problem of Legitimacy of Armed Intervention. Libyan Casus Belli]." *Mezhdunarodnaya Zhizn'*, no. 8 (2012): 107–18.

"Voyennaya politsiya VVO poluchila partiyu travmaticheskikh pistoletov 'Osa' [Military Police Received Traumatic Pistols 'OSA']." *RiaNovosti*, February 13, 2014, http://ria.ru/defense_safety/20140213/994703801.html [accessed May 10, 2018].

Walzer, Michael. "The Ethics of Warfare in the Jewish Tradition." *Philosophia* 40 (2012): 633–41.

Watts, Barry. *Six Decades of Guided Munitions and Battle Networks: Progress and Prospects*. Washington: Center for Strategic and Budgetary Assessments, 2007.

Webb, Walter. "The Frontier and the 400 Year Boom." In George Taylor (ed.), *The Turner Thesis: Concerning the Role of the Frontier in American History*. Lexington: D.C. Heath and Company, 1972.

Weigley, Russell. *The American Way of War*. Bloomington: Indiana University Press, 1973.

Weizman, Eyal. "Lawfare in Gaza: Legislative Attack." *Open Democracy*, https://www.opendemocracy.net/article/legislative-attack [accessed May 10, 2018].

Whittemore, Ryan. *Air-delivered Non-Lethal Weapons in Counterinsurgency Operations*. Maxwell Air Force Base: Air Command and Staff College, 2008.

Wilmoth, Gregory. "False-Failed Innovation." *Joint Force Quarterly* 22 (1999): 51–57.

Wistrich, Robert, and David Ohana (ed.). *The Shaping of Israeli Identity: Myth, Memory and Trauma*. London: Fank Cass, 1995.

"World Court Tells Israel to Tear Down Illegal Wall." *Guardian*, July 10, 2004, http://www.theguardian.com/world/2004/jul/10/israel3 [accessed May 10, 2018].

"X" [George Kennan]. "The Sources of Soviet Conduct." *Foreign Affairs* 25, no. 4 (1947): 566–82.

The XXI Century Encyclopaedia – Russian Arms and Technologies, Volume XII: Ordnance and Munition. Moscow: Arms and Technologies, 2006 (Russian/English).

The XXI Century Encyclopaedia – Russian Arms and Technologies, Volume XV: Security and Law Enforcement Facilities. Moscow: Arms and Technologies, 2007 (Russian/English).

Yadlin, Amos. "The Strategic Balance of Operation Protective Edge: Achieving the Strategic Goal Better, Faster, and at a Lower Cost." In Anat Kurz and Shlomo Brom (ed.), *The Lessons of Operation Protective Edge*. Tel-Aviv: Institute for National Security Studies, 2014.

Yair, Gad, and Sharona Odom-Weiss. "Israeli Diplomacy: The Effects of Cultural Trauma." *The Hague Journal of Diplomacy*, no. 9 (2014): 1–23.

"Yaladim palestinim ganvu tziyud tzvai mitank [Palestinian Children Stole Military Equipment from a Tank]." *Yedioth Ahronoth*, January 2, 2002, http://www.ynet.co.il/articles/0,7340,L-1489428,00.html [accessed May 10, 2018].

Yaniv, Avner. "A Question of Survival: The Military and Politics under Siege." In Avner Yaniv (ed.), *National Security and Democracy in Israel*. Boulder: Lynne Rienner, 1993.

——— (ed.). *National Security and Democracy in Israel* (Boulder: Lynne Rienner, 1993.

Zaitsev, D., et al. "Perspektivy primeneniya kompleksov oruzhiya neletal'nogo deystviya na osnove KVCH-izlucheniya v spetsial'nykh operatsiyakh po presecheniyu masovykh besporyadkov [The Prospects for the Application of Millimeter-Wave NLW in Operations for the Suppression of Mass Riots]." *Strategicheskaya Stabil'nost'*, no. 2(55) (2011): 54–59.

Zaitsev, D., et al. "Porazhayushchiye faktory oruzhiya neletal'nogo deystviya - fizicheskiye osobennosti, biologicheskiye effekty i taktika primeneniya [The Hitting Factors of Weapons with Non-Lethal Action: Physiological Features, Biological Effects and Employment Tactics]." *Strategicheskaya Stabil'nost'*, no. 3(64) (2013): 2–11.

Zaitsev, D., et al. "Rol' i mesto oruzhiya neletal'nogo deystviya v konfliktakh nizkoy intensivnosti [The Role and the Place of Weapons with Non-Lethal Action in the Low-Intensity Conflicts]." *Strategicheskaya Stabil'nost'*, no. 4(61) (2012): 27–35.

Zaytsev, D., D. Soskov, and A. Foteev. "Oruzhiye neletal'nogo deystviya na osnove izlucheniy - fizicheskiye osobennosti i perspektivy primeneniya [Weapons with Non-Lethal Action Based on Radiation: Physical Characteristics and Employment Perspective]." *Voennaia Misl'*, no. 4 (2013): 30–37.

Zemlyanko, A., and V. Rozhkov. "Poteri v lokal'nykh boyevykh deystviyakh [Losses in Local Armed Operations]." *Zashchita i Bezopasnost'*, no. 4(47) (2008): 18–19.

Zerubavel, Yael. "Bar Kokhba's Image in Modern Israeli Culture." In Peter Schafer (ed.), *The Bar Kokhba War Reconsidered: New Perspectives on the Second Jewish Revolt against Rome*. Tubingen: Mohr Siebeck, 2003.

———. "The Multivocality of a National Myth: Memory and Counter-Memories of Masada." In Robert Wistrich and David Ohana (ed.), *The Shaping of Israeli Identity: Myth, Memory and Trauma*. London: Fank Cass, 1995.

Zhurkin, V., M. Nosov, and D. Danilov. "Voprosy mezhdunarodnoy bezopasnosti [Questions of International Security]." In N. Shmelev (ed.), *Rossiya v mnogoobrazii tsivilizatsiy [Russia in the Diversity of Civilizations]*. Moscow: Ves' Mir, 2011.

Zinn, Howard. *Vietnam: The Logic of Withdrawal*. Boston: Beacon Press, 1967.

Zino, Aviram. "Ha'mishor ha'mishpati: keitzad ne'erechet Israel le'yom she'le'achar ha'mivtza be'haza [The Legal Dimension: How Israel Prepares Itself for the Day after the Operation in Gaza]." *MaarivOnline*, August 9, 2014, http://www.maariv.co.il/news/new.aspx?pn6Vq=E&0r9VQ=FMMFL [accessed May 10, 2018].

Index

About the Author

Dr. **Ofer Fridman** is lecturer in war studies and director of operations at the Kings Centre for Strategic Communications. Dr. Fridman joined the Department of War Studies in 2016 after completing his PhD in politics at University of Reading. His research focuses on the intersection between strategy, military thinking, and contemporary conflicts, including information confrontations, the role of civilian casualties, and military transformations. He has published many articles in different journals, including *RUSI Journal*, *Joint Force Quarterly*, *The Journal of Defence Strategic Communications*, *Comparative Strategy*, and others. He recently authored *Russian "Hybrid Warfare": Resurgence and Politicisation* and coedited *Hybrid Conflicts and Information Warfare* with Vitaly Kabernik and James C. Pearce. Before embarking on an academic career, he served for 15 years in the Israeli Defense Forces.

About the Author

www.ingramcontent.com/pod-product-compliance
Lightning Source LLC
LaVergne TN
LVHW101446200926
848483LV00001B/62
9781498586917